# Don't Breathe the Air

NUMBER SIXTEEN
*Environmental History Series*
Dan L. Flores, General Editor

# Don't Breathe the Air

*Air Pollution and U.S. Environmental Politics, 1945–1970*

SCOTT HAMILTON DEWEY

TEXAS A&M UNIVERSITY PRESS
*College Station*

The paper used in this book meets
the minimum requirements of the
American National Standard for Permanence
of Paper for Printed Library Materials, z39.48-1984.
Binding materials have been chosen for durability.

Library of Congress Cataloging-in-Publication Data

Dewey, Scott Hamilton.
    Don't breathe the air : air pollution and U. S.
environmental politics, 1945–1970 / Scott
Hamilton Dewey.—1st ed.
        p.   cm.—(Environmental history series ;
no. 16)
    Includes bibliographical references and index.
    ISBN 0-89096-914-0
    1.   Air—Pollution—United States—History.
2.   Environmentalism—United States—History.
3.   Environmental policy—United States—
History.  I. Title.  II. Series.
TD883.2. D48   2000
363.739'2'0973—dc21            99-088267

*To my parents,*
Charlotte and Donald Dewey

*& to the many unsung heroes*
*in the long fight*
*against air pollution*

# Contents

# Acknowledgments

Few if any long pieces of research and writing are ever completed without the help of many different individuals in many major or minor ways. While this particular project was often a relatively solitary one, and although I alone can claim credit for any shortcomings, I am indebted to a number of persons and institutions for assistance during the years when the project was taking shape.

I am grateful to the staff of Texas A&M University Press, Series Editor Dan Flores, copy editor Sally Antrobus, and the anonymous readers of the manuscript for their patience and perseverance in bringing this project to completion. My thanks, too, to journal editors Hal Rothman of *Environmental History*, John Boles of the *Journal of Southern History*, Kari Frederickson of the *Florida Historical Quarterly*, Doyce Nunis of the *Southern California Quarterly*, and various anonymous reviewers, who helped to improve portions of this research that appeared earlier in journal articles.

My dissertation adviser at Rice University, Allen Matusow, was also patient and long-suffering in handling an earlier manifestation of the present work, always remaining encouraging and keeping his sense of humor even when the project hit snags. I appreciate his help and advice as well as that from fellow long-suffering dissertation committee members John Boles and Stephen Klineberg. I am grateful to the history departments at Rice University and California State University, Los Angeles, for their generosity in helping to support a fledgling scholar. I am also deeply indebted to the hardworking staff of Rice University's Fondren Library, particularly those down in the Inter-Library Loan office, upon whom I leaned heavily for hard-to-find dissertations and reports.

I must also thank the archivists, records officers, and other staff members at various locations. Officials and staff of the National Archives and Records Administration were generous with their time and attention in handling my requests for documents little used and sometimes difficult to trace, both in Washington, D.C., and Suitland, Maryland. The staff at the lovely Henry E. Huntington Library in San Marino, California, were similarly courteous and considerate. I especially enjoyed my brief sojourn with the kind and helpful staff of the Edmund S. Muskie Archives at Bates College in Lewiston, Maine.

In this book I seek to tell, in part, the story of the many, mostly forgotten early citizen activists who fought in the battle for air pollution control prior to 1970, and it is to them that the book is dedicated, as the most fitting memorial I can offer. I am also indebted to the various scholars who pioneered the study of air pollution history years ago but often never got the recognition they deserved, such as Robert

Dale Grinder, Marvin Brienes, Shawn Bernstein, and others who are repeatedly cited in the present study. I hope this book is worthy of their valuable, pathbreaking work.

Although it is neither a person nor an institution exactly, I also wish to acknowledge Southern California and its striking geography and atmosphere for providing much of the initial inspiration for this project in a process that began over thirty years ago. To experience the fact that such a beautiful landscape frequently had such a hideous atmosphere above it was an object lesson in environmental studies for a child growing up in the San Gabriel Valley. During the summer and fall months of 1995–98, when much of the present work was written, the various moods of the Los Angeles area—even the bouts of smog or the smoky smell of the September forest-fire season—somehow brought back a flood of pleasant or wistful memories.

In the summer of 1998, Wilbur R. Jacobs, a distinguished scholar of environmental history, Native American history, and the history of the American West, was killed in a terrible automobile accident on the way home from his beloved Huntington Library. Though I wish I had said more when he was alive, I want to thank him for his friendship, his hospitality, and his kind encouragement during times of self-doubt. I will never forget you, Wilbur, and I will try to remember as much as I can of all the good advice you gave.

Above all, though, I wish to thank my parents, Charlotte and Donald Dewey. They in particular offered encouragement and kept their faith in me when others might have had their doubts. My mother, especially, helped to teach me love and respect for nature, the environment, living things, and living systems—a very great gift. I hope that this project can in some small way serve as a partial repayment for their many years of hard work and self-sacrifice on behalf of their children.

Thank you all, very much.

Scott Hamilton Dewey
*La Cañada, California*

Don't Breathe the Air

# chapter 1

# Introduction

In his 1965 song "Pollution," Tom Lehrer, a former Harvard mathematician turned piano-playing comedian and social and political satirist, reflected on environmental conditions in the United States with characteristically crackling wit. Humorously describing how America's cities were suffocating on pollution and garbage, Lehrer offered the punch line: "Don't drink the water and don't breathe the air."[1]

In so doing, Lehrer gave relatively early expression to public worries about pollution and environmental deterioration that would sweep the nation during the late 1960s, in a new movement that would come to be called environmentalism. Unlike the earlier conservation movement, which had won important victories in preserving and protecting unspoiled wilderness areas, wildlife, and other scenic treasures of North America during the 1950s and 1960s, environmentalism of the late 1960s and early 1970s focused on environmental problems that more directly threatened human health and survival. The scene of environmental battles also changed with the shift from wilderness conservation to fighting pollution, for the new sort of environmentalism mainly focused on urban, industrial zones where most people were, as opposed to untrammeled wilderness areas where people mostly were not.

Air pollution was one of the main rallying issues of the nationwide environmental movement when it exploded onto the American political scene during the late 1960s. Indeed, air pollution, along with the related problem of water pollution, largely defined the environmental movement at this crucial time, as the fight against pollution emerged from the shadow of traditional wilderness conservation to take center stage. Anti-pollution activists largely took for granted the major achievements of conservationists earlier in the 1960s and treated such activities as a slightly irrelevant sideshow in the primary battle against pollution. Young militants in particular tended to disparage traditional conservationists as "the birds and squirrels people." Such activists saw their fight against pollution as a whole new challenge to the American political and economic establishment, and they dismissed conservation as too sedate and unthreatening. Reflecting an attitude typical of the time, radical students in consumer advocate Ralph Nader's Study Group on Air Pollution affirmed that atmospheric contamination was the product of "concentrated and irresponsible corporate power" and contemptuously declared, "'Clean air buffs' who fail to recognize this fact of economic and political life had best begin organizing nature walks or collecting butterflies." For their part, members of conservation organizations were surprised and a little bewildered by the seemingly sudden emergence of the new environmentalism, a term that gained its current meaning only after the mid-1960s. The heightened interest in fighting pollution relative to traditional conservation also spread beyond activist circles. Citizens presented with lists of governmental programs in nationwide polls taken during the late 1960s consistently favored the expansion of federal air and water pollution control efforts over nearly all other governmental activities.[2]

Reflecting the salience of pollution concerns and the relative marginalizing of conservation at the end of the 1960s, policy analysts at the Rand Corporation in 1971 characterized environmentalism and conservation as initially two separate movements with separate agendas. "Broadly speaking," they observed, "the conservation movement in the United States has until recently been primarily concerned with preserving and protecting unique natural phenomena, particularly wilderness, while the present environmental movement was initially mainly concerned with pollution problems. These two movements have now largely merged, however, so that the environmental movement's concerns can be said to include both categories." Notably, according to these observers, the younger movement had subsumed the elder.[3]

During this period, air pollution generated serious public concern about human survival, giving the issue a sense of pressing urgency far beyond that in traditional conservation, and perhaps even more than that provoked by its closest competitor at causing environmental anxiety, water pollution. In nationwide opinion polls conducted in 1970, water pollution slightly outranked air pollution in issue salience, with 74 percent of those questioned viewing water pollution as a very serious or somewhat serious problem, as against only 69 percent for air pollution. However, in the politically influential urban centers where aerial contamination was most severe, it often leaped ahead of water pollution in polls. Moreover, for those worrying about pollution, polluted air was in some ways fundamentally more sinister and threatening due to its unpredictability and its capacity to kill without warning, as had happened most notoriously at Donora, Pennsylvania, in 1948; London, England, in 1952 and 1962; and New York City in 1966. Contaminated water, once released, typically flowed downstream predictably, but contaminated air could blow in any direction. Also, thanks to improvements in sewage treatment technology during the twentieth century, polluted water was easier to treat prior to release and sometimes could be caught and purified even after its release. Polluted air was generally harder to cleanse at the source and, once released, could never be recaptured. Some dreamy, impractical postwar technology buffs foresaw ready technological fixes for water pollution or water scarcity in the desalinization of ocean water, or even the harvesting of icebergs, along with traditional technological remedies such as dams and aqueducts. Air pollution offered no such options. The most dramatic incidents involving water pollution during the postwar era involved dying fish and marine life; the most dramatic incidents involving air pollution involved dead humans. For reasons such as these, water pollution remained somewhat more likely to be seen as an aesthetic concern or long-term risk to human survival (from gradually running out of clean water), while air pollution was more likely to conjure images of potential mass death or to be seen as an immediate threat, as in the various alarming, book-length exposés of the dangers of aerial contamination that appeared during the 1960s.[4]

In addition to the prominence of air pollution among environmentalists and the general public, it also was one of the key issues, perhaps even the single leading issue, driving federal environmental policy during the late 1960s. It is far from coincidental that air pollution was the first specific matter addressed by powerful new federal legislation of the 1970s, following the brief, general National Environmental Policy Act of 1969 and a string of major federal enactments and investigations regarding polluted air from 1963 onward. The Clean Air Act amendments of 1970 first formally acknowledged that pollution problems, previously left in the hands of state and local governments, were often by their very nature interstate, nationwide problems properly subject to federal authority. Under the new regulations, the federal government partially wrested the issue of air pollution from the grip of often hesitant, inactive states and localities and established federal power to intervene on the matter directly, setting national air quality standards that states and cities would have to meet to avoid facing federal compulsion. This "federalizing" of an environmental issue set the pattern for subsequent federal policy, including the even stricter Federal Water Pollution Control Act of 1972, the Endangered Species Act of 1973, and other enactments. Meanwhile, changes in federal policy and pressure from environmentalists triggered a flood of new state laws on air pollution during the late 1960s and early 1970s, as states rushed to bring their control programs into conformity with federal mandates.[5]

Since air pollution was so important in driving environmental policy and politics in the crucial, formative period around 1970, it is impossible to understand fully the origins of the modern environmental movement and postwar environmental policy without an awareness of the issue. Yet the history of air pollution and efforts to control it has been largely overlooked by historians, analysts, and critics of environmentalism and environmental policy. There are relatively few published sources offering significant historical background on the topic, most of them preoccupied solely with the passage of federal legislation, while the array of helpful but hard-to-find unpublished dissertations and theses on the subject is not that much wider. In particular, there is little discussion of air pollution control at the state and local levels, even though the issue remained at these levels almost entirely until the end of the 1960s, notwithstanding early anti-pollution activists' persistent efforts to elevate air pollution to the status of a national, federal issue. Thus, it was at the state and local levels that most of the history of aerial contamination in the United States occurred, and this was where the initial battles against it were fought. This early experience, in turn, would help shape the modern environmental movement and its ideas—including its partiality toward federal intervention. As such, to understand the overall history of air pollution control, it is necessary to have information regarding patterns and developments at the state and local level—information often lacking.

Partly due to the general lack of historical awareness of such a central topic, various misconceptions have emerged regarding the history of environmental policy, environmentalism, and air pollution itself. One common misconception arises from observers mistakenly equating the conservation movement with the whole environmental movement. Various critics of environmentalism from both ends of the political spectrum have focused on conservationists' preoccupation with wilderness preservation and outdoor recreation, or on the backgrounds and attitudes of early conservationists, to argue that environmentalism was and is primarily a campaign by selfish, white, male, middle-class elitists to preserve pristine playgrounds for themselves at the expense of ordinary, working-class Americans. Critics on the left blast conservationists for failing to concern themselves with the environmental problems of working people in the cities; they also dig up material such as the racialist philosophizing of early conservationist Madison Grant, or statements from conservationist icon John Muir himself expressing irritation over the ecological insensitivity of recent immigrants from southern and eastern Europe at the turn of the century, to charge environmentalists with racism. Similarly, critics on the right have charged conservationists with insensitivity to the needs of the poor for trying to lock up natural resources in wilderness preserves and failing to recognize the need to use resources for job creation and economic development. Some even brand conservationists as irrational, wilderness-worshipping fanatics opposed to human progress. All such critics blast conservationists for placing nonhuman interests above traditional humanistic goals.[6]

Whether or not such charges against conservationists are fair—and it is likely that most are overdrawn—the crucial fact remains that wilderness conservation was never the whole environmental movement but was one major root among others. Unlike the conservation movement, which primarily involved rural areas, the antipollution movement was not just about "nature" but about people, and it chiefly involved urban areas and environmental problems that directly concerned the poor as well as the affluent. As readers of the following chapters will note, the fight against dirty air prior to 1970 clearly was not limited to middle-class white males, as working people, people of color, and particularly women joined in the struggle to protect their communities from serious environmental threats. Although liberal critics may still charge some pollution fighters with elitism for working to protect only their neighborhoods and not those of poorer people across town, and though conservatives still charge environmentalists with elitism for supporting pollution controls that allegedly dampen economic growth and job creation, it is harder to dismiss the anti-pollution wing of environmentalism as a mere frivolous recreational self-indulgence for the affluent or as an anti-humanist agenda.[7]

Another common misconception concerns the federal government's role in environmental policy. Ever since the emergence of powerful federal environmental regulations imposing new standards and obligations on lower levels of government, industrialists, economists, and conservative theorists have complained of federal officials usurping state authority. A myth has even arisen in some quarters that enterprising federal officials, together with the news media, created and whipped up public sentiment over issues such as air and water pollution just to seize power from the states. During the 1980s and 1990s, conservative Republican lawmakers in particular hoped to eliminate various federal mandates and to restore to the states primary authority over the environment and other policy issues.[8]

Yet far from creating the issue of air pollution and then willfully seizing state and local authority to address it, the federal government long resisted taking direct action on the matter and was dragged into it against its will only by sustained public outcry from ordinary Americans frustrated with inactive state and local governments. As readers will see, federal authorities remained hesitant and lagged behind public opinion on the issue through the end of the 1960s. The question to answer, then, is not why the federal government got into controlling pollution but why citizens ultimately lost patience with control efforts at the local or state level and demanded federal intervention.

Along with these broad misconceptions about the history of environmentalism and environmental policy, there are a number of specific misconceptions about the history of air pollution that I address in the present study. Here are a few of the major ones:

*Air pollution just was not an issue before the Second World War.*

This is simply incorrect. In cities where the atmosphere was fouled by coal smoke, public complaints made air pollution an issue and repeatedly caused flurries of rudimentary legislation and control activity during the years before 1940. However, as with many other issues, demands for reform came and went in cycles. Usually, little progress in facing a difficult and dimly understood problem had been made before the cycle of public interest came to an end, whereupon any progress was soon eroded, and the problem then continued to worsen until the next cycle of public interest and demands for reform. Looking back on many decades of such limited progress on air pollution control, later observers might naturally conclude that it had never been an issue, but this was not the case.[9]

*Farsighted experts in the fields of science, technology, and public health saw the need and the possibility for cleaner air, but the public was not interested.*

This is partially incorrect. Some of the most important early air pollution control initiatives were led by public health crusaders, combustion engineers, or even major local businessmen. Their efforts typically drew the enthusiastic support of certain groups and the apathetic disinterest or active opposition of other elements of the populace, as with most social or political issues in America. However, there is no question that at least at certain points in the public interest cycle, significant numbers of citizens from various backgrounds rallied around anti-smoke crusaders in many cities.

The complaint of public apathy may frequently have come from the "experts" themselves, who saw their role as boldly trying to create an issue in the face of a frustratingly apathetic public. While perhaps true in certain cases, such claims should be taken with a grain of salt. There may have been some public-minded crusaders who were disappointed at being unable to persuade the general public to gallop off toward atmospheric cleanliness as rapidly as the reformers might have wished. Yet the majority of the self-proclaimed experts, the engineers and scientists, were usually closely linked to the very industries and corporations that caused so much of the pollution problem. Being thus "understanding" of the industrial polluters' viewpoint, the technical experts were usually unwilling to try to force industry to clean up at a rate any faster than it found comfortable, which often meant virtually no progress whatsoever. As such, technical experts often helped to slow control efforts over what citizen activists may have sought, and they often sought to shut the public out of control policy altogether.

In fairness to industry, it should also be pointed out that they faced difficult, costly, built-in problems, while for many decades, such limited legislation and control policy as existed regarding polluted air was almost exclusively directed at industrial polluters, due to another long-lasting misconception that *industries were the only significant sources of air pollution.* Reformers often overlooked the individually less dramatic but cumulatively significant contributions to air pollution from ordinary citizens, and if they were aware of these, they often felt there was no hope of changing individuals' behavior through legislation.

*Prior to the Second World War—indeed, prior to the 1960s—air pollution was viewed only as an aesthetic nuisance, not as a threat to health.*[10]

Again, this is largely incorrect. As readers of the following chapters will find over and over again, there was significant public concern, among ordinary citizens as well as public health advocates, regarding the health effects of air pollution, not only prior to the 1960s and the Second World War but even far back into the nineteenth century. In the case of London, England, one of the first places in the world to have a really severe, chronic air pollution problem, health worries dated back to the medieval era. Such concerns largely grew up together with the major use of coal, though they appeared even earlier than that. People had an instinctive sense that it was unhealthful to breathe thick smoke and airborne particles, particularly the vile-smelling emissions from burning coal, and scientific research by now has amply justified their suspicions.

Rather, the myth that smoke and air pollution were viewed only as an unsightly annoyance and not as a health threat developed largely because there were powerful forces in society that long sought—and still seek—to deny or minimize the potential health injury from contaminated air. These forces consisted mainly of industrial interests, historically the most powerful source of resistance to air pollution control measures, together with the technical experts and legislators over whom they often enjoyed a large measure of control. If air pollution was construed as a public health threat, then it naturally became a matter of much greater urgency as well as a potential basis for costly liability on the part of large polluters. If, however, dirty air was conceived merely as an unaesthetic nuisance, measures to control it could be undertaken at a much more leisurely pace. Even those industrialists and engineers who tended to downplay the health risks of air pollution knew full well from an early date that it was more than just an aesthetic problem, since smudgy, acrid coal smoke caused extensive, measurable economic damage to business and industry as well as to private citizens. Sometimes this economic damage was extensive and visible enough to lead to lawsuits or control campaigns specifically on economic grounds. Yet with air pollution's health effects largely denied and with economic injury usually spread relatively thinly over a great many separate individuals, demonstrated air pollution damage was seldom dramatic enough to generate sustained anger or activism in areas where the inhabitants became resigned to living with filthy air.

This long-successful denial of health impacts points up another important theme in the history of air pollution and wider environmental policy: the politicization of science. Part of the trouble for early smoke fighters claiming health damage from contaminated air was the relatively primitive state of both medical and general science. Air pollution is a subject of great scientific complexity in its chemical nature and interactions and its biological ramifications; it is also particularly hard to trace or capture experimentally due to the relative unpredictability and volatility of substances in a gaseous state. As a result, air pollution is still only imperfectly understood today, after decades of sophisticated research; prior to the Second World War, it was understood very little.

Yet, alongside these fundamental complexities, there remained the simple fact that most of the scientific expertise in the prewar days was controlled or influenced by private, corporate industrial interests that had a clear stake in the outcome of the scientific debate and in the denial of health effects. Those who controlled science had the power to shunt both scientific research and public awareness away from health effects and toward tamer considerations, at least to some degree; they also could offer hesitant public officials good, "scientifically based" excuses for taking little action and for giving air pollution less attention than public health advocates might have wished. Even leaving aside the issue of whether scientists were consciously or unconsciously swayed by their principal sources of research funds, grants, and employment, traditional interpretations of scientific professionalism and certitude also tended to have a quietistic effect on early environmental reform, for if scientists felt unable to declare conclusively that a practice was harmful, that practice generally would be allowed to continue. As such, any scientific uncertainty, regarding complex issues where there was often much room for uncertainty, always played into the hands of those allegedly causing harm to the environment or to public health, since the burden of proof was placed on their critics. For all the scientists' traditional self-image of perfect objectivity and apoliticism, their demands for near total certainty on questions involving the public health and welfare, rather than just for very strong probability, had inevitable political ramifications. As can be seen repeatedly in the following chapters, even the decision that more scientific research and information were necessary before a final decision could be made could have political implications, since this would become a favorite delaying tactic of those seeking to avoid pollution control costs.[11]

A final, partial misconception, closely related to the question of scientific understanding, is that *air pollution control in the prewar period was impossible due to the absence of adequate and available control equipment and technology*. There is no question that early control technology was limited in its availability and effectiveness, and this situation persisted for many years. Significant improvements in combustion efficiency for industrial equipment began to appear by the end of the nineteenth century and the beginning of the twentieth, but the introduction of newer equipment and replacement of older devices was very gradual; even after the end of the Second World War, many industries in the United States, the United Kingdom, and elsewhere were still using facilities dating to the time of the First World War or earlier and not much improved over nineteenth-century technology.

However, as experience proved time and time again, in certain fundamental ways, air pollution control was ultimately more a matter of political will than of mere technological adaptation. Advanced control equipment meant nothing if there was insufficient political will to require its installation and use. Both industries and ordinary citizens and homeowners usually hoped to wait until there were technological solutions that could solve their problems with almost no effort or cost, and even a relatively minor cost could be enough to make polluters resist purchasing control equipment as long as they could get away with it. Engineers' promises of greater fuel efficiency from improved combustion equipment and methods usually were not enough to persuade industries to come up with the initial capital outlay for new equipment if they could still limp along with the old, while certain sorts of control devices promised only costs and no direct economic benefits to industry whatsoever.

Meanwhile, there were always other options for any town that was really serious about air pollution control—behavioral changes. Centuries ago, some Britons were already aware that noxious coal smoke could be reduced through requiring the use of higher grades of coal and banning the dirtiest sorts, a technique Raymond R. Tucker would use successfully during the late 1930s when he was commissioner of smoke regulation for the City of St. Louis. Automatic coal-stoking equipment for industrial furnaces and boilers, which promised greater fuel efficiency and less smoke than traditional hand-firing, had already appeared at the beginning of the twentieth century, though Tucker again was the first to reduce smoke by requiring the use of such equipment, decades later. There were also obviously more draconian solutions available: shutting down particular noxious industries, for instance, or limiting the demographic and industrial growth of particular areas—options seldom if ever considered at a time when people presumed limitless economic and demographic growth even more unquestioningly than they do today. All of these potential solutions, however, carried economic and social costs that industrialists, municipal governments, and ordinary citizens were reluctant to pay. Probably the most fundamental theme in the history of air pollution control and general environmental policy has been that of political will for pollution abatement repeatedly faltering in the face of economic considerations and crises, since traditional economic doctrine marginalizes many environmental considerations as mere "externalities."[12]

Given such long-standing obstacles in the path of air pollution control efforts, I explore in the present study how the issue ever broke out of these traditional constraints to the degree that it has done. I seek to illuminate and correct various misconceptions regarding the history of air pollution, environmentalism, and environmental policy by analyzing the history of air pollution abatement efforts at the state and local levels and how these helped to create and influence both the environmental movement and federal environmental policy. The chief focus is on the United States from 1945 to 1970, the time and place in which modern environmentalism and environmental policy first came into being, though earlier patterns and precedents in the history of air pollution in the western world are traced briefly.

Since the United States includes fifty states and numerous localities, each with its own stories of aerial contamination, I analyze three separate and different case studies in depth, to reveal significant commonalities and variations in the patterns of air pollution control throughout the nation as a whole before 1970. The first case study is of Los Angeles and the State of California, the best known story after that of the federal government and arguably the place where modern air pollution control was born. Relatively speaking, Los Angeles was a success story that set important precedents for the future, for although Southern California remains plagued by smog, preventing such pollution from growing inexorably worse, even lethal, was a tremendous accomplishment in itself. However, automobiles were the main source of atmospheric contamination in Southern California, and the powerful, remote automotive industry proved mostly beyond the reach of Los Angeles County without help from higher authorities. The second case study is of New York City, a less well known and less successful story of a big city plagued by a serious air pollution problem from interstate as well as local sources. In its comparative lack of success at dealing with its atmospheric woes or those from neighboring New Jersey that crossed the state boundary and descended on it, New York was much more typical of the overall national pattern than was Los Angeles. Finally, there is the example of central Florida, where, contrary to common assumptions that air pollution was exclusively a big-city problem, a rural, agricultural area was plagued by harmful emissions from an influential regional industry that local, state, and federal authorities were long reluctant to confront. The frustrating problems faced by Los Angeles, New York City, and central Florida—including powerful, recalcitrant interstate industries effectively beyond the capacity of local or even state governments to control; air pollution crossing state lines from places with lax regulations; and the deep reluctance of state and local authorities to antagonize even intrastate industries by demanding emissions reductions—serve as examples of wider patterns existing in urban and rural areas throughout the United States prior to 1970, patterns that ultimately necessitated the direct intervention of the federal government in air pollution control policy. The last chapter summarizes these nationwide developments.

Various interwoven factors came together in the postwar era to shake the air pollution policy cycle out of its old, time-worn tracks. For instance, in one of the key studies of the origins of environmentalism, Samuel P. Hays has argued that environmentalism, including concern about air pollution, resulted from the overall transition of the United States from a producer-oriented culture, preoccupied with survival and acquisition of necessities before all else, to a consumer-oriented society, concerned with greater quality and enjoyment of life. Thus, for Hays, environmentalism merely represents a somewhat broader-reaching, enlightened form of consumerism, an extension of traditional human economic desires. While Hays's thesis may apply less well to clean air than to some of the other pleasant "environmental amenities" he describes, it is likely that the process he delineates, and the greater expectations of health and comfort that it created, had a role in elevating the issue of air pollution.[13]

A variety of other factors were also at work. The nature of government and the relationships between the different levels of government and the people were changing markedly in the postwar period, building on the foundation of a national government laid down earlier by Progressivism, the New Deal, and the nationalizing experience of the Second World War. As the federal government gradually extended its authority into formerly state or local matters, people throughout the nation became more aware of the federal government than ever before and came to expect more effort from government to solve human problems. The impulse of looking to higher authority when lower authority failed certainly was nothing new to the history of air pollution, since in a number of cases, citizens (mostly unsuccessfully) sought action from their state governments in the prewar years and would have gone to federal authorities, had there been any concerned with the problem to go to. After the war, American citizens increasingly did have somewhere else to turn when their state and local officials let them down. This new possibility doubtless helped lead to the new demands and militance about "rights" that were such an integral part of the environmental movement as well as of other more famous movements of the time. Similarly, the decades from the 1920s through the 1960s saw the emergence of a new national culture and national awareness, brought in part by a consumer culture that was becoming nationwide and by nationwide print and broadcast media. Americans were no longer quite so oblivious to their fellow citizens living hundreds or thousands of miles away, and unlike at the end of the nineteenth century, when people in each smoky city might think theirs was the only one, after the war, concerned citizens could better follow what people in other cities were doing to combat dirty air.[14]

Certainly a key pair of factors influencing life in the twentieth century—science and technology—also had a role in raising the issue of air pollution and in allowing greater progress to be made in understanding and confronting it. The nation's hugely expanded scientific capacity and technological prowess, which had leaped ahead rapidly during the Second World War, made it more feasible to analyze individual air pollutants and the reactions between them, including the invisible, gaseous pollutants as well as the visible solid or liquid ones. New control equipment, such as the electrostatic precipitator, first developed in the 1920s to trap particulate emissions, was introduced or improved. As fast as science and technology progressed, the nation's postwar aura of technological omnipotence led citizens' expectations to race ahead even faster, as shown by the common argument in letters to newspaper editors during the 1950s and 1960s that if we could build the atom bomb (later, build the hydrogen bomb, or put a man in space, or put a man on the moon), then certainly we could control air pollution. In both these ways—actual progress and the presumption that there could be even more—science and technology may have helped to fuel the drive for air pollution control in particular and environmentalism in general by expanding the sense of the possible.

On top of all these other factors, though, is the central fact of citizens mobilizing to protest what they came to see as intolerable conditions. Along with policy developments in Los Angeles, New York City, and central Florida, the present study focuses attention on the experience of the citizen activists who helped to build air pollution into a major national issue between 1945 and 1970—an essential part of the story of the emergence of environmentalism left out of most existing studies. For it was the people, whether spontaneous or manipulated, well-informed or ignorant, active or apathetic, who ultimately created the specific issue of air pollution and the wider issue of the environment—both matters that various levels of government long hesitated to address seriously. Since rhetoric is important in a study of public attitudes, values, and perceptions, I have sought to let ordinary citizens discuss the problem and describe their experiences in their own words as much as possible, to illuminate the process of the public discovering and trying to comprehend and confront a complex environmental issue.

Of course, the American people were never all ecological saints. As in other situations, in the battle against dirty air, ordinary Americans could sometimes be heroic, sometimes petty, sometimes steadfast, and sometimes fickle; they could cause major headaches for control officials, sometimes with good justification and sometimes for irrational or arbitrary reasons; they had a tendency to show the typically human but particularly American characteristic of wanting to have their cake and eat it, too, or in this case, of wanting to have clean air without having to sacrifice anything for it; they also long persisted in the characteristic American faith that there must be a quick technological fix to solve any problem effortlessly and painlessly.

Whatever their strengths and shortcomings, however, mobilized citizens were the essential ingredient in the struggle to break the air pollution issue out of well-worn ruts and elevate it to a new level of significance and governmental attention, without which the other factors meant little. More should be known about these early environmental activists, their complaints, and their legitimate grounds for complaint before they are casually tossed aside as mere selfish elitists or irrational fanatics opposed to progress. For while many environmental issues have recently been discussed or dismissed as mere amenities, or as sociocultural or linguistic constructs, breathable air is a necessity, and the threat the postwar public faced from air pollution was real and significant.[15]

Such unpleasant reality crashing in also helped cause a challenging or shaking of the traditional American presumption of endless expansion built on limitless resources and brought a partial opening of the American mind toward a newer, ecologically inspired sociocultural paradigm recognizing the interconnections between humans and other living systems. The new worldview also reshaped the environmental movement, combining traditional conservationist concerns over wilderness and wildlife with human-centered worries over pollution and public health in a new, more comprehensive whole as these two different wings of environmentalism increasingly accepted the legitimacy of each other's issues. Although most Americans, and even most environmentalists, have never made it all the way to full ecological or environmental awareness, they made major strides along that path between 1945 and 1970, in large part because of their experience of struggling with air pollution.

Reflecting on a dramatic, revolutionary movement of an earlier time, the great nineteenth-century abolitionist and civil rights leader Frederick Douglass observed: "Power concedes nothing without a demand."[16] The same might be said for bureaucratic inertia and public apathy. The following chapters trace how the public demand for clean air in the United States challenged these obstacles and took the particular shape it did during the years before 1970.

# chapter 2

# Recurring Cycles

*A Brief History of Air Pollution and Control Efforts
in Britain and the United States before 1945*

Many people today might assume that air pollution is a post-1900 phenomenon, even chiefly a post–Second World War phenomenon. However, human-caused air pollution has harmed health and caused other damage since ancient times. Even prehistoric peoples suffered from it. Lung tissue samples from frozen or mummified remains of ancient humans often show anthracosis, or blackening of the lungs, a condition later to become especially common in smoky industrial centers of the late nineteenth and early twentieth centuries, such as Pittsburgh. Such archeological evidence suggests that chronic indoor air pollution was part of life for many early humans who lived in simple, one-room abodes with a hole in the roof for a chimney. In the coldest sites of early human habitation, such as the Russian steppe, where even holes in the roofs were often dispensed with, damage to lungs, the eyes, and the bronchial tract would have been still worse.[1]

Just as human-caused air pollution goes far back in history, so, too, do some of the basic themes and cycles that generally characterized air pollution control policy through the time of the Second World War. Where people congregated in large numbers, burning fuel for domestic and commercial purposes, smoke problems gradually developed. The continuing increase of human population and activity might bring such pollution to a serious level. By that time, however, the problem was already firmly built into the social and economic fabric of the communities in question, such that it would involve hardship and dislocation to mitigate it significantly. Some individuals complained of harm to property and health from contaminated air, and occasionally laws were passed to abate the nuisance, but after such brief flurries of complaints and reform efforts, the air pollution usually continued to grow, uncontrolled, due to the sheer force of human desire for increasing production, re-production, and consumption. Those who had the means could perhaps move to less crowded, less polluted surroundings; most others just learned to live with atmo-spheric filth. People continued to complain, usually about the most obvious, nox-ious sources of aerial contamination, such as particularly smoky or foul-smelling trades and industries, but the economic desires for the products of these trades over-whelmed environmental concerns. Meanwhile, ordinary folk seldom if ever recog-nized their own contribution to the problem through their own fireplaces, stoves, and chimneys, which were less dramatic than industrial polluters but more numer-ous and cumulatively very befouling to the air. This general situation would prevail through the middle of the twentieth century.

Writers in ancient Rome already complained of air pollution there. During the Medieval Era, declining urbanization brought less urban air pollution throughout Europe. However, a major new problem emerged with the expanded use of coal—the dirty, smoky fossil fuel that would be the principal frustration of air pollution foes through the 1950s. The example of London, a large medieval town that would become the greatest city of early modern Western Europe as well as the first major city wholly reliant on coal for its fuel needs, offers a particularly good early case study of the developmental trajectory of a major metropolitan air pollution prob-lem. Londoners' experience with air pollution from coal over the seven centuries from the 1200s onward would set patterns many other cities around the world would later follow.[2]

Major fuel-using industries first came to rely on coal, which was shipped down the English coast from northern seaports to London starting in the 1200s. The new fuel, commonly called "sea-coales," was unpopular from the beginning. People saw it as "unnatural," ungodly, even satanic, and they long resisted using it in their homes. In the face of such rooted hostility, in earlier centuries only the more energy-inten-sive trades used coal, and such trades became stigmatized for their coal smoke, soot, and stench.[3]

Regardless of the stench, though, coal was cheap and abundant at a time when London was rapidly depleting its supply of accessible firewood. As the population grew and the price of firewood rose, poorer Londoners gradually turned to coal for their heating needs, first on a seasonal basis when wood was scarcest and later year-round. By the late 1500s, despite lingering distaste, the lower classes of London mainly burned coal, and access to firewood became a class distinction and privilege of the wealthy. By the mid-1600s, virtually all Londoners, rich and poor, nobles and com-moners, burned coal exclusively. Coal made possible the rapid growth of London during the sixteenth and seventeenth centuries and became inextricably woven into the socioeconomic fabric of the city.[4]

English coal was mostly dirty, smoky, bituminous or "soft" coal. By the end of the thirteenth century, London already had enough tradesmen using this cheap fuel to make many citizens complain. However, England's earliest known air pollution incident happened not in London but to the north, where the coal was mined. In 1257, Queen Eleanor, wife of Henry III, was so sickened by the stench of coal smoke at Nottingham Castle that she fled to preserve her health. Historian Peter Brimblecombe notes that similar fears of danger to health "permeate almost all medieval complaints about air pollution." These health concerns grew out of the prevailing medieval medical belief in miasmas, a concept derived from ancient Greece holding that foul-smelling things such as swamp vapors or soft coal fumes were dangerous, unhealthful, and causes of disease in themselves. The miasmatic theory of disease held sway in Western medicine through the late nineteenth century, when scientists offered convincing evidence for the newer germ theory of disease that has been the dominant paradigm in medical science ever since. Regardless of the scientific inaccuracy of miasmatic theory, though, Queen Eleanor and countless other nameless sufferers were right to worry about coal smoke harming their health, and these noxious fumes would generate endless concern and legitimate complaint down to the present day.[5]

By the end of the thirteenth century, the growing smoke problem in London brought perhaps the earliest air pollution control effort in the Western world. Following public complaints, in 1285 a commission was created in London to investigate the problems resulting from coal use, followed by a 1306 royal proclamation banning the use of sea coal, allegedly under pain of death—a regulation that was largely ignored and left unenforced, like many that followed.[6]

This and other early control efforts were to no avail. The desire of ever more Londoners for ever more fuel, building materials, and other products left the city few alternatives, and coal became firmly established, first as an industrial and then as a domestic energy source. This did not stop the authorities from trying to satisfy angry citizens, though. The city experimented with various other control methods, such as requiring higher chimneys; limiting the hours of the day in which furnaces could be operated or the months of the year when coal could be used; zoning to keep brick kilns away from sensitive fruit trees; and later, in the fourteenth and fifteenth centuries, taxing coal and regulating its shipment. All of these were early versions of abatement techniques that would be tried during the twentieth century; all were ultimately ineffective. By the reign of Elizabeth I, London was burning 50,000 metric tons of soft coal each year, still mostly in industry, and the queen herself was "greatly grieved and annoyed with the taste and smoke of sea-coales." As nearly all householders burned coal by the mid-1600s, the city's smoke problem only got worse.[7]

London's deteriorating smoke situation stimulated a flurry of early research and control efforts during the seventeenth century. Enterprising inventors experimented with charring coal or combining it with other substances to form processed fuels with less noxious or acidic emissions—techniques still often tried during the nineteenth and twentieth centuries. Pioneering meteorologists and epidemiologists recorded major early air pollution incidents and traced links between coal smoke and elevated death rates, while other early scientists studied the ways coal smoke damaged health and property. In 1661, John Evelyn, England's most celebrated early air

pollution reformer, proposed an ambitious program of industrial relocation to preserve public health and drafted legislation for Parliament to effect this. Like many
later environmental initiatives, Evelyn's plan was defeated by traditional economic
considerations and apathy. He also erred in focusing only on industrial emissions
and ignoring the vast cumulative pollution from Londoners' homes, a misconception that would persist into the 1950s.[8]

As London grew in size and density, the local smoke problem only continued to
worsen, and it powerfully affected the city's culture. By the end of the eighteenth
century, Londoners burned more than a million tons of coal each year, cloaking
London in a perennial smoke pall visible from many miles away and giving it the
corresponding nickname "the big smoke." The sooty filth in London's air, and subsequently on sidewalks and streets, made it impossible to keep clothing clean, so
citizens wore black or drab clothing, a "protective coloration" that would later appear in grimy industrial cities throughout Europe and North America. Indoors, paint-
ings and wall-hangings, books, maps, and other documents all were eaten away by
corrosive coal smoke, while outside, the pervasive smoke corroded iron, brick, and
marble, blackened structures, and eradicated certain species of plants from London's
parks. To dodge some of the undesirable effects of the smoke pall, wealthier Londoners tended to move westward (upwind) toward neighborhoods such as Ken-
sington, leaving the more polluted East End to the city's poor, who remain there
today. This gravitation of affluence upwind of the smokestacks would also be an
often repeated pattern in continental Europe and the United States.[9]

During the nineteenth century, Londoners would also learn that not all coal was
the same. Back in 1595, a Welsh visitor recommended the relatively clean-burning,
hard anthracite coal of Wales to Londoners as a solution for the smoke scourge. Like
any other fossil fuel, anthracite gives off some carbon monoxide when burned, due
to imperfect combustion; it also releases some sulfur dioxide and other compounds
due to sulfur and other impurities present in the coal. However, relative to ordinary,
low-grade bituminous coal, high-grade hard coal has much less of the sulfur that
makes coal smoke acrid and acidic and fewer of the noncombustible pyrites and
silicates that produce more particulate pollution such as smoke, dust, and fly-ash.
Anthracite is thus much cleaner-burning than bituminous coal. In the 1800s, public
protest and legal action forced some London businesses to use anthracite to abate
the nuisances they caused. However, such hard coal was slightly more expensive than
soft coal and required different handling, so custom, economics, and geology dic-
tated that most of London and England would continue to use the dirty, smoky,
bituminous variety—a problem later shared by other smoky cities such as Pittsburgh
and St. Louis. This limited tinkering with fuel substitution in the absence of sufficient
political will to make it widespread and permanent would be another recurring theme
in other industrializing nations.[10]

As London's smoke scourge continued to worsen, it brought more dramatic incidents along with gradual overall deterioration. Meteorological fluctuations brought

periodic becalmed days when the wind did not sweep away the city's smoke and fog, allowing them both to accumulate. Even in the late 1600s, "Great Stinking Fogs" brought horse and wagon collisions due to near zero visibility. As the city grew and smoke increased, such episodes of air pollution and fog grew costlier and deadlier, in terms of both general mortality and accidents, from railway collisions to smoke-blinded Londoners falling in the river Thames and drowning. The late nineteenth and early twentieth centuries brought numerous such incidents. In 1905, recognizing that what London experienced was no mere normal, innocuous fog, Dr. Harold A. Des Voeux, a London physician, first coined the new name "smog," a combination of smoke and fog, to describe the unpleasant new phenomenon.[11]

By the 1800s, though, London was far from being the only major English city with a serious air pollution problem. The Industrial Revolution—with its coal-powered steam engines, factory machinery, and locomotives producing great wealth together with overrapid urbanization, tremendous crowding, and chaotic squalor—came to cities in the north of England much faster and more intensively than to London. This produced frightful smoke palls in such places even before the nineteenth century. Brimblecombe calls the industrial north "an environmental disaster" by the late 1700s, while historian Carlos Flick notes how in northern industrial towns by the late nineteenth century, "grim humor held that generations of people in these regions had come to believe that in nature the sky was gray and vegetation was black." Although record keeping in these areas was generally sporadic compared with London before 1900, it is likely that damage to public health and property was often worse than in the metropolis.[12]

As the steam engine drew attention to smoke in cities throughout Britain, Parliament renewed its control efforts, and various acts to abate smoke nuisances in particular cities or nationwide were introduced in the House of Commons during the nineteenth century. In 1819, the first of a series of parliamentary committees was appointed to study how steam engines could be made "less prejudicial to public health and comfort." This committee, like others that followed, found that smoke could indeed be prevented or reduced, though not without costs. The committee's investigations led to a bill to require all steam engines to consume their own smoke. The bill passed in 1821, but it was largely ineffectual, like most that came after. The 1840s brought further enactments but little lasting progress, while industrial lobbyists successfully killed or weakened various bills, setting a pattern for their counterparts in other industrial nations.[13]

The single most important British air pollution control law of the nineteenth century was the Alkali Act of 1863 to control severely damaging hydrochloric acid emissions from the alkali (sodium carbonate) industry, which had laid waste to significant portions of northern England. This law for the first time allowed trained government inspectors to enter factories, not to protect workers but to preserve the atmosphere and surrounding property, and together with subsequent enactments, it did bring some actual, enduring progress by forcing chemical manufacturers to use existing, affordable techniques to abate particular emissions. Since the Alkali Inspectorate's authority covered all noxious industrial emissions *except* ordinary smoke, though, their accomplishments did not affect general smoke levels.[14]

Private citizens and civic groups also attempted to take action on smoke in the 1800s. For instance, in London from the 1880s through the end of the century, the Fog and Smoke Committee sought to publicize smoke issues and educate the public about proper firing practices. Late in 1881, this organization staged a major Smoke Abatement Exhibition, showcasing improvements in both industrial and domestic combustion technology, though it did not persuade many Londoners to give up their cherished, smoky fires in open fireplaces, fires that sent up to 90 percent of the energy value of their coal fruitlessly up the chimney. Some private citizens fought smoke in other ways, as concerned journalists followed the air pollution story and neighborhood groups and individuals took action to clean up their surroundings by installing improved equipment, reporting chronic violators to the authorities, and taking legal action. Some factory owners also made voluntary efforts to reduce their emissions.[15]

Yet actual reduction of the smoke nuisance in English cities remained spotty and limited. In 1913, the agitation of various local and national smoke abatement organizations led to the introduction of yet another general smoke control bill in Parliament and another study committee, though the committee's investigation was interrupted by the First World War, which also torpedoed smoke abatement efforts in the United States. After the war, in a pattern also typical on the other side of the Atlantic, British industrialists and ordinary citizens continued fruitlessly trading blame and refusing to accept responsibility for their share of air pollution, while authorities mostly failed to enforce any existing weak regulations and sporadic governmental panels of inquiry bemoaned the lack of progress. Britain's atmospheric contamination remained uncontrolled into the 1950s, leading to London's Great Killer Smog of December, 1952, when stagnant weather conditions trapped aerial filth over the city, suffocating an estimated four thousand Londoners. This frightful incident, still one of the worst air pollution disasters of all time, finally triggered serious reform in Britain's Clean Air Act of 1956.

Britain was not the only nation of Europe to have an air pollution problem prior to the twentieth century, though its smoke was traditionally by far the worst. During the eighteenth and nineteenth centuries, many European and American visitors viewed the smoke in English cities with fascinated horror. Yet smoke problems, and unenforced prohibitions against pollution or coal use, also had a long history in continental Europe. Already by 1348, a municipal proclamation in Zwickau, Saxony, forbade metalworkers to use coal. As in England, such laws did not stop coal use or smoke. Yet other European nations had more firewood than England and had less accessible coal deposits, and so took longer to industrialize. Also, London was by far the largest city in Europe after the late Middle Ages, so no other European city had the same concentration of traditional smoke sources. While medieval manufacturing towns near coal deposits tended to have noxious smoke problems earlier, generally this was not seen as a major problem on the Continent, so early abatement measures were rarer.[16] Other Europeans also burned fuel more efficiently, and less smokily, than the wasteful English.[17]

Yet most nations of Western Europe still had no effective aerial emissions controls by the early twentieth century. This left them ill prepared to confront the results of their recent, rapid industrialization as their air pollution problems mounted from the mid-1800s onward. The severity of aerial contamination on the Continent prior to the Second World War was most starkly revealed in the Meuse Valley incident of December, 1930, in Belgium, when sixty local citizens died and many more became ill as stagnant weather conditions trapped and concentrated contaminated air within an industrialized river valley for a number of days. The Meuse Valley disaster, which got international attention and news coverage and helped stimulate neighboring France's 1932 control enactment, was one of the earliest clear demonstrations that acute air pollution conditions could actually kill people. As was typical of early air pollution policy, however, this warning was largely passed over as a freak incident.[18]

One industrializing western nation that showed even less concern about fuel efficiency than Great Britain was the United States. The sense of limitless abundance that shaped and distinguished the American psyche in so many ways was also reflected in American patterns of energy use. During colonial times, settlers from Europe and even indigenous Americans burned huge quantities of wood per capita, by European standards. The Native American tribes generally were able to avoid overtaxing their woodlands, however, by maintaining relatively small, stable populations and by periodically rotating their dwelling sites to places where the forests had regrown. European colonists took no such pains, and wherever they settled, they rapidly eliminated forests, using them for firewood and construction materials and clearing them for cropland for their ever-expanding villages and towns.[19]

The new Americans could get away with their reckless use of natural resources for a long time, though, because by European standards, their supply of resources was enormous. America's forests stretched almost continuously from the Atlantic Coast to the Mississippi River, offering a seemingly limitless store of firewood to the relatively few early settlers. Consequently, through the end of the American Civil War, when England had been relying on coal for centuries, most American homes, railroads, and steamboats still relied on firewood, which was cut and consumed in prodigious amounts.[20]

Yet changes were afoot that would bring to America the sort of atmospheric problems seen in English cities. Already by the end of the eighteenth century, the larger cities and towns on the East Coast of North America were beginning to outstrip their local, accessible supplies of firewood. This situation was first evident in Boston and the rest of relatively populous southern New England; before too long, New York City, Philadelphia, and other population centers of the Northeast experienced the same effect as London had centuries earlier. Firewood in the urban hinterlands ceased to be cheap and abundant and required transportation from ever greater distances. At the same time, the Industrial Revolution was beginning to take root in parts of New England and the Mid-Atlantic states at the beginning of the nineteenth century. While most early textile mills relied on waterwheels for their power, suitable water-power sites were limited and subject to seasonal fluctuations in river flow. American industry, too, required a more abundant and flexible energy source to expand.[21]

Thus, by the early 1800s, Americans in the fast-developing northeastern states basically faced a choice between two alternatives—either find a new energy source or cease urban and industrial growth. Given Americans' particular antipathy to accepting limits, the Northeast soon followed England down the path toward reliance on coal. Fortunately for citizens in the Northeast, their most accessible coal supply was the anthracite fields of central Pennsylvania, with their higher-grade, less polluting fuel. Prior to the twentieth century, when soft coal use rose, northeasterners never knew the sort of smoke that Londoners had long endured, or on which Americans in other cities would soon be gagging.[22]

Farther west, where Americans burned bituminous coal, serious smoke problems emerged during the nineteenth century, and many industrial cities went through a process of gradual discovery similar to that in London, though on a smaller scale and in a shorter period of time. Many aspects of the sequence of events in the English capital were played out all over again in the American heartland, including urbanization; industrialization; depletion of local firewood; substitution of soft coal; complaints about property damage; suspicion of unhealthful effects; the search for new technology; the dodging of responsibility; and sporadic activism stifled by the combined weight of overall public ignorance and apathy, traditional economic assumptions, and the sheer ponderous inertia of business as usual. Most early American smoke fighters never knew that their local struggles were merely repeat performances.

Pittsburgh, the city that would become the capital of the nation's iron and steel industry and a symbol of America's industrial might, was precocious in its troubles with smoke. Already an iron-making center during the eighteenth century, it quickly depleted its local supply of firewood and turned to soft coal for various purposes. Consequently, by 1791, when the manufacturing town still had less than four hundred inhabitants, an observer complained that Pittsburgh was "kept in so much smoke as to affect the skin of the inhabitants." By 1804, some local politicians worried that the smoke might grow so severe as to stifle the city's development. By then, Pittsburgh was already gaining a wider reputation for its dirty air. By the 1860s, it was commonly known as "the Smoky City," a nickname it would keep through the mid-1940s as its atmospheric problem worsened.[23]

Other industrial cities of the Midwest, frequently situated like Pittsburgh in narrow river valleys that helped trap smoke and fumes, were already developing smoke problems during the first half of the nineteenth century. Another midwestern pioneer of urbanization, industrialization, and air pollution, Cincinnati, was frequently smoky by 1815. By the early 1820s, St. Louis, Pittsburgh's chief rival for smokiness, had already burned up its local forests and turned to using soft coal found nearby in southern Illinois. Major air pollution inevitably followed. In 1822, civic leaders in St. Louis embarked on an unsuccessful campaign to teach local citizens how to burn coal with less smoke. In 1823, the *Missouri Republican* described smoke so dense "as to render it necessary to use candles at mid-day," while an anonymous letter to the editor during the 1840s warned of the city becoming "an emporium for disease" from all the smoke. The city's smoke problem steadily worsened until the 1930s.[24]

Chicago, the biggest and best-known midwestern industrial city of all, did not have as long a history as these other cities, but it made up for lost time in developing a severe smoke problem during the half-century after 1833. Chicago never had the same ideal preconditions for a chronic smoke scourge that Pittsburgh and Cincinnati had: it was not located in a narrow, steep valley prone to temperature inversions that periodically put a lid over the top of the valley, keeping pollutants from escaping and dissipating in the open atmosphere. Indeed, Chicago, the "Windy City," situated on a flat plain near Lake Michigan, was extremely well ventilated. Rather, Chicago's smoke problem grew out of the sheer volume of industrial and domestic emissions gushing into the air day and night as its citizens, factories, steamboats, and railroad locomotives lavishly consumed soft, smoky coal. By the 1880s, despite its wind, Chicago had a nearly continuous smoke pall covering the whole horizon. The smoke shocked visitors from cleaner areas and often made it almost impossible even to see across the street.[25]

Chicago's smoke problem had grown serious enough by the last quarter of the nineteenth century that in 1881, after citizens' reform groups had worked on the problem for several years, the city enacted what is generally acknowledged as the first serious smoke law in America. This law declared dense smoke a nuisance per se, meaning that plaintiffs would no longer have to overcome various obstacles and conditions to proving a nuisance under traditional common law; the Chicago ordinance also authorized a municipal inspection force to patrol for smoke violations. Contrary to the early pattern in some other cities, such as St. Louis and Pittsburgh, the Illinois Supreme Court upheld this ordinance in 1892 in a precedent-setting decision holding that a city could declare dense smoke a nuisance per se if state laws granted them the power to do so. Thereafter, many other cities passed similar laws and sought similar legal backing from their state governments. However, neither the largely unenforced original Chicago ordinance nor others that followed elsewhere were able to eliminate the cities' growing smoke problem.[26]

As in England, smoke caused serious damage in America's industrial centers. During the first decade of the twentieth century, many cities, including Minneapolis and Louisville as well as Pittsburgh, St. Louis, and Chicago, experienced repeated darkness at noon from smoke palls that limited visibility to less than a city block. The thick haze caused traffic accidents, injuries, and deaths, and also endangered shipping on the Great Lakes. Urban foresters in St. Louis and Cleveland reported many of their city's trees killed by gaseous emissions, while buildings, bridges, marble statuary, merchandise, and other goods and materials were blackened by soot and corroded by sulfuric acid in the atmosphere.[27]

Citizens also complained bitterly about the smoke that ruined their freshly washed laundry, invaded their homes, and begrimed their dwellings inside and out. While not until much later would America's booming industrial towns enjoy the extensive epidemiological analysis of mortality records London had had since the 1600s to prove elevated death rates and other effects, citizens nevertheless worried and complained over damage to health. By the early 1900s, physicians warned of the dire physiological and psychological effects of the smoke pall blocking healthful sunlight and keeping people indoors in poorly ventilated quarters, while medical experts linked smoke to tuberculosis and psychological depression that in turn produced more alcoholism, irritability, and juvenile delinquency.[28]

In the face of persistent and shockingly visible smoke problems, citizens and local governments throughout America's industrial heartland struggled to find effective means of confronting the problem without giving up either soft coal or urban and industrial growth. By 1912, twenty-three of the twenty-eight American cities of more than 200,000 inhabitants had smoke abatement bureaus, while thirty-one smaller cities also had smoke ordinances or inspectors. Laws covering wider jurisdictions appeared as well. In 1913, citizens in New York State passed a smoke law covering all of Albany County, while already in 1897 Ohio had an early if limited statewide law restricting emissions from boilers.[29]

As in England, such laws came as a result of public complaint and citizen mobilization. Citizen agitators varied considerably, though. Robert Dale Grinder, in his pioneering study of early smoke control efforts in American cities, identifies three main varieties of smoke fighters: local business leaders concerned about the economic implications of smoke damage to a city's property, merchandise, and reputation; engineers, who viewed smoke as a technical matter to be solved by more efficient combustion in proper equipment; and citizen activists, who usually worried more about health effects and favored strict laws and enforcement. In the early stages of a city's battle against air pollution, these different sorts of smoke fighters often could cooperate effectively. Yet their differences in outlook also could produce conflict.[30]

The citizen activists typically were hard-liners on smoke control: they wanted to see definite results promptly. As such, once smoke control programs were in operation, smoke fighters of this sort gravitated toward what Grinder calls the "prosecutionist" wing of the anti-smoke movement—those favoring strict enforcement of smoke laws and harsh penalties for violators—as against the "educationists," who hoped to work more gently through good advice and persuasion. Activists or prosecutionists could come from many sources, from concerned doctors and crusading journalists to local Socialists.[31]

Perhaps the largest, most important component of the prosecutionist wing of the smoke control movement in many cities, though, was women's organizations. Such activist women's groups were usually middle or upper middle class in orientation; notable examples include the Women's Club of Cincinnati, the Wednesday Club of St. Louis, and the Ladies' Health Protective Association of Pittsburgh. Working-class wives and mothers might have favored smoke abatement as well, and they almost certainly hated the smoke that ruined their laundry, befouled their homes, and threatened their families' health, but most of them probably did not have the time or energy left over from their daily tasks to attend meetings or organize clubs. Middle-class women, however, were active and earnest in fighting smoke despite the fact that fellow members of the comfortable classes were often the ones generating smoke and profiting from it. Grinder attributes this female activism to a sense of mission born of the turn-of-the-century ideology of "civic motherhood" or "municipal house-keeping": women perceived the smoke nuisance as endangering the health of their families, harming the cleanliness, beauty, and morality of their communities, and thus threatening their traditional female role as protectors of these qualities.[32]

The other wing of the anti-smoke movement was much tamer. The educationists steadfastly favored the traditional, libertarian American approach of education and voluntary persuasion to correct local smoke nuisances, rather than compulsion by municipal authorities. The educationists typically worried about economics rather than illness. Local business leaders and civic boosters who sought smoke controls, for instance, viewed smoke as a threat to growth rather than to health. Such smoke fighters offered statistics showing that St. Louis had suffered six million dollars in smoke damage in 1906, or that smoke had cost Cleveland six million, Cincinnati eight million, and Chicago fifty million dollars in 1911 alone. Because of such worries, local business interests and associations were active in pushing for smoke control in many cities. Thus, the Citizens' Smoke Abatement Association of St. Louis was mostly a businessmen's organization "designed to aid the trade as well as the salubrity of St. Louis." Chicago's Society for the Prevention of Smoke, formed in 1892, was also led by major local business figures, including some of the chief planners of the city's famous 1893 Columbian Exposition. Similarly, in Pittsburgh, Andrew Carnegie, Henry Clay Frick, and George Westinghouse all shared with the local chamber of commerce an interest in smoke control. Such major industrialists and smokemakers hoped that the smoke nuisance could be addressed through clever and inexpensive technology that would not interrupt the pursuit of profits unduly.[33]

The other principal educationists were the engineers. While they frequently spoke in terms of efficiency, the engineers generally shared the industrialists' economic preoccupation. Even more than their employers, engineers were certain that the smoke problem could be banished by proper equipment and firing methods—their stock in trade. Thus, engineers recommended the use of steam jets, fire-brick arches, automatic stokers, downdraft furnaces, and other technology then available, along with careful fueling and boiler ratios. While recognizing that such improvements might entail immediate costs, engineers argued that the improvements were well worth the expense and would pay for themselves in the long run through improved fuel economy, though such arguments seldom overcame businessmen's stronger desire to keep down upfront costs.[34]

Obviously, the differences in orientation between the prosecutionist and educationist wings of the anti-smoke movement always threatened conflict, and local leaders had to work hard to hold the movement together so that it could confront the many unrepentant polluters and apathetic citizens in any one city. Many citizens and civic leaders tacitly accepted traditional industrialists' arguments equating smoke with prosperity. This traditional view of air pollution as a necessary concomitant of economic growth often persists a century later.[35]

Even when majorities of citizens favored local smoke control laws, unenlightened and unashamed manufacturers could pull strings and manipulate those city councilors they controlled to prevent or delay regulation, while major business leaders earnestly committed to smoke control were often stubbornly resisted by other businessmen, including smaller, more marginal operators such as tugboat owners. In Pittsburgh and St. Louis, industrial interests were able to insert in the smoke laws weakening qualifications that held smoke violators safe from prosecution if there were no known means or devices to prevent their emissions; other American cities, like English cities, saw similar demonstrations of industrial lobbies' power. Then, once new regulatory agencies were in place, smoke control officials—whatever they previously might have said about protecting the public interest—usually were brought around to seeing things the industrialists' way, due to traditional economic factors, the industrialists' political influence and frequent resistance to reform, and the sheer intractability of the problem of air pollution. Consequently, most control authorities offered advice but little enforcement, emphasized economic factors while ignoring health risks, and remained largely ineffectual.[36]

After World War II, the educationist-prosecutionist battle would be fought all over again in many American cities, usually with no memory of earlier campaigns. Also, unlike in Britain, where smoke had already become a matter of concern to the national government during the nineteenth century, under the decentralized federal system of American government, air pollution would remain almost entirely a worry of local—often not even state—governments through the 1950s. Although cities had gradually grown more aware of one another's smoke control efforts by the second decade of the twentieth century, most national awareness of air pollution evaporated with the general collapse of the early anti-smoke movement during World War I, when smoke came to symbolize expanded production and patriotic support of the war effort. Thereafter, persistent localism, like educationism and disregard of health impacts, would remain a primary stumbling block to effective control of air pollution in the United States through the 1960s.[37]

America's early anti-smoke campaigns also established other unfortunate but persistent patterns. As with their English antecedents, early American smoke fighters stumbled upon, and proved unable to resolve, the problem of apportioning blame between ordinary citizens and industry. Throughout the early smoke abatement campaigns, control activists continued to insist incorrectly that industrial polluters caused the whole problem, and any laws passed typically concerned only industry, while industrialists pointed their fingers at the smoke from domestic combustion. Such fruitless blame-trading would continue long after the Second World War. Turn-of-the-century smoke fighters, even avid prosecutionists, also showed the typically American presumption that there must be a relatively cheap, easy technological quick fix for any air pollution problem, and that such problems could be fixed without incurring major social or economic costs. When such costs arose, especially during economic downturns, public support for smoke control quickly evaporated. Such a response would also prove typical of the post–World War II years.[38]

American cities outside the industrial heartland also confronted early air pollution problems, though usually not such severe ones up to the 1930s; East Coast cities still mostly used cleaner-burning anthracite. Some communities west of the industrial heartland had a different advantage: natural gas, the cleanest-burning of the fossil fuels. The largest city on the West Coast at the turn of the century, San Francisco, relied almost entirely on natural gas, and its growing rival to the south, Los Angeles, similarly grew up on local gas and fuel oil. Such cities never had used much coal and thus seldom experienced the sort of persistent smoke pall found in the Midwest, although Los Angeles was already showing symptoms of its peculiar atmospheric difficulties before the turn of the century. The same was largely true of cities in the oil belt around Texas and Oklahoma, such as Houston, Dallas, and Oklahoma City, where the stench of oil refinery emissions replaced coal smoke.[39]

Elsewhere in the West, though, especially in the Rocky Mountains, cities gained access to natural gas more slowly and long relied on coal for domestic and industrial purposes. By 1900, Denver already had a noticeable problem with coal smoke. Salt Lake City shared with Denver and other Rocky Mountain towns both a reliance on coal and a heavy concentration of nearby mining and smelting operations, so the Utah capital also suffered from coal smoke. By 1912, women in Salt Lake City had organized like women elsewhere to help fight smoke, and in 1919 the city conducted a joint study of the problem with the U.S. Bureau of Mines, resulting in an early smoke abatement ordinance that same year and in monitoring of local airborne particulates and sootfall by 1926. The original law covered only smoke and fly-ash; not until 1941 would the city gain a more general ordinance controlling all sorts of air pollution. By then, from about 1938 on, Salt Lake City had been connected to natural gas supplies, which helped reduce sootfall levels nearly 75 percent during the next two decades.[40]

Smelters were among the most noxious polluters of the air in early industrial America, and wherever the mining frontier went, the smelters naturally followed: California's great mercury mine at New Almaden, the Comstock Lode in Nevada, the various gold- and silver-mining towns of Colorado, and the great copper mines of Bisbee, Arizona, and Butte, Montana, to name a few examples. Whenever local trees were not used up for shaft timbers in the mines, they were burned to smelt the ore, often leaving mining areas deforested even before operators turned to coal for their fuel. The smelting of mountains of ore frequently extirpated vegetation for miles around, as the smelters threw out tons and tons of sulfur oxides, which killed or injured plants before falling to the ground as acid rain, acidifying the soil so that many plants could not regrow. The smoke from the smelters also carried off tons of other impurities burned out of the metal ore, such as arsenic and cyanide compounds along with various toxic heavy metals, which further poisoned the soil and the atmosphere local citizens breathed. Lumber towns in the Mountain West and the Pacific Northwest also endured heavy smoke from the lumber industry's burning of great mounds of waste sawdust.[41]

Although the South had relatively little industry, what it had was dirty and seriously contaminated the air. Among the very worst examples of smelter pollution were the operations in Tennessee's Great Copper Basin, which caused deforestation and crop damage extending into neighboring Georgia and brought the first major lawsuit over interstate air pollution. The case reached the U.S. Supreme Court in 1905 but ultimately was left mostly unresolved after years of litigation. The South had other air pollution problems as well, which were similarly left to fester. Most of the industrial centers of the South relied on soft coal and were located in mountain or river valleys that collected the smoke, so they developed local smoke nuisances much like those in northern industrial cities. If the response of the northern cities was ineffectual, that of southern cities appears to have been almost nonexistent. A visiting air pollution fighter found serious smoke problems in Chattanooga, Nashville, Knoxville, Memphis, Louisville, Birmingham, and Atlanta by the 1940s, but only Atlanta was preparing to take any real action. If the South lagged, though, it was not by much, for the 1920s and 1930s generally saw relatively little smoke control action at either the state or local level in other American regions.[42]

The federal government at this time basically left atmospheric policy to local governments. Besides a few cooperative technical surveys of particular cities conducted by the Bureau of Mines, and this agency's more ambitious study of lead smelter emissions and resulting impacts on soil, vegetation, livestock, and human health in Selby, California, around 1915, the U.S. Weather Bureau watched for smoke in certain cities and kept some limited records. Later, during the Great Depression, the federal government provided money for the Works Progress Administration and other New Deal programs to help conduct some of the first systematic urban air pollution surveys in various American cities, measuring smoke and sootfall and recording meteorological data. But beyond these fairly rudimentary efforts, the federal government basically had no significant, continuous role in air pollution control until the 1950s, and no major role until after 1963. However, during the prewar period, the federal government did participate in two projects with important ramifications for the future. The first of these was the joint study and arbitration tribunal regarding the Trail Smelter Case; the second was the early study of the health effects of leaded gasoline.[43]

The Trail Smelter Case of 1926 to 1941 was a precedent-setting foray into international environmental law, the first instance of international arbitration and agreement over damaging air pollution that crossed national borders. The case involved emissions from a zinc smelter in Trail, British Columbia, that damaged crops and timber around Northport in Washington State after 1924. In 1927, the Trail affair was put on the agenda of the International Joint Commission, which adjudicates issues and disputes between the United States and Canada. The case was not considered until 1928, and significant activity did not begin until 1933, after repeated goading from Washington's two U.S. senators. Finally, in August, 1935, the two nations signed a Treaty of Arbitration creating a binational tribunal to study the problem, determine damages, and adjudicate just compensation. By 1938, the tribunal concluded that there *had* been damage, and in 1941, after conducting "probably the most thorough study ever made of any area subject to atmospheric pollution by industrial smoke," it found Canada responsible for the smelter's conduct, a new precedent in international and environmental law. Precise damages were never determined, however, and the whole matter was forgotten after the bombing of Pearl Harbor later in 1941.[44]

The other most notable early federal effort, regarding possible health effects from widespread use of leaded gasoline, was ultimately left unresolved as well. In the early 1920s, the General Motors Corporation (GM) started catering to consumerism by producing flashier, more powerful new cars that required higher-octane fuel to perform well. In 1922, GM researchers found that tetraethyl lead (TEL) made an excellent anti-knock gasoline additive for higher-compression, higher-powered engines; in 1924, GM created the Ethyl Corporation to produce and market their new invention. Yet public health researchers warned of possible dangers to humans from atmospheric lead, and the U.S. surgeon general and the federal Public Health Service (PHS) requested proof of TEL's safety. GM ultimately agreed to a joint federal-corporate study in which GM would largely control the research.[45]

In April, 1925, unsatisfied with the results of the earlier study, the surgeon general convened a conference of experts to discuss the toxicity of tetraethyl lead. Despite intense industry propaganda as to safety, enough strong doubts remained at the end of the conference that the Ethyl Corporation voluntarily stopped production and distribution of leaded gasoline. This self-imposed ban was supposed to last until all public health fears were laid to rest by further research conducted by a blue-ribbon panel of scientists and public health experts from government, industry, and academia appointed at the end of the surgeon general's conference. However, Ethyl restarted production after only a seven-month interim study in which the expert panel decided that there were "at present no good grounds for prohibiting the use of ethyl gasoline" provided that the concentration of tetraethyl lead in motor fuel was no more than three cubic centimeters per gallon. Although the expert panel noted lingering uncertainties and urged continued government study of health risks in pursuit of some proper, final conclusions, all subsequent research through the late 1950s was conducted and tightly controlled by the Ethyl Corporation, which unsurprisingly found no basis for health worries even into the 1970s. Under a 1926 voluntary agreement with the PHS, the Ethyl Corporation and the DuPont Chemical Company freely marketed gasoline with up to three cubic centimeters of TEL per gallon until 1958, when they sought to increase the limit to four cubic centimeters per gallon to satisfy a new generation of still higher-powered automobiles. Only in the mid-1960s were TEL's potential dangers rediscovered.[46]

The tetraethyl lead affair set important precedents for future air pollution control efforts and general environmental policy in America. First, it established or reinforced the practice of leaving to corporations themselves the responsibility of assessing the toxicity and riskiness of their potentially dangerous products, a situation that persists even today. Second, corporations would receive the benefit of any doubt about the safety of their products and could go on producing them in the absence of conclusive proof of their harmfulness; those who questioned product safety shouldered the burden of legal and scientific proof. Third, the early tendency toward the politicization of science reflected in the TEL affair—the instrumental harnessing of science toward specific legalistic, economic, and political ends and the conversion of scientific debate into a plaything for lawyers and propagandists—would set another troublesome precedent for the twentieth century.[47]

Both notable federal forays into air pollution control prior to the Second World War also show the continuing development of yet another traditional pattern: the tendency to study air pollution to death while letting business proceed as usual. In both cases, research efforts served to defuse political challenges and to allow powerful economic interests to go about their business. In early cases such as these, the decision to take no final action but only to pursue fuller understanding might have been sincere: scientific knowledge was unquestionably lacking. Yet studies finding only the need for further study became a standard part of the history of air pollution control in America at the local, state, and federal levels. After some point, there is no doubt that many such thorough, detailed studies were used deliberately to prevent or delay control efforts until public agitation died down, while allowing major polluters to show apparent good citizenship. With an endlessly complicated, technical subject like air pollution, in which humans seem eternally doomed to incomplete knowledge, the "further study" tactic to prevent or delay more effective controls is often highly effective.

By the eve of the Second World War, although some scientific understanding had been gained and some improved combustion techniques had been developed, air pollution control in America had advanced relatively little over the efforts of preceding centuries. Coal smoke mostly continued to billow out of domestic chimneys and industrial smokestacks uncontrolled, and even available control techniques often were not used. Industrialists and ordinary citizens were reluctant to put themselves to extra expense or effort to purchase more efficient combustion equipment or higher-grade, costlier fuel; municipal authorities generally lacked the will or political support to compel them to do so; and much of the public was too apathetic to care one way or the other. There was some renewed interest in smoke control in the United States during the 1930s, and urban smoke problems may also have improved somewhat during the Great Depression due to the sharp drop in industrial activity, though most would have agreed that this cure was worse than the disease. Basically, air quality in coal-burning industrial cities in the United States and elsewhere only continued to deteriorate through the end of World War II—with one notable exception.

The exception to the rule was the city of St. Louis, Missouri, traditionally among the very worst of America's smoky industrial cities. By the late 1930s, it had surpassed even Pittsburgh, the Smoky City, in its atmospheric filth. City officials resolved to do something about the terrible problem. In August, 1934, Mayor Bernard F. Dickman made Washington University mechanical engineering professor Raymond R. Tucker his personal secretary with the specific mission of coordinating efforts to "clarify the air." In February, 1935, Tucker led a citizens' advisory commission in recommending that the burning of dirty, sulfurous soft coal from nearby southern Illinois be forbidden except in furnaces and boilers with efficient automatic stoking equipment. By December of 1936, draft ordinances to create a new Division of Smoke Regulation in the city's Department of Public Safety and to regulate the quality of fuel burned in the city were submitted to the city aldermen, while medical reports determined that local smoke was detrimental to health by indirectly contributing to various respiratory diseases, besides killing plants and causing serious economic damage.[48]

Coal interests in nearby southern Illinois were furious at the suggestion that their dirty bituminous coal might be shut out of the St. Louis market, and they threatened retaliation. They even made an early effort at environmental blackmail, taking out full-page advertisements in the major St. Louis newspapers inviting local industries to relocate to southern Illinois, where they would never have to fear any "inhibitive smoke ordinances" with "impractical or prohibitory stipulations." When the city's new smoke control ordinance passed in February, 1937, southern Illinois interests called for a boycott of St. Louis products and sought an injunction against the new law in federal district court as an interference with interstate commerce; the judge rejected this reasoning.[49]

In October, 1937, Tucker was sworn in as the new "Commissioner of Smoke Regulation" and began a long campaign to educate the public about the necessity of certain policies that might cause short-term expense and hardship. He also set about creating a highly trained, professional staff of investigators to enforce the new regulations strictly but fairly and to bring credibility to the whole control program. Over time, Tucker would get from the Board of Aldermen a series of further regulations, limiting the ash, sulfur, and volatile-element content of fuel burned in the city and requiring permits for automatic equipment in which lower-grade coal could be burned. Eventually, by 1940, he even forced the most powerful and recalcitrant local polluter, the Terminal Railroad, to clean up its locomotives' emissions either by using higher-grade fuel or by converting to electric or diesel locomotives. Yet progress on smoke was slow, mandatory automatic stokers or higher grades of fuel were expensive, the costs of cleanup fell harder on the city's poor, and Tucker drew sharp criticism from journalists, citizens, and industrialists alike during the 1930s, while Mayor Dickman lost his job. Nevertheless, by 1940, the city's air was unmistakably clearer, and Tucker then became a public hero who was able to help St. Louis successfully resist industry pressure to abandon its control program during World War II.[50]

The example of Tucker in St. Louis was yet another reminder that a successful air pollution control program needed more than just education and exhortation to voluntary compliance or dreams of painless technological quick fixes. Serious atmospheric cleanup required strict enforcement and determination to persist in the effort despite inevitable costs and complaints. It required changes in the behavior of both citizens and industry. Contrary to the wishes of many Americans, meaningful air pollution control was not just a matter of science and technology for engineers and technicians to solve; it was also a difficult social, political, economic, and legal matter. Some control officials would learn this lesson during the next few decades, but many others would long persist in ignoring it and would repeatedly knuckle under to the forces resisting cleanup.

Other American cities were not so determined or successful at reducing emissions before World War II. In July, 1941, Pittsburgh passed a tough new smoke control ordinance patterned after St. Louis', but the war temporarily derailed this campaign. The Smoky City's highly successful atmospheric cleanup, which drew nationwide attention and praise, was delayed until after the war, first until October 1, 1946, then until October 1, 1947, following a barrage of the same sort of threats and warnings about relocations and competitive disadvantages St. Louis had heard earlier. As with Tucker's crusade, Pittsburgh required use of better grades of fuel or approved automatic stoking equipment and backed up their regulations with enforcement. This approach soon brought visible results. The "Cinderalla City" also benefited from the gradual replacement of soft coal by natural gas, a fuel substitution that would help many cities reduce air pollution after the war.[51]

Various more recently industrialized nations of the world outside of Western Europe and the United States, notably the Soviet Union and its neighbors in Eastern Europe but also Japan, Mexico, Argentina, Canada, and Australia, were developing air pollution problems by the 1940s and 1950s, if not sooner. Typically, such nations took their first tentative steps to study or control emissions only well after the Second World War, by which time some of these problems were already serious and firmly rooted. Throughout the twentieth century, in an unfortunate cycle of eternal recurrence, and despite the negative examples of earlier industrial nations, industrialization in the developing world has almost always been accompanied by the sort of severe air pollution and other gross environmental abuses that England pioneered long before.[52]

In sum, little progress was made on controlling atmospheric contamination between the days of ancient Rome and the Second World War. Control efforts were long hampered by inadequate technology and primitive understanding of the nature of pollutants and the atmosphere. Most of all, though, early air pollution control measures were stalled by economic and social forces that dominated politics and law. Business as usual—the individual cost-controlling and profit-seeking desires of both corporations and ordinary citizens, together with sheer apathy and inertia—usually won out in this recurring cycle, and the common resource of the air suffered ever further exploitation and degradation. All the research and bold promises of quick solutions from professional experts and technicians were not sufficient to handle the problem, especially since these experts usually were closely allied with the very forces resisting change. Further research usually just meant further delay, and the lessons learned from one cycle of interest in air pollution control were largely forgotten by the time the cycle began again, while knowledge and experience long remained more local than universal.

In the face of such a long-standing pattern of repetition, the case of St. Louis in the 1930s offers a rare and striking example of the cycle being broken and at least some of the difficult social and economic aspects of the problem being faced. While St. Louis was only one local success story, it was nevertheless an important example pointing the way to the future. The chapters that follow trace the story of the traditional recurring cycles of air pollution control policy being broken on a wider scale—though in other ways still persisting—up to 1970.

# I

# Los Angeles

# chapter 3

# Trouble in Paradise

*The Discovery of Smog*

During the post–Second World War years, the principal example of traditional cycles of air pollution control being broken, or shifted to a whole new level of governmental attention, is the story of the city of Los Angeles and the state of California. It was in Los Angeles that complex, modern urban air pollution, as opposed to the simpler, more straightforward smoke nuisances of St. Louis and Pittsburgh, was first recognized and confronted. Many of the key pioneering scientific studies and control initiatives that would later shape emissions control efforts throughout the nation and the world were first undertaken in Los Angeles. The myriad different sources contributing to a modern urban air pollution problem were gradually but methodically unveiled, revealing the degree to which air pollution was built into all levels of consumer society. These discoveries would lead Los Angeles into a long crusade to abate all these different emissions sources. The city's campaign against local, stationary sources, although a tough fight, would prove determined and successful enough to make the western metropolis the clear national leader in air pollution science and control techniques. However, the city's less successful, sometimes bitter fight against mobile sources of air pollution—chiefly automobiles—would expose the limits of traditional local control approaches and would ultimately entangle remote authorities and interests such as the state government in Sacramento, the federal government in Washington, D.C., and the major domestic automobile manufacturers in Detroit.

It is deeply ironic that Southern California should have become synonymous with chronic air pollution. During the latter half of the nineteenth century, the region was famous for its healthful atmosphere as well as its gentle climate. According to local boosters, such factors made all Los Angeles a "veritable sanitarium," and sanitariums were built throughout the surrounding area as victims of tuberculosis and other health complaints in cramped, polluted eastern cities fled by the thousands to the fast-growing western metropolis. The city of Pasadena, at the base of the San Gabriel mountain range to the northeast of Los Angeles, drew particular acclaim for its "sapphire skies overarching orange and lemon groves, with a climate all-the-year-round not dreamed of elsewhere . . . where July is as agreeable as December, March as September." Other enthusiasts struggled to come up with superlatives adequate to describe Pasadena's climate: "Glorious! Delicious! Incomparable! Paradisiacal!!" Given these attributes, together with the unending efforts of local boosters and real estate interests to attract new residents, Los Angeles County generally experienced skyrocketing demographic growth from the 1870s through the 1940s and beyond as immigrants from other states flooded into this purported new Eden.[1]

The City of the Angels, however, had a history of local air pollution well before the 1940s, when the so-called smog first began to appear. In 1542 when Spanish explorer Juan Rodriguez Cabrillo discovered the area that would become Los Angeles, he called nearby San Pedro Bay "La Bahia de los Fumos," the "Bay of Smokes," due to the peculiar way smoke from the campfires of indigenous communities collected and hung motionless in the air over their villages. In 1868, when only five thousand people lived in the area, residents for five days experienced an "atmosphere . . . so filled with smoke as to confine the vision within a small circumference," then still an "unusual atmospheric phenomenon" that led to "manifold surmises, conjectures, speculations and rumors" among bewildered local inhabitants.[2]

The region's incipient atmospheric problem only increased as urbanization and industry came to Southern California around the turn of the century. By 1901, Angelenos already spoke of a local "smoke nuisance," while in 1903, an editorial in the *Los Angeles Herald* described an air pollution episode that darkened the sky and was thought to be a solar eclipse until citizens recognized that it was actually clouds of smoke accumulated from their homes, businesses, factories, steamships, and railroad locomotives. As the newspaper editors observed, "It was like meeting a railroad train in a tunnel." Such conditions remained sporadic, however.[3]

This recurrent and sometimes acute air pollution was disconcerting for a city and region that had long justly prided itself on its climate, atmosphere, and scenery, and citizens moved to nip the problem in the bud. Los Angeles' first local smoke control law was passed in 1905, followed by further measures in 1907, 1908, 1911, 1912, and 1930, while noxious odors and oil-burning orchard heaters, or "smudge pots," became targets of regulation in 1930 and 1931, respectively. From 1907 onward, smoke inspectors for the City of Los Angeles investigated causes of public complaints. Neighboring San Bernardino County created a countywide air pollution control district in 1936, more than a decade before Los Angeles County, although this agency remained largely inactive through the 1960s. There was also sporadic civil litigation over air pollution from the 1880s through the 1930s. Efforts such as these had some minor beneficial impact, and Los Angeles also benefited from oil strikes along the Southern Californian coast and the opening of major natural gas fields in California's

Central Valley. During the late nineteenth century oil and gas permanently replaced imported coal as the region's primary energy sources. However, the Los Angeles area's control efforts were failing to keep pace with its tremendous rate of demographic and industrial growth and corresponding increase in aerial contamination even before the 1940s, when the real trouble began.[4]

What Southern Californians would gradually learn, to their dismay, was that their benign, blessed climate became a curse when handled improperly. The prevailing high-pressure atmospheric conditions in the Los Angeles basin, which allow so much cloudless sunshine through so many months of the year and which moderate the wind and generally keep the temperature pleasantly unchanging, also produce a meteorological phenomenon known as temperature inversion, which persists through many months of the year. Temperature inversion is a reversal of normal atmospheric conditions, under which the air becomes progressively cooler with increasing distance from the earth's surface. Under such ordinary conditions, the air mixes freely, and combustion byproducts leaving chimneys, smokestacks, and tailpipes at high temperatures will tend to rise and gradually dissipate amidst the vast sea of cooler gases that comprise the atmosphere. With temperature inversion, however, cooler air—from over the ocean, in the case of Los Angeles—slides underneath a layer of warmer air that was heated near the surface of the regional landscape the previous day, then rose and was trapped by the prevailing high-pressure system. Under such conditions, normal mixing of air does not take place, and hot pollution emissions only rise normally through the cooler layer of air trapped underneath the warm layer of air. When they meet the upper layer of warmer air, pollutants cease to rise and begin to spread out laterally at a low altitude (usually between one thousand and three thousand feet). The warm air effectively acts as a lid atop the Los Angeles basin, confining aerial contaminants within the relatively small pocket of air trapped underneath.

The mountain ranges that ring the Los Angeles area to the east and north and give rise to the term *LA basin* limit the ability of the trapped pollutants to mix and dissipate laterally, while the prevailing gentle westerly winds from the Pacific Ocean prevent pollution from moving westward. Thus, all the combustion byproducts from various human activities tend to accumulate throughout the day, pushed northeastward against the San Gabriel Mountains and other lesser ranges. This produces high concentrations of air pollution throughout the area, but particularly in foothill communities in the San Gabriel Valley such as Pasadena, once known as the place "where the goodness of life is greatest," before the inversion generally breaks up and allows pollutants to dissipate overnight. Among other natural limiting factors that people had defied to create the fantasyland of Southern California—notably the lack of water—Los Angeles turned out not to have sufficient ready air resources for the giant and prodigal industrial metropolis it was becoming.[5]

The phenomenon that would become one of the defining features of Los Angeles and would call into question many of the ebullient earlier claims of local boosters—

photochemical smog—first appeared during the 1940s. By 1939, local meteorologists were becoming aware of a deterioration in visibility throughout the Los Angeles basin that could not be attributed to natural causes. During the summer of 1940, office workers downtown briefly experienced a nearly incapacitating siege of eye-stinging, throat-irritating air pollution showing the characteristics that would later typify the LA smog. There were subsequent brief episodes during the fall of 1940 and the early summer of 1941, but these were spaced and sporadic enough that few citizens saw any developing pattern in them. Angelenos were surprised all over again by further attacks of the mysterious, eye-stinging haze in December of 1941 and at various points during 1942. These incidents coincided with America's entry into World War II and led some local citizens to suspect that they were some sort of Japanese chemical attack.[6]

By 1943, citizens and local authorities recognized the new aerial affliction as a recurring phenomenon and took the first tentative steps to confront it. Public complaints about the problem began flowing in from residents in Los Angeles as well as Pasadena, Burbank, and other nearby communities downwind of the major city. In answer to these complaints, the Los Angeles City Council ordered an investigation of the "peculiar atmospheric condition," while the Los Angeles County Board of Supervisors instructed the County Health Officer to investigate and determine the cause of the "nuisance." Public suspicions generally focused on the new industries that had come to the area throughout the early decades of the twentieth century and that leaped ahead especially during the defense buildup of the late 1930s and early 1940s, such as the various new aircraft factories, steel mills, electric generating stations, and oil refineries dotting the local landscape. In particular, Los Angeles residents initially blamed a foul-smelling, smoky synthetic rubber plant on Aliso Street near downtown, and the federal government's wartime Rubber Reserve Corporation that had built it, for causing the area's new, distinctive air pollution. As was highly typical of early abatement efforts in virtually all American cities that undertook them, both ordinary residents and public officials initially felt there must be just one particular factory or type of industry that was causing the whole problem. However, after the Aliso Street butadiene plant was temporarily shut down and fitted with such control equipment as was then available, the new air pollution problem continued unabated. From this point onward, it would gradually dawn on Los Angeles authorities and citizens alike that their unpleasant new atmospheric problem was far more complex than they had initially supposed, and that there was no single, simple, direct remedy for it.[7]

During the latter years of the war, city and county officials in Los Angeles laid the foundation for an air pollution control program. Their early efforts were frequently fraught with frustration and confusion as they grappled with a dimly understood and scientifically complicated problem that was already coming to be perceived as "the most baffling, the most unprecedented, and the most widely distributed affliction ever to harass the county," as long-time county supervisor John Anson Ford recounted.

On October 13, 1943, in response to complaints and petitions from angry citizens throughout the area, the Los Angeles County Board of Supervisors created the Los Angeles County Smoke and Fumes Commission (SFC) and appointed three local scientists and two members of the general public to study "the general problem of smoke and noxious fumes" in the region. During its brief and troubled existence, like the early smoke control organizations in other cities and counties throughout the United States, the SFC was never adequately funded, and board members felt they were only "window dressing" to quiet the public outcry over smog. The new agency also occasionally suffered from internal dissension, as the nonspecialist members sought to take action on the basis of existing knowledge, while the scientists and engineers on the board, in a manner that would prove highly typical of scientific and technical experts throughout the nation, generally insisted that ever more research was needed before control programs could be established—a view industrial polluters invariably shared. The commission did contribute significantly to the limited knowledge about the area's atmospheric scourge, which by 1944 had come to be known (inaccurately) as smog, as they struggled to draft a countywide air pollution control ordinance that would address the problem seriously but would still be palatable to the various independent jurisdictions within the county. However, official foot-dragging within established county departments, together with the resistance of the more heavily industrial incorporated cities and personal animosity between certain county supervisors and committee members, prevented the SFC from accomplishing much before it was disbanded at the end of January, 1945.[8]

The committee's early efforts were not entirely for naught, however. In November, 1944, the City of Los Angeles adopted a new ordinance creating a Bureau of Air Pollution Control based on the county Smoke and Fumes Commission's earlier suggestions, while in February, 1945, the county board of supervisors unanimously passed a nearly identical proposal to help foster a mutually compatible joint city-county control effort. Both jurisdictions established positions for control officers commanded to investigate "the origin and cause of all known sources of air pollution" and to issue policy recommendations. In February, 1945, the city appointed Major Harry E. Kunkel as program head; in May, the county selected Isador A. Deutch, a twenty-year veteran of Chicago's Department of Smoke Inspection and Abatement, who hoped to help make Los Angeles the "cleanest industrial community in the world." For all their effort and optimism, though, Kunkel and Deutch's respective recommendations for new legislation and agency organization were obstructed through most of 1945 by resistance from local business interests and by official apathy. Meanwhile, the city and county health departments in which air pollution control authority remained ensconced conducted some modest efforts to control sources such as the railroads, diesel trucks, the lumber industry, and the oil refineries through education and gentle persuasion. As usual with such early efforts to gain voluntary cooperation, these efforts accomplished little atmospheric cleanup, with the exception of the diesel truckers, who were at least somewhat receptive to the city's "educational program" and calls for voluntary cooperation, and the railroads, which were then converting nationwide from traditional coal- or oil-burning steam locomotives to newer, cleaner, more efficient diesel engines for mostly economic, not environmental, reasons.[9]

By 1946, when revamped city and county control programs began operation, the Los Angeles smog problem had only grown worse. That summer produced the most severe air pollution attacks yet seen in the area, repeatedly stinging residents' eyes and noses and reducing visibility downtown to less than a mile. The city's affliction even drew media attention throughout the nation for the first time. Meanwhile, Kunkel and Deutch were left limping along with totally inadequate budgets and inspection staffs, while elected officials continued to show little interest in the problem, and many of the region's major polluting industries and the independent, incorporated cities that housed them stubbornly refused to cooperate with Los Angeles city or county authorities. Above all, the sheer quantity of air pollution continued to expand hugely as new residents and industries poured into the Los Angeles area after the war at an even faster rate than before. The problem was entirely out of control.[10]

Sustained public outcry ultimately forced officials to take the matter of air pollution more seriously, and from the summer of 1946 to the summer of 1947, local authorities joined concerned citizens in pushing for the creation of a special countywide air pollution control district with adequate power to control the problem and compel compliance from the myriad autonomous jurisdictions within the sprawling greater Los Angeles metropolitan area. This action was taken only after the observed failure of efforts to gain voluntary cooperation from pollution sources and incorporated cities. Along with angry citizens in foothills communities such as Pasadena, the move toward a sterner regional control effort was supported by the influential local real estate and hospitality businesses, which began to look beyond the exciting short-term prospect of rapid industrial development to wonder about the long-term, economy-dampening impact of the smog that was gradually besmirching the area's name and making it the butt of jokes throughout the rest of the nation. As is common in America, the understanding that major economic interests were at stake, not just the health and comfort of individual citizens, helped give the issue greater attention and legitimacy in political circles.[11]

Local smog fighters were joined by the influential *Los Angeles Times,* which launched a concerted campaign to promote atmospheric cleanup through numerous articles and hard-hitting editorials, and as in other American cities, this press support was a prerequisite for gaining a more serious air pollution control program. The *Times* also arranged for a visit by the most celebrated smoke controller in America of that time, former St. Louis smoke control chief Raymond R. Tucker, who surveyed the local situation and made a full report offering recommendations for further control efforts. While Tucker's report was mostly a recapitulation of the lessons learned from smoke control efforts in eastern and midwestern coal-burning cities, Tucker helped to emphasize key organizational and jurisdictional requirements, and the visitor's national stature, together with the *Times's* unabashed hype, brought even greater public attention to the issue. Meanwhile, other local papers in and around Los Angeles similarly publicized the need for air pollution control.[12]

Of the twenty-three specific recommendations in the Tucker report, probably the most controversial was one calling for "necessary State legislation" to create a countywide air pollution control district for the Los Angeles area. Polluting industries and their allies among the officialdom of the region's many incorporated areas invoked a century-long tradition of "home rule" in California government in arguing against the county gaining control authority over independent jurisdictions. Such entities continued to favor a libertarian, voluntary, cooperative approach doomed to ineffectuality through lack of any higher authority or compulsion. In particular, they sought to have the various incorporated cities represented on the board of the proposed new control district, rather than surrendering regulatory power to the county board of supervisors. However, as most cities in the area were by then suffering from smog, and citizens throughout the area were angrily demanding action, even most independent cities wanted some pollution control agency and were not ready to prevent the creation of one altogether through their opposition to particular details in the county's proposal. Thus, after extensive discussions and minor concessions, the backers of Assembly Bill 1 finally gained the support of most of the cities.[13]

California business interests presented a greater obstacle to passage of A.B. 1 than the incorporated areas. Because it was politically untenable to be forthrightly against air pollution control, given the charged political atmosphere in Southern California at the time, the railroads, lumber companies, and oil refineries formed a coalition to suffocate the proposed bill behind the scenes. For example, some anonymous party, probably representing industrial interests, mounted a systematic campaign to kill the proposal by a mass mailing of deliberately misleading pamphlets to state legislators and the general public. Later, in May, 1947, after pro-control forces had finally gained the cooperation of the state's powerful railroad lobby and the bill had passed the state assembly nearly unanimously, oil and lumber lobbyists plotted to kill the measure while it was reviewed by a committee of the state senate. The oil men especially objected to a proposed provision requiring permits for construction or modification of facilities to ensure their compliance with emissions regulations, an essential part of any control program but an unwanted new infringement on industry's accustomed liberty, while the lumber industry wanted freedom to continue burning mountains of waste sawdust. Pro-control forces threatened the two uncooperative industries with a firestorm of negative publicity for their intransigence, and the *Los Angeles Times* made a down payment on this promise, issuing a ringing denunciation of the industries' "selfish opposition" to a measure promoting the public good. Finally, after difficult last-minute negotiations, first the oil then the lumber industry capitulated, and A.B. 1 passed the state senate unanimously on June 2 and was signed into law by California governor Earl Warren soon thereafter. In October, 1947, after considering hiring Raymond R. Tucker but failing to meet his terms, the Los Angeles County Board of Supervisors appointed Dr. Louis C. McCabe as head of the new Los Angeles County Air Pollution Control District (APCD), and the existing city and county programs were absorbed by the new agency.[14]

McCabe, the well-respected former head of the United States Bureau of Mines' combustible fuels division, then the lead federal agency dealing with air pollution, immediately set about trying to organize the new APCD into an effective regulatory agency. With a budget of $178,000, the Los Angeles APCD was already the best funded air pollution control agency in the nation. McCabe's staff of forty-seven included twenty-two inspectors to ferret out problems and violations and handle public complaints, along with twelve engineers to study local pollution problems and offer technical advice to industry. McCabe, like other engineers and scientists of his day, was a firm believer in the power of science and technology to solve any air pollution problem, and under his leadership, the district embarked on a major program of research into the nature and causes of air pollution, expanding a system of research contracts with local universities started under the earlier, separate city and county control programs. However, the APCD remained mostly oriented toward practical results, not pure science and theory. In the absence of conclusive evidence on the myriad aspects of the endlessly complicated topic of air pollution chemistry, McCabe instituted a policy of seeking to control all known significant air pollutants and sources as much as was technologically feasible and economically reasonable, rather than waiting for the slow development of knowledge about which were the most dangerous and damaging.[15]

While the APCD's initial approach to research and regulation was reasonable enough, given the overall lack of information about a problem demanding immediate action, the district's policy of taking action before all the facts were in led to industry complaints and political difficulties. Industry representatives complained that the APCD had no conclusive proof that *their* industry was the one causing smog and so had no right to bother them until this relationship was established, showing the traditional but fallacious assumption that there must be only one significant cause of a major urban air pollution problem. In particular, the oil industry and its regional trade association, the Western Oil and Gas Association, argued that emission standards should be set only after a comprehensive (or exhaustive) investigation of the nature of local air pollution, not before, and that proposed limits would upset the local economy and throw people out of work while possibly accomplishing little to improve the atmospheric situation. This call for a more carefully focused control program based on complete research and perfect understanding of the problem was superficially logical but disingenuous, since in fact, research and understanding on the question never would be perfect or complete (and still are not), and industries throughout the nation often used calls for further research as a way of delaying and obstructing any actual control efforts.[16]

Meanwhile, state legislators on the Assembly Committee on Air and Water Pollution, among other critics of the APCD's approach, wondered why the district was basing control standards upon what was merely doable rather than on what standards would safeguard the public health. APCD officials responded that the public health basis would be preferable if necessary information as to the health effects of dirty air existed, but it did not, and the county had neither the time nor the money to answer all the medical questions before taking control action. The district's research program, even if small relative to the task before it, was already by far the most ambitious such effort in the world. Yet some citizens worried that the control agency was setting its standards of technological and economic feasibility too low and so sheltering marginal industrial operations from the just consequences of their pollution and general inefficiency. In sum, McCabe's shotgun approach to air pollution control—seeking to make *some* progress promptly on each of a variety of emission sources by using available technology and demanding some level of cleanup from all polluters—may have been the only practical, workable way to start cleaning up the atmosphere in Los Angeles in the late 1940s, given the inadequate state of knowledge about the problem together with public pressure for quick results. Still, the apparent inefficiency of such an approach, relative to theoretical perfection, led to criticism from many quarters, especially as the APCD's early abatement efforts led to little visible reduction in smog.[17]

McCabe never had to bear the brunt of this criticism, however, since he returned to the U.S. Bureau of Mines in July, 1949, when the city's honeymoon with its new control agency was ending and public criticism was beginning to heat up. Rather, his successor, Colonel Gordon P. Larson, formerly with the U.S. Army Corps of Engineers, got stuck leading the agency through five difficult, disharmonious years, including investigations by the Los Angeles City Council and the county grand jury and charges of "inefficiency and chicanery," before he was forced out of office in late 1954.[18]

In the meantime, the APCD made notable progress in various areas. Following the example of other American cities that had taken action to control "smoke nuisances," the APCD focused on large, stationary industrial emission sources as the likely cause of most of the smog problem. By 1949, grey iron foundries had been controlled as far as available technology allowed, and the APCD rightly rejected specious claims from industry representatives that the cost of installing control equipment would price them out of the market. By 1951, open-hearth steel mills were similarly controlled, and oil storage tanks were required for the first time to have floating lids to prevent the evaporation of hydrocarbons. The heavy particulate emissions—black smoke and soot—that typified air pollution problems in coal-burning eastern and midwestern cities were almost totally eliminated in Los Angeles during the early 1950s through strict enforcement of a ban on visible emissions. Through such accomplishments, Los Angeles was already becoming the national and world leader in air pollution control at this time.[19]

Under McCabe, the LA APCD also embarked on an ambitious program to control sulfur dioxide emissions from local oil refineries. In the late 1940s, sulfur dioxide was still the only invisible, gaseous air contaminant drawing much attention from control experts due to earlier damaging or fatal incidents involving the gas, and it was the only one they had much experience handling, so McCabe naturally but wrongly suspected sulfur dioxide of being a major component of Los Angeles' distinctive smog. Over the stubborn resistance of the oil companies, he launched, and Larson and others later continued, an uncompromising campaign to control the refineries' sulfur emissions. This helped control a potential pollution problem almost entirely and reinforced the district's reputation for toughness but fairness. Later, in 1952, after earlier successful efforts to quell the open burning of rubbish in dumps, Larson began the process of banning the filthy, inefficient backyard garbage incinerators that most Angelenos relied on as a cheap means of solid waste disposal. While this proposal caused a firestorm of protest from incinerator manufacturers and segments of the public that eventually helped to depose Larson as director of the APCD, the backyard incinerators ultimately were banned in 1957, permanently ending a noxious and much-protested source of local air pollution.[20]

Although actions such as the control of particulate emissions, sulfur dioxide limits, and the backyard incinerator ban largely eliminated significant sources of damaging aerial contamination, they did not put an end to smog, as officials had blithely suggested they soon might. For this reason, they also led to increased public criticism and industrial opposition, as oil refinery executives and ex–backyard incinerator owners alike could complain that they had been singled out unfairly and put to considerable expense for nothing; the smog clearly was still there. Again, this was a faulty line of reasoning, based on the flawed premise that a complex modern urban air pollution problem had only one basic ingredient that was the only thing in need of control. But this argument was popular nonetheless. Large numbers of ordinary citizens blamed the oil refineries as the single major cause, industrialists in turn pointed their fingers at private citizens, and both groups chastised the APCD. Such ignorant but strident criticism made difficulties for the agency as it sought to discover what other sources of air pollution were most directly responsible for causing the city's trademark affliction. Indeed, it was only the increasing severity of the smog that maintained overall support for the APCD's control efforts from the general public and the county board of supervisors.[21]

The answer regarding the *principal* cause of smog—the automobile—came as a surprise and was long resisted by industrialists and individual citizens alike. Conventional wisdom, and even most expert opinion, long held Americans' favorite consumer durable blameless, though some investigators suggested cars as a culprit by the mid-1940s. Already in 1943 and 1944, visible smoke from heavy diesel- and gasoline-powered trucks drew extensive public complaint and official attention. Later, Isador A. Deutch of the County Office of Air Pollution Control had realized that complaints of eye irritation came mostly from areas of high automobile traffic density, such as downtown Los Angeles, and he suspected that aldehydes and other incompletely burned hydrocarbons in the motor exhaust might be a significant part of the area's smog problem, though he did not know precisely how these substances contributed to the haze or eye irritation characteristic of the LA smog. In 1945, Deutch requested that citizens voluntarily limit driving for one day during the smog season to see whether this affected smog formation, though he chiefly suspected "cars burning excessive amounts of oil," reflecting the traditional focus on visible emissions. In 1946, county smog researchers reported that "under certain conditions," particularly "slow or interrupted traffic movement necessitating periods of deceleration and idling," motor vehicles could emit "relatively large quantities of partially oxidized hydrocarbons" capable of causing eye irritation "if present in sufficient concentration." That same year, the city's Bureau of Air Pollution Control under Harry E. Kunkel analyzed auto fumes at the long, congested Second Street tunnel leading into downtown, where particularly severe eye irritation had been pinpointed.[22]

Yet Deutch and Kunkel were discouraged from actively pursuing this possible angle by such notable scientists or air pollution experts as A. A. Millikan, Nobel prize winner and president of the California Institute of Technology, and Raymond R. Tucker himself. In 1947, Deutch, still harboring suspicions, met with the engineers and directors of the General Motors Corporation's research divisions for chemistry and industrial hygiene in Detroit and recommended research into new engine design and other means of reducing automotive emissions. The GM staff responded that "they had never considered the automobile as capable of producing irritating gases in objectionable amounts." In his famed report on the Los Angeles air pollution situation, Raymond R. Tucker declared "Automobiles Absolved from Most Blame," while Louis McCabe, after taking control of the new APCD in 1947, mostly ended Deutch's early studies of the automobile's contribution to the problem. In 1949, McCabe further declared that while a mere layman standing next to a bus might feel certain that motor vehicles were the cause of all pollution in the area, and while old and poorly maintained automobiles should not be allowed to pollute the atmosphere, "neither should folklore be encouraged that will place the onus of metropolitan area atmospheric pollution on the automobile, without proof." The experts had spoken, and the automobile was off the hook—for the time being.[23]

Yet suspicions lingered regarding the contribution from cars. In November, 1949, at a football game at the University of California at Berkeley, some three hundred miles to the north of Los Angeles, thousands of spectators suffered "intense eye irritation." Although meteorologists found that the San Francisco Bay area had been experiencing a Los Angeles–type temperature inversion during that whole week, people only experienced the eye-stinging characteristic of the LA smog on the day of the football game, when the crowded parking lot turned into a massive traffic jam producing unusually concentrated vehicular exhaust.[24]

Closer to home, truck farmers in the area around Los Angeles from Long Beach to the San Fernando Valley had begun experiencing inexplicable major damage to leafy vegetable crops such as spinach and lettuce from 1944 onward, to such an extent that growers suffered nearly half a million dollars in crop damage in 1949 alone;

many farmers and commercial flower growers were forced to relocate or go out of business. Researchers such as Dr. John T. Middleton of the University of California agricultural research station in Riverside and plant pathologist Dr. Frits Went of the California Institute of Technology (Cal Tech) began to study this problem. While scientists' experiments were able to confirm that the damage was caused by air pollution, they could not determine what exactly was producing the damage, though some suspected sulfur dioxide or airborne fluorides.[25]

The mystery was solved during the late 1940s by a Dutch-born biochemist at Cal Tech, Dr. Arie Haagen-Smit. From 1948 to 1949, he used equipment and techniques from totally unrelated previous experiments on the complex chemistry of flavors to determine that Los Angeles' air contained organic peroxides that could only have come from petroleum but that were not released in the form in which they were found by any known pollution source. This discovery, together with the observed correlation of eye irritation with sunlight and growing suspicions that smog was a byproduct of chemical reactions among other, primary pollutants, led to the photochemical theory of smog formation that later gained acceptance. Working with his Cal Tech colleague Went on an APCD research contract, Haagen-Smit later combined ozone with gasoline vapor to produce a cloud of pollution that looked and smelled like the LA smog and also caused the same characteristic leaf damage to experimental plants. He then combined olefins, volatile chemicals from gasoline, with nitrogen dioxide, a combustion byproduct of high-temperature, high-compression engines such as those in America's postwar automobiles, and exposed the gaseous mixture to ultraviolet light, imitating the effect of sunlight. This experiment produced ozone, which damaged Dr. Went's experimental plants as expected. Later research found another byproduct of the reaction of unburned gasoline hydrocarbons with nitrogen dioxide in the presence of sunlight, peroxyacetyl nitrate, or PAN, to be a major cause of the typical lachrymation (eye-stinging, tear-producing) response caused by the LA smog. At last, some of the most distinctive and unpleasant aspects of Los Angeles' air pollution apparently had been figured out.[26]

Though Haagen-Smit's research conclusions in November, 1950, had convinced the leadership of the Los Angeles APCD by the following year to focus on controlling such sources of unburned hydrocarbons as motor vehicles, refineries, and garbage incinerators, the scientist had a harder time gaining wider acceptance for his photochemical hypothesis. Haagen-Smit's early articles discussing his findings were rejected by journal referees and other experts, while the Western Oil and Gas Association funded research by the Stanford Research Institute suggesting that home heating, rubbish burning, and countless other sources, not the oil industry and its products, were responsible for the LA smog. But APCD director Larson unswervingly persisted in his efforts to control refinery emissions, now including evaporative losses of hydrocarbons as well as sulfur oxides, and by the fall of 1954, such emissions had been cut to such a degree that they could no longer be a significant source of smog precursors in the Los Angeles basin. Meanwhile, an independent organization of

Southern Californian businessmen, civic leaders, and local governmental officials had formed the Southern California Air Pollution Foundation in November, 1953, to study and find solutions for the local atmospheric problem. Their research, done under contract by the Midwest Research Institute of Kansas City, Missouri, eventually confirmed Haagen-Smit's findings, agreed with the APCD that refinery evaporative losses and garbage incineration were contributing factors, but concluded that automobiles were the major source of the problem. Ultimately, following a demonstration by Haagen-Smit, even scientists at the Stanford Research Institute were able to duplicate his results. By about 1955, scientific understanding had come to accept Haagen-Smit's photochemical hypothesis for smog formation; by the late fifties the control of refineries together with the banning of backyard incinerators in 1957 left no major, uncontrolled source of the unburned hydrocarbons producing photochemical smog save one—the automobile.[27]

Even once the relevant science was well established, it was still no easy matter politically to affix primary blame for the nastiest aspects of the local air pollution problem on automobiles. Southern Californians, even more than other Americans, loved their cars and were already highly dependent upon them, Los Angeles being the first of the great, sprawling sunbelt suburb-cities to be built around the personal automobile. One wit described the city as "a group of suburbs in search of a center." Some suspicious citizens, not eager to own up to their own culpability for the local atmospheric affliction, suspected Haagen-Smit and Larson of helping "the interests" cover up their role by blaming the "little man's automobile and incinerator"—at the same time that the oil industry was attacking the APCD's control program from the opposite angle.[28]

Most of all, the automobile industry long refused to accept or admit that they were a major part of the problem. Already in 1951, Larson was seeking to impress upon the auto makers their responsibility, a claim the industry rejected. In 1952, the APCD director began sending the agency's research reports on automotive pollution to the major auto makers, and after a year of pleading, he got their promise to study the alleged problem. Also in 1953, Los Angeles County Supervisor Kenneth F. Hahn, who would become the county's leading politician crusading against smog, began sending letters to all the major auto makers, asking them to take action to control their products' emissions. A Ford Motor Company representative insisted that auto exhausts "dissipated in the atmosphere quickly and do not present an air-pollution problem," while John M. Campbell, director of General Motors' research laboratories division, was typical in arguing: "As far as we are aware, Los Angeles is the only community having this particular complaint [eye-stinging smog] . . . indicating that perhaps some other factors than automobile exhaust gases may be contributing to this problem." In personal discussions with Larson, Campbell pointed to backyard incinerators as the cause of the problem. Late in 1953, Larson went to Detroit to try to impress representatives of the auto makers with the seriousness of the matter and to invite them to visit Los Angeles to see for themselves; he called on them to begin promptly working harder to solve the problem, and he warned them that early in 1954, the APCD would start aggressively publicizing the automobile's contribution to Los Angeles' atmospheric woes, since the matter could not wait any longer. Within weeks of Larson's visit, the Automobile Manufacturers Association, the principal trade organization of American auto makers, created a new Vehicle Combustion Products Subcommittee within their Engineering Advisory Committee, mostly to try to debunk claims that the car was a significant contributing factor. Later, in January, 1954, Campbell and other technical experts from Detroit came to review the Los Angeles situation and again publicly questioned their vehicles' role in causing smog, though Campbell privately admitted misgivings about his company's products. It would be years before the Big Four auto makers of the day (then including the American Motor Corporation) would publicly own up to the automobile's contribution to the LA smog, and even then, through the early 1960s, they insisted this was something that happened only in Southern California.[29]

While the battle over automotive air pollution was brewing, the APCD had other troubles. By 1954, Gordon Larson, who was generally conceded to be honest, conscientious, and technically proficient but also plodding, politically insensitive, and overly understanding of industrial polluters in the eyes of some, had lost the confidence of the Los Angeles County Board of Supervisors. Following an investigation into why the APCD had not made more progress in controlling local air pollution in late 1953, the supervisors in early 1954 moved to demote Larson from the position of general director of the APCD, keeping him as technical director in recognition of the significant scientific research and engineering progress made during his watch. One supervisor, Herbert C. Legg, even went so far as to accuse Larson and some of his subordinates of maladministration and unprofessional conduct, although extensive further inquiry completely exonerated Larson of all charges. By late 1954, though, as a serious smog siege gripped the region and local citizens angrily asked why the problem was not fixed already, various county supervisors and the new mayor of Los Angeles, Norris Poulson, who had been elected largely on the air pollution issue, were again seeking to find a scapegoat to take the blame for the continuing smog, and they again turned to Larson. Larson was demoted to director of the APCD's newly created Research and Engineering Division, and he left the county's service soon thereafter.[30]

Following the turbulent Larson years came an era of relatively good feelings between the APCD, the public, the supervisors, and to some extent even local industry. Unlike Larson, the APCD itself had always maintained the support of the county board of supervisors, and it had weathered and won out against legal challenges to its various regulatory provisions. Whatever their shortcomings as politicians or diplomats, McCabe and Larson had gotten the fledgling agency established as a permanent fixture on the political and legal landscape of Southern California. Their extensive research efforts enabled them to justify the actions they took with significant scientific evidence, and their "firm but fair" approach to demanding cleanup from all sources they uncovered, large and small, industrial and domestic alike, ultimately won the overall respect of Angelenos. Moreover, the APCD's conscientious and largely successful efforts to control one source after another—first steel mills and foundries, then oil refineries, then backyard incinerators, then other more minor sources, all of them contributing to the overall air pollution problem in the Los Angeles basin—gradually weeded out potential suspects as to what was still causing the persistent photochemical smog that was the most salient and hated feature of the region's troubled atmosphere. While earlier suspicions as to the "principal cause" of smog may have proven unfounded, the effective control of such emissions was nevertheless a significant and beneficial achievement and helped clarify subsequent policy and scientific understanding about air pollution—as well as removing from the atmosphere millions of pounds a year of harmful pollutants—in a gradual process of discovery that policy analysts James E. Krier and Edmund Ursin have called "exfoliation." The efficiency of the agency's regulatory work never approached the level of theoretical perfection, but the APCD, through its combination of relatively extensive research, strict enforcement, and overall public support for smog abatement, did mount a practical, effective control effort that brought results, particularly in comparison to the relatively feeble attempts of most other cities.[31]

The administrative changes of 1954–55 that drove away Gordon Larson also brought in new directors with better political and public relations skills, if less technical expertise, than the scientists and engineers who had previously run the district. Captain Louis J. Fuller, a thirty-year veteran of the Los Angeles Police Department, was made chief enforcement officer for the APCD, while S. Smith Griswold, who had worked for twenty years in county administration besides serving in the military during World War II and the Korean War, became overall chief control officer. Griswold and Fuller also received a major increase in funding and manpower, with the budget growing to nearly two and a half million dollars—more than the spending of all other American jurisdictions on air pollution control put together—for the 1955–56 fiscal year, while staff rose from 165 authorized positions in 1954 to 280 the next year and 467 the year after that. The new management team also vastly increased public relations efforts to keep people informed of their activities and achievements, something the more taciturn, technocratic Larson had not emphasized.[32]

Above all, the success of Griswold and Fuller in policing polluters in the Los Angeles basin revealed a lesson that other jurisdictions in the nation sometimes stumbled upon but rarely learned: that beyond a certain point, air pollution control was primarily a matter of public policy, politics, and law, not merely a science and engineering project. Fuller, in his enforcement efforts, was not excessively hamstrung by scientific niceties; he merely enforced the law that was given to him and was less likely than Larson or other scientists and engineers in other American cities to be excessively patient and understanding of the real technical difficulties industrial polluters sometimes faced. Local polluters simply learned that they could not count on continually gaining the forbearance of the regulatory agency, as was so typical of other cities' control agencies and for which Larson had been faulted, so they instead learned to comply. This brought genuine results, and made Los Angeles the clear world leader in air pollution control for more than a decade. Griswold and Fuller of course benefited from the solid scientific foundation for control efforts that McCabe and Larson had begun to lay down, but they also maintained and extended a research program that the rest of the world borrowed from and imitated. All in all, Griswold and Fuller regained public confidence in the APCD for more than a dozen years.[33]

As they had promised would happen, visible APCD enforcement efforts went up markedly under Fuller and Griswold. In 1953, Larson reported to the county board of supervisors regarding 27,958 enforcement actions taken over the previous five years of the agency's existence. The tally of cease and desist orders served was 14,190, with 4,868 of these resulting in hearings at the APCD office, while 7,320 equipment operation permits had been granted and 498 denied. The APCD Hearing Board had considered 747 variances for additional time to install acceptable equipment. Relatively few persistent violators were taken to court, however: Larson noted only 309 criminal actions and 26 civil actions over five years. While this was already a relatively impressive record of enforcement by national standards, Fuller and Griswold stiffened enforcement and made much freer use of the threat of court proceedings. In mid-1955, the district boasted of its team of sixty-eight inspectors on the prowl twenty-four hours a day, seven days a week, and noted how in the first six months of 1955 they had issued more than eleven hundred court citations to persistent violators—more than in the previous seven years of the APCD's operation, and with a better than 90 percent conviction rate. Inspectors received special training at the agency "Smoke School" to be able to recognize violations of the visible emissions limits without special equipment, but they were also armed with early Polaroid cameras, which produced instant evidence to be used in court proceedings. District pamphlets offered photographs of black-and-white APCD squad cars, equipped with two-way radios, racing off to apprehend violators and of grim-faced, uniformed inspectors writing up citations—like something out of the popular police drama *Dragnet* then on television.[34]

However, there was more to the revamped APCD than just strong-arm tactics. Variances were still available in cases of economic hardship, though not on as liberal terms as critics alleged during Larson's time. Recognizing that "the war against smog cannot succeed without voluntary cooperation from industry and the public," the district continued and expanded Larson's policy of trying to help industry find solutions to its pollution problems while also offering much more advice and information to the public through the new Public Information and Education Division. Realizing the need to plan for the future in an area changing and expanding as rapidly as Southern California, the district also launched a new program "to control potential future sources of pollution" by restricting "new industrial growth to those operations which can be controlled effectively through air pollution control devices and equipment" and by the "development of zoning based on the studies of pollution concentration and wind movement." In a city, state, and nation in which industries had so often been allowed to build and operate without many limits, such notions of zoning and other environmental restrictions on new industrial development were relatively innovative.[35]

In 1955, impelled by the terrible smog sieges of the early 1950s, Los Angeles County also created the world's first air pollution alert system. County Supervisor Kenneth F. Hahn had recommended this in December, 1954, but County Counsel Harold W. Kennedy reminded him that such a system could not be put in place before there were established standards for air pollution danger levels corresponding to the varying phases of any emergency alert system. Under "Regulation 7—Emergencies" of June 22, 1955, the APCD ultimately set 100 parts per million (ppm) of carbon monoxide, 3 ppm of nitrous oxides or sulfur oxides, or 0.5 ppm of ozone as the levels, any one of which would trigger a first-stage smog alert, including warnings broadcast to the public and a request for citizens and industries to curtail their atmospheric emissions voluntarily. With a second-stage smog alert, at levels of 200 ppm carbon monoxide, 5 ppm nitrous oxides or sulfur oxides, or 1.0 ppm ozone, all incineration was to stop, and all emission sources operating under a variance were similarly supposed to cease. A third-stage smog alert, at levels of 300/10/10/1.5 ppm, was a truly major and potentially life-threatening emergency under which all industrial and automotive pollution was banned. To monitor these levels, the district established fifteen continuous air sampling stations around the region, noting that the program marked "the first time anywhere that the lower atmosphere has been investigated on such a broad scale, and continuous surveillance of pollutant levels maintained." LA's basic smog alert system was later adopted by other major cities, such as San Francisco and New York City.[36]

Through the late 1950s and 1960s, Los Angeles air pollution control authorities continued their epic battle against local, stationary emissions sources, often using innovative new approaches that set an example for the rest of the nation. In May of 1956, the United States Supreme Court dismissed an appeal by the Union Oil Company of California, which for years had been claiming that the APCD's first rule prohibiting dense visible emissions was unconstitutional because it was "so vague, uncertain, indefinite and unintelligible that it failed to ascribe an ascertainable standard of guilt"—a legal tactic tried in many other cities by local industries seeking to derail air quality legislation. The Supreme Court found no substantial federal question involved, and the APCD won another round against the local oil industry. As noted, garbage incinerators were banned in 1957, and Angelenos grumblingly grew accustomed to paying for the wastes from their relatively opulent consumer lifestyle to be hauled away to distant landfills. Later, during the 1960s, Los Angeles County led the way in studying and seeking to regulate aircraft emissions.[37]

More strikingly, in 1957, the district began a major new effort to control industrial polluters, leading to the passage by county supervisors of what became known as "Rule 62" on November 12, 1958. This regulation required all steam power plants in the region to burn natural gas rather than cheaper, more sulfurous and polluting heavy fuel oil between May 1 and October 1, when smog was worst and when local residents were demanding less of the higher-grade fuel for domestic heating purposes. Local industrial representatives charged that this requirement to use costlier natural gas would be an economic disaster for them; they also argued that sulfur dioxide from burning fuel oil was not a significant contributor to photochemical smog. The industrial interests, led by the Western Oil and Gas Association, sought to establish an alternative policy whereby the industries would shift to use of natural gas only when serious smog was forecast and only on a voluntary basis—a position characteristic of industries throughout the nation during the early postwar decades. In response, the Los Angeles County counsel argued that smog and weather forecasts had proven hopelessly inaccurate as a basis for regulating emissions and that it was unfair for major industries to control their pollution only on a temporary, voluntary basis when local citizens were required to do so year-round, as with the ban on garbage incineration. The district prevailed with the support of the county board of supervisors, and in March, 1961, it extended the ban on high-sulfur fuel to seven months rather than six. The county and the APCD also ultimately won in litigation challenging the constitutionality of Rule 62, though this dragged on until 1966, and the oil interests continued to threaten an appeal until 1968. Once more, the APCD had stubbornly persisted in its drive to control stationary sources, although it was long hampered by the Federal Power Commission, the agency regulating interstate pipelines, which was reluctant to allot more natural gas to Southern California for nontraditional environmental purposes that the federal agency found difficult to comprehend.[38]

The APCD took yet another innovative and controversial step toward controlling air pollution from stationary sources in 1966 with the promulgation and approval of Rule 66 to regulate emissions from industrial solvents, paint, and any other "architectural coatings," including ordinary house paint. This regulation, which the APCD called "perhaps the most sophisticated and technically complex air pollution regulation that has ever been attempted," was expected to eliminate some two hundred tons a day of the existing six hundred tons a day of mostly hydrocarbon emissions from paint and solvents in the Los Angeles area, or nearly 15 percent of the remaining cumulative daily emissions of stationary sources under the district's jurisdiction. The rule limited most users of industrial solvents to emissions of only forty pounds a day and banned certain more volatile solvents altogether, targeting the worst third of total emissions from this overall source. Rule 66, a novel approach to controlling industrial and domestic emissions and quite different from more traditional regulations relating to obvious major "point sources" such as blast furnaces or smokestacks, was drafted after months of consultation with twenty major industrial groups and trade organizations to seek cooperation and to prevent charges of arbitrariness.[39]

Nevertheless, two years later, a group of major chemical companies and the Chemical Industry Council of Southern California sought to gain an exemption for one cheap, widely used, and highly volatile solvent banned under Rule 66, trichloroethylene (TCE). When asked why they had accepted the regulation without complaint earlier, an industry representative privately observed that at the time, opposition was considered "impolitic," but two years later, the industry felt it was safe to attempt to challenge and undercut the new rule. Louis J. Fuller, overall director of the APCD after S. Smith Griswold left to join the federal air pollution control program in 1965, blasted the four companies for not keeping faith and for playing "the well-known game called 'Let's Have Another Study.'" He continued angrily, "This kind of stalling action may have been successful back East where the factories of the solvent makers are located . . . but it never has been successful in Los Angeles County." The county board of supervisors was similarly unreceptive to the chemical industry's effort to bend the rules, and as usual, backed up the APCD and roundly rebuffed the industrial interests.[40]

As a result of this sort of attitude on the part of the Los Angeles County Air Pollution Control District and the county board of supervisors, Los Angeles County accomplished much in terms of air pollution control from the 1940s to the 1960s. Through most of this period, Los Angeles led not only all the other cities and states of the nation but even the federal government and the other nations of the world in its research and enforcement efforts to fight dirty air. By 1966, the APCD had created more than one hundred rules and regulations covering all sorts of stationary sources, and it was estimated that the county control program prevented from entering the atmosphere 5,350 tons a day of harmful pollutants that might have done so had there been no control effort. They had also prosecuted nearly thirty thousand violators of the air pollution regulations over a ten-year period, at a time when most cities were still groping for "the mirage of 'voluntary compliance.'"[41]

As a result of its clear national leadership in air pollution science and enforcement techniques, Los Angeles came to have an influence far beyond the spread of its own smog pall. By the later 1950s and through the 1960s, control officials and scientists from Los Angeles and the state of California increasingly went forth almost like missionaries—unhappy polluters might have said like imperialists—to spread the message of real air pollution control to more benighted areas of the nation, just as Raymond R. Tucker of St. Louis had earlier sought to do regarding smoke abatement. For example, after he lost hope of seeing created in California a powerful, comprehensive state environmental authority, which he had hoped to lead, S. Smith Griswold retired from the Los Angeles Air Pollution Control District in late 1965 to go to Washington, D.C., to serve as a top abatement official in the federal air pollution control program. John T. Middleton, plant pathologist at the University of California's air pollution research center at Riverside, later became head of the entire federal control effort from 1966 to 1970. Benjamin Linsky, the first director of the San Francisco Bay Area Air Pollution Control District—the first multicounty control authority in the nation, started in 1956—later served as the top control official in West Virginia and was a fixture on the national air pollution lecture circuit. These are only a few salient examples of a wider pattern present among lower-ranking air pollution officials and scientists as well. Since much of the early federal research effort was devoted to funding grants and experiments out in California, many federal control officials also got their practical experience dealing with the problems of Los Angeles and California. With or without the actual transfer of personnel, Los Angeles, as the most famous air pollution battleground, also taught by example those other cities or states ready to hear the lesson. Arthur J. Benline, top air pollution control official in New York City during the early 1960s, candidly acknowledged this leadership when he announced in 1962 that his agency would begin studying volatile emissions from paint spraying based on research done in Los Angeles: "We copy them. . . . Their budget is five times ours. They investigate—we check."[42]

Yet for all of Los Angeles' monumental exertions and achievements to control local stationary emissions sources, by far the greatest and most pervasive source of pollution in the county was not stationary and proved to be frustratingly beyond the reach of local authority. The 5,350 tons a day of pollution that county officials were preventing were just over a quarter of the total daily emissions that would have been there. Of the other seven-tenths of the total, or an estimated 14,225 tons of air pollution a day still causing damage to structures, materials, and vegetation and potential injury to human health, up to 80 or 90 per cent came from motor vehicles. Los Angeles officials, unable to solve the problem all alone, would find themselves hammering on the doors of Sacramento, Washington, D.C., and Detroit, seeking help on the issue.[43] Progress was anything but swift.

# chapter 4

# Smog Town vs. the Motor City

*Taming the Automobile*

By the late 1950s, Los Angeles air pollution control officials had some reason to crow about their successes. True, the western metropolis still had smog regularly, but not as severely as before. Real progress had been made in cleaning up various stationary sources of aerial emissions, from oil refineries to steel mills to backyard garbage incinerators, in what was coming to be widely recognized as the most ambitious and successful atmospheric cleanup effort the world had yet seen. LA Air Pollution Control District officials proudly pointed this out to readers of local newspapers in the Los Angeles metropolitan region in November, 1957: how over the preceding ten years, industry, government, and ordinary citizens had shelled out an estimated $70 million on the anti-smog campaign, not counting the $48 million worth of garbage incinerators that had been outlawed or the additional $20 million a year Angelenos would have to pay for garbage pickup and disposal. Of this figure of $70 million, nearly $50 million had been paid by local industry to purchase new control equipment, while another $12.5 million went for research by public and private agencies outside the APCD; the district alone had spent nearly $10 million from 1948 through the end of 1957.

This monumental effort—and this astronomical budget—was preventing an estimated fifteen hundred to two thousand tons of dangerous air pollutants from entering the city's air every day from stationary sources that had been controlled almost to the limit of technical ability. Robert L. Chass, the district's director of engineering, testified that local conditions would have been "infinitely worse" without these controls, even "so high as to render the atmosphere dangerous to human health and safety." With some cooperation from local meteorological conditions, the fifteen first-stage smog alerts during the smog season of 1955 had been reduced to ten in 1956 and only one in 1957. The APCD credited itself and all the other industries, agencies, jurisdictions, and ordinary citizens of Los Angeles with pulling the situation back from the brink of disaster. As the district's information officers reminded readers, Angelenos were embarked on an adventure of historic significance:

> Los Angeles County has the world's biggest smog headache, and certainly its most notorious. It also has the world's most complete air pollution control program, and its most effective. The deservedly acclaimed efforts made in cities such as Pittsburgh and St. Louis are a bare beginning compared to what has been done here. . . . Ours is the toughest air pollution problem in the world, and we are blazing the trail in learning to control it. The fruits of a victory will be not only a more enjoyable future for us, but a valuable contribution to world science as well.[1]

However, the title of the district's self-congratulatory declaration of partial victory identified the remaining fly in the ointment: "APCD Completes Job—All except Auto Exhaust." With an almost audible bureaucratic sigh of relief, the agency placed nearly all remaining blame on the auto manufacturers, observing: "Now, however, we may have reached the practical limit of what we can do for ourselves— the hope for future improvement rests squarely with the automobile industry in Detroit. Until a cure is found for the auto exhaust we can look for little reduction of our present smog level." While the district noted that it was helping in the development of auto exhaust control devices through scientific research and laboratory testing of any new developments, "achieving the actual solution" was being left "properly in the hands of private enterprise," including the major auto makers and other manufacturers or research laboratories.[2]

The remaining uncontrolled source was itself a huge and growing problem. Throughout the postwar period, Southern California was one of the fastest growing regions in the nation in terms of population, and this meant a rapid increase in the number of cars, particularly in California. In 1954, the Los Angeles basin, the 1,629-square-mile portion of Los Angeles and Orange counties most afflicted with air pollution, had a population of just over five million—more than all but eight of the fifty states of the union, and as many as the total population of the entire Rocky Mountain West. Just between 1950 and 1954, the LA basin had grown by more than 750,000 inhabitants, or around 18 percent, more than the total population of Pittsburgh; and this followed the decade of the 1940s when Los Angeles County had greater total population growth than any entire state save California itself.[3]

The number of automobiles and trucks increased even faster, nearly tripling from an estimated 878,000 in 1930 to 2,361,000 by 1954, the greatest concentration of motor vehicles in the world, half again more than in the five crowded boroughs of New York City and more than were owned in any foreign nation save the United Kingdom, Canada, and France. Southern California's car-crazy culture correspondingly brought the lowest per capita rate of public transportation usage of any American city, along with the lowest number of persons per vehicle, 2.8, as against New York City's 7.1. The number of persons per motor vehicle in Southern California dropped steadily toward two during the 1960s, while the total number of cars rose with the booming population. Meanwhile, gasoline consumption exploded, rising from an estimated 1.8 million gallons a day in 1936 to nearly 4.8 million gallons a day in 1954, a level that was expected to increase to more than 6 million gallons a day by 1960. This tremendous expansion in demand for motor fuel was happening at a time when automobile engines blew out at least 7 to 10 percent of their gasoline intake as wholly unburned hydrocarbons. In other words, in 1954, Angelenos' autos and other vehicles were dumping more than a quarter of a million gallons of gasoline directly into the atmosphere every day, a vast flood of smog precursors. Without major changes in vehicle technology or public transportation systems, the steadily proliferating chrome monsters of the LA freeways threatened to undo whatever good the APCD had done and to push the city back toward the brink of atmospheric disaster.[4]

As noted in the previous chapter, the discovery of the major role of automobiles in producing air pollution was a slow and sometimes painful process, due to the unjustified but firmly rooted presumption of the innocence of motor vehicles in that regard. Gradually, though, with Arie Haagen-Smit's successful solving of the photochemical puzzle and the growing acceptance of the demonstrated link between motor vehicles and smog, even the automobile industry began to accept some responsibility for the atmospheric affliction of Los Angeles, if not yet for any other place on the globe. County Supervisor Hahn's letter-writing campaign and APCD director Larson's personal visit to Detroit led the Automobile Manufacturers Association to establish its new Vehicle Combustion Products Subcommittee in late 1953 and to send ten of its top engineers out to Los Angeles to observe and study local auto emissions in early 1954, followed by a "one-million-dollar-a-year industry research program" to investigate the problem. While this paled next to the industry's outlay for a handful of top salaries or the gratuitous cosmetic changes in body design every year, it did represent some commitment.[5]

However "dead serious" the auto industry may have claimed to be, there were those who felt a little pressure could not hurt. In 1953, County Supervisor Hahn asked County Counsel Harold W. Kennedy for a legal opinion regarding the county's power to require exhaust control devices on all vehicles sold in the county. Kennedy responded that such devices could be made mandatory as soon as such a device was available and adequate, but that prior to then, any requirement for such a device would be "arbitrary, capricious and void." While the county supervisors could justify prohibiting the use of motor vehicles altogether around the city center if true emergency conditions existed, the city and county did not face such conditions yet. The exhaust control requirement would have to wait until technology advanced, and the best Los Angeles County could do was to try to figure out a way to spur such technical progress.[6]

So began the pursuit of emissions control devices for Los Angeles automobiles, an undertaking that proved more costly and complicated than most participants could have imagined at the outset. The County of Los Angeles faced many obstacles in its drive for the vehicular emissions controls that it needed. Beyond the lack of available technology, there was the additional problem that the obvious likely source of such technological development, the auto industry, had no real incentive to spend too much time or effort on the problem, since the unhappy county initially had nothing beyond moral suasion and empty threats with which to compel them to do so. The state of California represented slightly more than 10 percent of the total auto market in the United States, then still entirely dominated by American manufacturers, and the Los Angeles area accounted for at least half of that; yet with the sprawling city wholeheartedly committed to a transportation system based on the personal automobile, Angelenos were left hoping that the auto industry would introduce necessary technological improvements out of public-spiritedness. At the very time when the region's air pollution was mounting to unacceptable levels, local citizens were cheerfully going along with the automobile industry and allied interests in tearing up and discarding the metropolitan area's old light-rail system of public mass transit and replacing it with ever more cars, roads, and freeways. Angelenos depended on the automobile more than Detroit depended on Angelenos, and the auto industry knew that car-crazy Californians were hardly going to give up their favorite consumer durables, regardless of whether the industry promptly came out with workable control devices.

Thus, however much the city might complain, it was in Detroit's interest to have only a modest, relatively plodding effort in the direction of vehicular exhaust controls. Although by 1954 the auto industry already accepted that automobiles were the largest single source of hydrocarbons in the LA atmosphere and were major contributors to smog, throughout the rest of the 1950s they constantly found need for further "confirmatory work" to ascertain precisely the nature of the automotive contribution and whether they were in fact the primary sources of smog—yet another example of industrialists' favored trick of delay through endless research. In 1954, the auto industry began an informal agreement to pool research on exhaust control devices, formalizing this as a cross-licensing arrangement the following year. While this was explained—and celebrated by journalists—as a way of speeding progress by sharing the fruits of the best engineering minds of the industry, critics would later allege that this agreement served to delay technical advances by ensuring that no individual manufacturer could race ahead of the others. One way or another, the pace of progress was painfully slow.[7]

Facing a united front of powerful corporations in remote jurisdictions, with a stranglehold on most automotive engineering expertise and a limited desire to address the problem of vehicular emissions, Los Angeles County had few alternatives. It could appeal to the conscience of the industry and hope for some results; it could appeal to higher authority in hopes that this could force action as the county could not; or it could try to break the industry's control over the relevant technology. Each of these methods would be tried, with varying degrees of success, over the next decade and beyond.

Although moral suasion applied to mighty international corporations might seem naive to readers in the post-1970 period, the early Cold War years were a time of great pride—almost religious faith—in American free enterprise and its economic institutions as well as in science and technology, and in the wake of the Second World War and the atom bomb and other scientific marvels, many Americans generally tended to assume that the great corporations could and would readily harness technology to fix any human problem. The efforts of Los Angeles County officials to convince the major auto makers to produce exhaust control devices through the late 1950s reflect this assumption. The tone of Kenneth Hahn's early letters to the auto makers is not so much combative as expectant—wondering when they would do what was presumably easy for them to do. When further years elapsed with no visible progress, county officials began to push a little harder. For instance, in late 1957, S. Smith Griswold, head of the APCD, suggested that if the auto makers themselves were not in the business of manufacturing catalysts for purifying vehicular exhausts, then they should offer a prize of a million dollars to any outsider who could develop a suitable catalyst. Only with the greatest reluctance would Griswold finally agree under questioning that he felt the industry was dragging its feet. At this point, LA control officials still hoped to get the industry to help voluntarily. It was only later, after further years of industry research producing few tangible results, that frustration set in, and the tone of the Angelenos became more blunt and confrontational.[8]

Meanwhile, scientists and engineers in Los Angeles, Detroit, and elsewhere gradually expanded their understanding of the technical aspects of the auto exhaust problem. There was much room for improvement. Through the 1950s, for instance, automotive engineers actually believed that new V-8 engines would offer not only more complete combustion but better fuel economy as well, and it was not until 1960 that researchers determined conclusively that both pollution and fuel efficiency were in large measure functions of vehicle weight and engine size, such that six-cylinder autos were fundamentally less polluting and more fuel-efficient than eight-cylinder cars. Also, researchers initially thought that the majority of the tailpipe emissions resulted when cars slowed down, so they guessed that "control could be achieved by a device that would cut off the flow of gasoline to the carburetor during deceleration." Although they came up with a vacuum control device to prevent such flooding of the carburetor, this was not the answer.[9]

By the late 1950s, the auto makers had settled upon four possible approaches for controlling exhaust emissions: a low-temperature chemical catalytic converter, a high-temperature catalytic converter, a direct flame afterburner, or a set of careful engine adjustments. Both sorts of catalytic converter would be chambers in which engine exhaust would be passed over chemical catalysts that would help react byproducts of incomplete combustion, such as unburned hydrocarbons and carbon monoxide, and would transform them into less harmful compounds. The lower-temperature catalytic converter was more desirable, since it would not need special, high-grade, heat-resistant materials to house the catalysts and would pose less risk of injury to anyone who might brush against it on a recently parked car; but effective low-temperature catalysts were more difficult to find, while carbon monoxide was resistant to chemical conversion except at high temperatures. A direct flame afterburner offered the highest temperatures and the simplest concept but was potentially the most dangerous design for people who might accidentally touch the intensely hot device soon after a car was parked; it also would require very heat-resistant materials, which raised the cost. From early on, auto makers most favored the idea of engine modifications and careful tune-ups to gain optimal combustion from a traditional engine, for this involved the least change in basic design and promised to be cheapest for auto manufacturers. It also meshed with their idea that automotive emissions would not be much of a problem if only owners would maintain their vehicles properly—a not entirely disingenuous effort to pass on responsibility for vehicular pollution to consumers, and a matter on which California, a state with no regular vehicle inspection program and with common sloppy maintenance habits, made an easy target.[10]

But it had taken the auto industry years even to come up with these approaches, let alone any workable devices. As it became increasingly evident that gentle persuasion alone would not be enough to prod the auto makers toward more serious efforts to solve LA's vehicular emissions problem, county officials increasingly sought help from higher levels of government that might be able to apply more pressure on Detroit than they could. Complaints from Los Angeles and other places around the nation helped bring the first federal legislation on the air pollution issue in 1955, but the early federal program was so modest in scope and so exclusively preoccupied with research, not results, that for years it had little impact on the LA smog situation. This being so, activist Angelenos went to the next best place: the California State Capitol in Sacramento.[11]

The state government had gotten pulled into Los Angeles' pollution politics back in 1947 with the passage of Assembly Bill 1 authorizing creation of the Los Angeles County Air Pollution Control District. Yet this was mostly a formality to give the state's stamp of approval to local actions and involved no further commitment on the part of the State of California. Policy remained entirely local, although the California state legislature established committees to review the matter. The smog siege and resulting public uproar of October, 1954, again briefly injected the LA smog into state politics: Governor Goodwin J. Knight felt compelled to cease campaigning, briefly, to hold emergency hearings into the air pollution situation in Los Angeles. Earlier, in 1953, the Los Angeles City Council had noted in frustration that "smog does not respect city boundaries," that "we can't put a fence around Los Angeles and expect to keep other counties' smog out of here," and that therefore, "the state should bear all or a major portion of the cost of cleaning up the atmosphere." Other cries for a multicounty, regional control approach had also been heard—but largely ignored—by the state legislature. At that stage, Governor Knight waffled on the issue of state responsibility before ultimately declaring air pollution to be an "essentially local" matter. After the smog siege of 1954, however, Knight ordered "an all-out smog war" making full use of "all the state's technical resources," including the research laboratories of the various University of California campuses, in pursuit of a "practical and inexpensive" automotive exhaust control device. Knight also ordered a special emergency grant of a hundred thousand dollars for the California Department of Public Health to combat the health risks from dirty air. The resulting report from this agency still affirmed that air pollution remained primarily a local responsibility, but one with which the state could and should help through research and technical assistance. In accordance with this new acceptance of partial responsibility, under a state law of 1955, the state Department of Health created a new Bureau of Air Sanitation to investigate air pollution and possible countermeasures. In this way, an atmosphere of crisis at a critical point in the political cycle helped elevate LA's problem to the level of a state concern.[12]

During the late 1950s, further bills to expand state authority and responsibility for abating air pollution failed to pass, but pressure for increased state intervention continued to grow. In 1958, APCD officials formally requested state legislation and cooperation to help control automotive exhaust, not just in LA County but throughout the state. On the last day of 1958, the Los Angeles City Council even passed a resolution requesting that the state legislature enact a law prohibiting the sale of automobiles in California after January 1, 1961, unless they were fitted with smog prevention equipment, while Los Angeles mayor Norris Poulson, who had been elected largely on the smog issue back in 1953, called for "vigilante committees" to be formed to put pressure on the auto makers and public officials to do something about vehicular pollution. Additional state laws proposed in 1959 would have expanded California's role in the testing, certifying, and mandating of vehicular emissions control devices, but these were rejected, partly on the specious but then still common assumption that Los Angeles' photochemical smog was a strictly local phenomenon that would never trouble any other place in the world. However, Senate Bill 117, a law directing the state Department of Public Health to study and set recommended statewide air quality standards, did pass that year.[13]

The major leap forward in California's participation in air pollution abatement came in 1960 with the passage of Assembly Bill 17, the California Motor Vehicle Pollution Control Act. This law established within the state Department of Health a Motor Vehicle Pollution Control Board (MVPCB) composed of representatives from four state agencies—the Departments of Public Health, Motor Vehicles, and Agriculture and the California Highway Patrol—along with nine members from the general public appointed by the governor to represent the "interests of various affected groups throughout the state . . . to the fullest extent possible." This board was to set criteria for vehicular exhaust control devices based on what they determined to be reasonable costs of purchase and installation and adequate durability and reliability. The board was also responsible to test such devices when available and to certify any that met its cost and performance criteria and the emissions standards earlier set by the Department of Public Health under the 1959 law: 275 parts per million of hydrocarbons and 1.5 percent carbon monoxide by volume in the exhaust plume. As the law was written, exhaust control devices would become mandatory only after the MVPCB had certified at least two, to prevent legislating a monopoly. New vehicles could no longer be registered without a certified control device one year after the board's certification of the second device for new cars, while used cars, treated as a separate category, could not be registered in the name of a new owner without certified controls one year after two certified devices for used cars existed. It would be unlawful to drive vehicles that were unregistered or unequipped with a mandated device. However, in a concession to rural California counties, where people viewed air pollution as someone else's problem and resented having to pay for the Angelenos' affliction, counties were allowed to opt out of the requirement for used vehicles. The provision for new cars was statewide, however, and rightly so; as anti-pollution advocates argued, Los Angeles County had no way to prevent cars from rural counties from entering LA County and spewing filth there uncontrolled—though this of course did not help as regards out-of-state tourists' cars. Although the law contained many of the regulatory provisions promoted by Los Angeles County officials in earlier, unsuccessful bills of the late 1950s, A.B. 17 failed to include any provision for mandatory vehicle inspections, an item stubbornly resisted by rural counties and local and regional automobile clubs. Nevertheless, it was a monumental step, throwing the weight of the State of California behind mandatory auto emissions controls—as soon as means became available. APCD officials happily predicted that smog would be rare in three to four years, gone in five or six.[14]

Yet there still were no certified exhaust control devices. Although in October, 1959, APCD director Griswold enthusiastically announced the development of a simple direct flame afterburner with only one moving part, costing less than $125 installed and 80 to 90 percent efficient in controlling hydrocarbons, the auto industry was less enthusiastic. Around the same time, the General Motors Corporation announced plans to study auto exhaust to try to disprove links to cancer or other diseases, asking, "Why spend millions of dollars to develop equipment to control invisible exhaust fumes if they are not injurious to health?" Earlier in the year, the major American auto makers had emphasized lingering scientific uncertainty, calling for more research rather than control action. Ford was working on a low-temperature catalytic converter with about a 60 to 70 percent rate of efficiency against hydrocarbons, while General Motors' hot catalytic converter claimed up to 90 percent efficiency in removing unburned gasoline from the exhaust stream, and the Chrysler corporation claimed good results with an experimental flame afterburner, but none of these devices was anywhere near the marketing stage in terms of cost and durability. Further reports of progress only sporadically broke the overall silence. At the end of 1959, the Los Angeles County Board of Supervisors unanimously adopted a resolution recommending January 1, 1961, as the deadline for smog control devices to be required in California and laying out some cost and performance standards for the contemplated devices: cost of less than a hundred dollars, efficiency of at least 80 percent in removing major pollutants, and a life span of at least 25,000 miles. But no such device existed, nor would one be ready by early 1961.[15]

Yet the 1960 state law served another purpose: it mandated a major market for whoever—inside or outside the auto industry—could first offer exhaust control devices that met the state board's qualifications. As such, California had declared that Los Angeles County and the other metropolitan areas of the state would no longer be entirely at the mercy of the auto industry to set the pace for developing control equipment. Going far beyond S. Smith Griswold's earlier idea of a million-dollar prize for the first workable catalytic converter, the state was offering a potentially much greater reward: a share of the profits from supplying mandated control equipment to one of the largest regional automobile markets in the world. While critics had earlier complained that no outside industries were going to go to the expense of research and development for exhaust control products that might never be required, the 1960 law made the economic incentive a seeming certainty.

The Motor Vehicle Pollution Control Act stimulated major new interest and excitement about the development of emissions control technology. Previously, when the state had passed its 1959 law requiring the promulgation of health-protective air quality standards, some state officials recognized that they were whistling in the dark with their dramatic gesture. The presumption then was that whatever standards the state might set, they would simply have to wait until the major auto makers could satisfy them, for as one official remarked, "We're not about to abolish automobiles in California." After the 1960 law, suddenly new alternatives seemed available as companies outside the auto industry eagerly entered the race for an answer to automotive pollution. In August, 1960, the Business Section of the *New York Times* was being only partially hyperbolic when it trumpeted: "There's a new 'gold' rush under way in California. . . . The eventual strike could run into the hundreds of millions of dollars." The article offered statistics to make inventors and manufacturers salivate: the largest number of automobiles of any state, seven and a half million of them, with more than 85 percent of those—all new cars and any used cars in metropolitan counties—needing smog equipment. Analysts pointed out that the California market alone could mean as much as $700 million in sales, but that this "would be only the beginning of a burgeoning new anti-smog-device industry" as other cities and states followed California's lead. The article justified its sensationalism by offering a list of nearly one hundred different companies working on the problem.[16]

Along with the smaller competitors were some substantial concerns outside the auto industry. Already in May of 1960, Union Carbide Corporation claimed to have developed a catalytic smog-control system capable of eliminating eight to nine tenths of all hydrocarbon and carbon monoxide emissions. In October, the Universal Oil Products Company announced that it would introduce a catalytic muffler called the

"Purzaust" system, and the following February, Universal Oil was the first contender to apply for the testing of its product. In May, 1961, W. R. Grace and Company announced that they, too, were "in the race" to develop a workable chemical catalytic converter and had tested theirs for 25,000 miles already; J. Peter Grace, company president, informed stockholders that their product, if approved, would be "a very substantial business." The following year, still other devices were developed by independent manufacturers outside the auto industry, including one from Minnesota Mining and Manufacturing and another from Oxy-Catalyst, Incorporated, of Wayne, Pennsylvania, where Gordon P. Larson went after leaving the APCD. The California MVPCB was expected to certify at least two before the end of 1961. Ultimately, none of the seven exhaust control systems received for testing by late 1962 was accepted due to high cost or inadequate life span, though improved versions of some of these would be among those certified in 1964.[17]

Meanwhile, another issue had arisen regarding automotive emissions not from the tailpipe. In December of 1959, American auto makers came up with a different, cheaper way to reduce total automotive emissions by recycling engine crankcase gases. These were gases that escaped from the cylinders during compression and that had traditionally been vented directly to the outside air without passing through the exhaust pipe. However, from the 1920s onward, systems to recirculate these escaped gases into the cylinders existed. Crankcase or "blowby" gases accounted for up to a quarter of the total hydrocarbon emissions of a traditional, uncontrolled motor vehicle, and they could be handled through existing, well-understood, and relatively inexpensive technology. Perhaps as a way of showing progress and easing the pressure for actual exhaust controls, the five largest U.S. auto makers offered to include control devices known as "positive crankcase ventilators" or "blowby recirculating systems" on all 1961 model year cars sold in California.[18]

This new attention to blowby gases led the MVPCB to study crankcase emissions. In December of 1960, they released new emissions standards, and the following September, they certified the first blowby control device for new cars. A second such device was certified in April of 1962, which would have triggered the state law and made the devices mandatory. However, prior to that point, in December, 1961, the auto industry, recognizing that the simple crankcase devices would soon be mandatory in California and were being demanded elsewhere, announced the voluntary installation of such equipment on all 1963 model cars sold throughout the nation. The legal requirement still applied to 1964 models from foreign manufacturers, however, and California then as now was the nation's largest consumer of foreign cars, so in late 1963, foreign makers were still rushing to get their crankcase systems approved. It was more difficult to fit blowby recyclers on used cars, but in December of 1962, the State of California certified two of these, meaning that any used cars changing ownership after January 1, 1964, could not be registered without an approved crankcase device. The state later went even further and began a program to require crankcase devices on used cars whether they changed hands or not.[19]

The American auto makers' decision to install crankcase blowby control devices voluntarily on all 1963 model cars was not a matter of sheer generosity, for they were beginning to face pressure from a new direction—the federal government. During the late 1950s, Paul F. Schenck, a Republican congressman from Ohio and a former high school shop teacher, became concerned about the danger of air pollution from automobiles. In 1957 he proposed a bill that would have prohibited the use or sale in interstate commerce of any motor vehicle that discharged hydrocarbons in a quantity found dangerous to human health. This first federal bill regarding automotive emissions died in committee in 1958, but in 1959, Schenck proposed a similar measure that was watered down to become a law to promote federal research into the problem of auto exhaust, and this passed the House of Representatives in 1959 and the Senate the following year.[20]

Meanwhile, the auto industry was also feeling increasing heat from the executive branch of the federal government. At the end of the 1950s during the Eisenhower Administration, Secretary of Health, Education, and Welfare Arthur S. Flemming showed interest in and concern about air pollution in general and sought a more active federal role than that allowed under the weak, tentative first federal air pollution control act of 1955—even including federal enforcement powers. While the rest of the federal government was mostly against such ideas, in late 1958 the Public Health Service (PHS) within the Department of Health, Education, and Welfare (HEW) began studies of automobile exhaust, reportedly studies "of a type and on a scale never before attempted," at the Robert A. Taft Sanitary Engineering Center in Cincinnati, Ohio. Flemming also publicly urged the auto industry to "intensify its activities" in that direction. Later, after the industry announced its offer to put crankcase controls on all 1961 model autos sold in California, Flemming asked why they should not be used everywhere in the nation, expressed concern over worsening levels of vehicular pollution, and intimated that the federal government would be forced to take further steps if "indifference persisted in the curbing of poisonous gases." Although auto industry representatives insisted that air pollution conditions varied widely all through the country, such that what worked in California's climate might not work elsewhere, Abraham Ribicoff, HEW secretary in the Kennedy Administration and Flemming's replacement, was also concerned about air pollution and was unimpressed with the industry's argument. In August of 1961, he effectively gave the auto makers an ultimatum: install crankcase ventilators on all 1964 model cars or face federal compulsion. Although the industry still protested that there was "no evidence to indicate that it would serve a useful purpose for communities" outside California, Ribicoff dismissed this idea, declaring that vehicular air pollution from unburned gasoline was "a problem all over the United States, not just the cities, but in the suburbs and in the farm areas." Although federal pressure on the auto industry in the early 1960s remained muted, it would continue to grow steadily during the decade.[21]

Besides the lesser matter of crankcase gas controls, during the early to mid-1960s, the issue of exhaust controls heated up into a major controversy. As of March, 1962, General Motors engineers were able to report the development of a "functionally and structurally satisfactory" catalytic converter, but one that still could not measure up to the California MVPCB's new, relaxed minimum life span of 12,000 miles. Other GM engineers were still working on other approaches, such as a proposal to pipe air to a point near the hot engine exhaust ports to help oxidize the engine's emissions more completely. Chrysler engineers were developing an elaborate system of minor adjustments to the carburetor, distributor, and spark timing system to try to control the emission of pollutants, especially during deceleration. All these various approaches reportedly were still years away from introduction as standard features.[22]

Yet officials in California were growing impatient after waiting years for a solution. In particular, critics of state policy in Los Angeles County began clamoring for the MVPCB to do more, faster, on the problem. One major complaint had to do with the way in which the MVPCB ascertained whether devices submitted for testing satisfied its 1961 performance criteria. Originally, when the board specified that devices must be good for 12,000 miles, this meant that the emissions of test cars must not exceed hydrocarbon or carbon monoxide limits at any point during those 12,000 miles. These standards actually were not applied strictly to every test vehicle: rather, recognizing varying emissions performance among the various cars of a test fleet, sometimes even of cars of the same make and model, the board's laboratory staff instead averaged the overall portion of the vehicle test fleet using a particular device and then declared that a device had failed to meet the board's criteria if the *average* emissions performance of all these vehicles rose above the limits at any time during the 12,000-mile test period. Those seeking more rapid action to install vehicular exhaust control devices urged that this system of averaging be extended still further, such that control equipment could be approved if periods of control performance better than the specified levels (usually earlier in the device's and vehicle's life) balanced periods of control performance failing to meet the specified standards, such that average performance of the test vehicles using a device remained at the desired level of control for at least 12,000 miles. In January of 1964, the new, more lenient averaging policy was officially adopted by the California MVPCB, both as a way to silence critics of the board's slow pace of progress and to give the auto industry a push in its efforts to develop control equipment by making it easier for independent manufacturers to get certification of devices that the auto makers would then have to adopt to continue selling automobiles in California. This important policy shift, together with an earlier legislative pronouncement allowing the board to consider systemic changes or engine adjustments as well as separately installed "tack-on" devices such as catalytic converters as legitimate smog control devices, significantly widened the range of possibility for early action to control automotive emissions.[23]

Finally, action came swiftly. In March, 1964, the Big Four auto makers pledged that they would comply with California's standards, but only on 1967 model cars, claiming they could not do it any faster. However, in June of that year, the California MVPCB went ahead and certified four anti-smog devices from independent manufacturers—three catalytic converters and one flame afterburner—which had been rejected under the old guidelines but satisfied the new ones. This meant that such devices would be required on any cars sold in California after June, 1965—the 1966 model year. The auto makers were dismayed at this, for they had been accustomed to effective control of the pace of policy developments and were horrified at the prospect of losing a good part of the exhaust control market to outsiders or being forced to purchase competitors' products. The auto makers bleated about the economics of their situation: they claimed an industry price elasticity of 1.0, meaning that a certain percentage increase in price would cause an equal percentage decline in sales. On this basis, the industry during the 1960s stubbornly resisted adding pollution control or passenger safety devices, though they spent billions of dollars on cosmetic changes year after year.[24]

After so many years of stalling, however, historians James E. Krier and Edmund Ursin note that "the auto companies proved miraculously equal to the new challenge." In August, 1964, reversing their position stated the previous March, the auto makers announced that they had developed engine modification systems of their own, offering performance superior to that of the four approved tack-on devices, and that these would be ready for introduction on 1966 models. The auto makers promptly sought and received certification for their emissions control approaches, and none of the certified devices from independent manufacturers ever found their way onto new vehicles from Detroit. The controls that the Big Three suddenly pulled out of their hats like magicians' rabbits were the very same systems they had announced back in 1962: Chrysler's set of engine, carburetor, and timing adjustments, called the "Clean Air Package," and from General Motors and the Ford Motor Company an inexpensive air injection system to burn off impurities leaving the engine—a system that critics alleged had been known and available for use in the 1950s. The simple air injection system cost only fifty dollars, less than any of the independents' contraptions; Chrysler's Clean Air Package cost only twenty-five dollars initially but required careful, regular maintenance. The industry had gotten its way and had limited the dislocation to its price-demand curve.[25]

The best the independent pioneers of automotive exhaust control devices could hope for was a share of the market for control devices for used cars, a market that ultimately never developed as it was supposed to have. In December, 1964, the MVPCB considered an exhaust control device for used cars that would have cost $150 to install and $50 a year to maintain. Wary of a negative public reaction such as was then brewing over a requirement for crankcase devices on used cars, the board members requested public input on whether they would be willing to pay such costs. The public, including even the Los Angeles County Board of Supervisors, offered a thunderous rejection, revealing once again the frequent fickleness of the public in wanting smog control but not wanting to have to pay for it. In 1965, the issue re-emerged due to an apparently inadvertent change in the phrasing of state law that seemed to eliminate the earlier two-device certification system as well as limiting the acceptable cost of future control devices to no more than sixty-five dollars. One of the four exhaust control devices from independent manufacturers certified in 1964 for new cars had also been found acceptable for used cars at that time, and the following year, in the wake of the change in state law, the manufacturer announced that it could meet the new cost limit and that since a second certified device was no longer required, the provision for mandatory control of used cars' emissions should go into effect. However, the board, wary of generating more controversy and public ill will by forcing a single device on motorists, went against the opinion of its legislative counsel and declared that the old two-device policy remained in effect.[26]

This decision, together with the auto makers' successful sleight of hand and other instances of bumbling or inconsistent policy on the part of the state board, caused most independent manufacturers to lose interest in the California vehicular emissions control market altogether. The independent concerns already claimed to have lost many millions of dollars in research and development costs working on new-car controls; suddenly the state agency seemed to be making the used-car market inaccessible and uncertain as well. Harold Lipchik, vice-president of the Chromalloy American Corporation, the company that had developed the first used-car exhaust control device to win certification, later noted that in 1964, his company curtailed its research program into exhaust controls after detecting "a complete change in attitude towards us on the part of the MVPCB staff, which by coincidence occurred right after the automobile manufacturers agreed to equip new cars with control devices." K. T. Norris of Norris-Thermador Corporation, producer of another certified device, testified before the California Senate's Highways and Transportation Committee that it would take an investment of nearly six million dollars just to get his company in position to supply used-car attachments, and he declared: "We lack confidence that provisions requiring application of exhaust control devices on used cars will ever become operative." He continued bitterly, "Years ago, some of my Detroit friends argued that our politicians would never require the average motorist to make a substantial investment to reduce smog. I was naive enough to argue with them." Noting a record of backtracking, changed requirements, and broken promises on the part of the MVPCB even with regard to crankcase ventilators, he concluded by observing that although his company had sold nearly four hundred thousand crankcase devices, they were still in the red for several hundred thousand dollars for research and development costs on this project alone. Although a new bill concerning this issue was passed in 1966, it raised required performance standards to a life span of at least 50,000 miles and annual maintenance costs of no more than fifteen dollars a year. Independent manufacturers were not impressed by this bait, and progress on used car exhaust controls was stalled for nearly five years.[27]

By 1966, the days of the California Motor Vehicle Pollution Control Board were numbered. In addition to the six years of slow and often limited progress on controlling vehicular emissions, and the various policy stumbles and public relations flaps that had cost the agency public good will, the state agency got into a feud with the Los Angeles County Air Pollution Control District. This animosity took root fairly soon after the creation of the MVPCB, back in the early 1960s, when the APCD tried to prevent the state board from certifying the first crankcase ventilator because the district had information showing that the device did not meet specified standards. The APCD proved to be right in its suspicions; the device was defective and was taken off the market a year after its introduction. Further friction developed between the two agencies as Los Angeles County officials pushed for more and faster action on vehicular exhausts, while critics felt the state board "seemed more bent on cooperation with and conciliation of the manufacturers." In December, 1963, at the urging of the APCD, the Los Angeles County Board of Supervisors voted to purchase only Chryslers for county fleet vehicles, since district scientists and engineers had determined that Chrysler's Clean Air Package adjustments were the only system then available from major auto makers that truly reduced emissions. The county's action represented an effort to put pressure on the other manufacturers and to drive a wedge into the auto makers' common front, which seemed to be obstructing and delaying progress on exhaust controls. A few months later, by contrast, the MVPCB was urging the state Department of Public Health not to proceed with proposed reductions in acceptable emissions levels, arguing that this would discourage the auto industry in its efforts to meet existing standards. When state public health officials did go ahead in mid-1964 with their plans to set hydrocarbon emissions standards at only 130 ppm and carbon monoxide at 1 percent by volume, the MVPCB did not put these new standards into effect until August, 1966, to apply only to 1967 model cars and beyond. Although the board's action could arguably have been to protect and encourage independent device manufacturers as well as the auto makers, this action, together with many others, furthered the public impression that the MVPCB was just a faithful ally of the motor industry and led to increased sniping from smog-ridden Los Angeles County.[28]

This simmering debate between the two major air pollution control authorities in California exploded into total war at the end of 1966. In November that year, APCD director Louis J. Fuller branded the state board's exhaust emissions control policy a flat-out failure. He offered data indicating that of the 1966 model cars from Detroit fitted with state-certified new exhaust control devices and systems, 37 percent failed to meet the older state standards of 275 ppm hydrocarbon/1.5 percent carbon monoxide before having gone 2,000 miles, 63 percent were already exceeding standards after 2,000 miles, and by 20,000 miles, 85 percent of the "controlled" autos failed to meet the emissions standards. Although it was expected under the post-1964 expanded averaging policy that many cars would underperform toward the end of the stipulated life span for a control device, the fact that so many were failing to meet the standards when they were still almost new indicated that there was little hope that the entire 1966 model year overall could hope to meet state emissions standards even under the generous averaging policy. This revelation led to a whole new vitriolic debate about the legitimacy of the state board's averaging policies as well as to bitter exchanges between state and county officials. In February, 1967, Fuller blasted the MVPCB's existing averaging policy as a "clear evasion" of the law needing immediate replacement. MVPCB officials responded that the present policy was the only workable way of removing some automotive pollution from the air and that it was accomplishing a great deal, although current devices admittedly were not perfect. Board members complained that the APCD used the MVPCB as a political scapegoat, heaping all blame for air pollution on the state board and the auto industry to prevent having to address a stationary source problem that was again beginning to mount in Los Angeles County; the board challenged the vehement Angelenos for first angrily demanding the early establishment of an admittedly imperfect control policy to get some sort of control program under way, then "screaming" when this imperfect program brought imperfect results. The state board also tried to shift some of the criticism to Detroit, warning the auto makers that 1968 model vehicles would not be sold in California if the products continued failing to meet standards and deteriorating in performance with increasing mileage. The auto makers were not alarmed by this, responding that they were confident that they could meet the standards and that the retooling for the 1968 models was already finished; they also probably recognized an empty threat when they heard one.[29]

However justifiable some of their positions may have been, the averaging controversy of early 1967 was one public discrediting too many for the troubled California MVPCB. To their credit, they had made pioneering efforts to explore, comprehend, and cope with vehicular emissions, and under their policies, they had already brought a degree of actual emissions control—at least on new cars, if not used ones—never before attempted anywhere else in the world, though still not enough to stay ahead of the growing vehicular emissions problem in Los Angeles County. As to the board's shortcomings, in addition to hesitance and outright mistakes in policy, the agency suffered from having been conceived under one set of assumptions but still being around after these assumptions changed. When the MVPCB was created in 1960, most of the state of California and virtually all of the rest of the nation shared the assumption that the purpose of such an agency would be to work closely with the polluting industry in a cooperative, not confrontational, mode. It was presumed that the auto industry could and would readily do its civic duty and clean up its products once its research scientists and engineers, with help from government and outside private researchers, came up with a quick, easy, and painless solution to the problem. Los Angeles County officials had already had enough experience dealing with industry to have a dawning awareness that such a vision of ready cooperation and voluntary action from the private sector was often fanciful, but most other Americans had not. So the MVPCB was created to be a deliberately moderate agency, with the majority of its members having no expertise in automotive air pollution and representing a variety of sectors, many of which had little interest in emissions control, such as the automobile clubs or the trucking industry. Most board members were competent and conscientious and gained considerable general knowledge of the issues they confronted, but on specifics of automotive engineering, they were forced to rely on their more expert staff, who in turn relied on close interaction with the auto industry.[30]

As such, the board actively sought industry cooperation, talked of teamwork and a division of labor with the industry, let auto company laboratories do its testing, and occasionally enjoyed the free technical advice and assistance of industry engineers who worked for the board while receiving their salaries from Detroit. Prior to 1964, when they adopted a more confrontational stance to assuage public complaints, board members repeatedly commended the auto makers for their "complete cooperation"; thereafter, the board made only a few lukewarm efforts to challenge the industry with which they had long been so cozy. In 1965, MVPCB executive officer and technical expert Donald Jensen left his $19,000-a-year position in California for a $50,000-a-year job as an engineer at Ford Motor Company, while a chief engineer on the MVPCB staff took two trips to Europe at foreign auto makers' expense. Such close, even revolving-door relations between industry and government often were taken for granted during the 1950s and early 1960s as necessary for progress, but these conditions began to raise eyebrows by the mid-1960s as painfully slow progress and perceptions of industry resistance increasingly led citizens of California and the nation generally to start demanding a more confrontational relationship between regulatory agencies and industry. With regard to vehicular emissions, by the mid-1960s, many Californians and other Americans had come to see the major auto makers as the problem, not the solution, and the MVPCB's close relationship to the now suspect industry brought suspicion on itself as well.[31]

Consequently, in August 1967, state legislators passed the Mulford-Carroll Act, S.B. 490, abolishing the old MVPCB and replacing it with a new California Air Resources Board (ARB). Unlike the old agency, strictly limited to vehicular emissions, the new board would have comprehensive authority over stationary as well as mobile sources throughout the state. Its fourteen members included the heads of the state Departments of Public Health, Motor Vehicles, Agriculture, Conservation (a new department, reflecting evolving priorities during the 1960s), and the California Highway Patrol as well as nine other appointees with "demonstrated interest and proven ability in the field of air pollution control." The latter qualification represented a move away from the lay board of the MVPCB toward greater technical expertise, although the requirement was phrased vaguely enough to allow the inclusion of citizen smog fighters without technical training. Like the better-known federal Air Quality Act of 1967, the Mulford-Carroll Act gave the ARB the mission of adopting a more regional approach to air pollution control, establishing air quality basins and setting ambient air quality standards for these, but otherwise leaving control of stationary sources in the hands of local authorities as much as possible and threatening to intervene directly only in the event of local nonfeasance—a partial extension of state authority over localities. Also, in recognition of some of the practices that had brought trouble to the MVPCB, the new law gave the new agency greater resources to conduct independent research and coordinate its operations. Critics of the measure, including Louis Fuller and, ironically, California Governor Ronald Reagan, had sought a more powerful, comprehensive agency, but special interests helped to pare the agency down to one much more modest in scope, though better than the old MVPCB. After Fuller rejected the position, the new ARB chairmanship was given to another distinguished senior statesman of air pollution control from Los Angeles County, Dr. Arie Haagen-Smit.[32]

At the end of its brief but troubled existence, the MVPCB did get to share in one final hurrah for California air pollution control policy. This grew out of another process that was transforming American environmental policy during the 1960s: the gradual but growing intervention of the federal government in environmental affairs, particularly motor vehicle emissions.

Since passage of the Schenck Act establishing a minor federal role in automotive air pollution research, the matter of vehicular emissions and general air pollution had been considered periodically in congressional hearings before the House Commerce Committee's Subcommittee on Traffic Safety, later renamed the House Subcommittee on Health and Safety, chaired by Alabama Democrat Kenneth A. Roberts. Through the late 1950s and early 1960s, Roberts, a moderate-conservative southern Democrat who generally disfavored federal intrusion into state and local affairs, regularly found no compelling reason to expand the modest, limited federal air pollution research program begun in 1955. In particular, Roberts rejected the notion of federal enforcement powers—an aversion shared by the various industrial interests that regularly spoke against an increased federal role. Yet serious air pollution attacks in London, England, during the winter of 1962 to 1963 helped Roberts to change his mind, removing an obstacle to the passage of more ambitious federal control laws in the House of Representatives.[33]

That same year, there was a flurry of new activity on air pollution in the U.S. Senate. Four separate bills were offered calling for the establishment of federal air quality criteria to serve as guidelines for states and localities in planning and instituting their own air pollution control programs, a proposal officially supported by thirty-three states and the District of Columbia. To handle this new attention to environmental issues, the Senate Public Works Committee created a new Subcommittee on Air and Water Pollution headed by Maine's freshman Democratic Senator, Edmund S. Muskie. In early September, 1963, with the prodding of President John F. Kennedy, who was eager to pass a new federal air pollution control bill that year, Muskie and his subcommittee held hearings on air pollution and determined that it was a nationwide, worsening problem that could be helped through federal assistance with research and funding for state and local control efforts. In December, 1963, after Kennedy's assassination, the new President, Lyndon B. Johnson, took advantage of the national sense of shock to bypass continuing industry opposition and ram the 1963 Clean Air Act through Congress as a fitting memorial to the slain president.[34]

After passage of the 1963 law, which created a significantly expanded but still modest, limited, research-oriented federal program allowing the federal government to offer some help with funding but no enforcement role, the Muskie subcommittee held field hearings in various cities to see how well the new law met the needs of all these different localities. The testimony cumulatively pointed to the need for auto exhaust controls and regional air pollution control agencies as the two most important gaps to be filled in the nation's campaign against contaminated air. The subcommittee recommended California's auto exhaust control efforts as a model for the rest of the nation, since auto exhaust was found to comprise up to half the nation's total air pollution, and Muskie found the auto industry's vision of vehicular emissions controls varying state by state or locality by locality to be redundant and senseless. In January, 1965, Muskie introduced a bill to strengthen the 1963 Clean Air Act by calling for the establishment of federal vehicular emissions standards paralleling California's and for increased spending on federal research into various aspects of automotive pollution. Around the same time, President Johnson and HEW Secretary Anthony S. Celebreeze also called for expanded federal efforts against air pollution.[35]

The U.S. Conference of Mayors, the National Association of Counties, and even the AFL-CIO strongly favored Muskie's proposed new bill, but the auto industry did not, instead calling for further studies of the problem. Although by this time the auto makers had agreed to meet California's pollution standards on 1966 model cars, in April 1965 they still claimed that there was insufficient evidence that cars were a problem anywhere else to justify introducing nationwide controls and that "the taxpayer would get far more for his pollution dollar by spending it to expand controls on pollution from industrial plants and homes instead." However, the auto manufacturers nevertheless accepted the ultimate inevitability of the further extension of vehicular controls and promised that they could be ready by the 1968 model year; as one unnamed senior automotive engineer whined, "It's clear the politicians think this is for home and motherhood and it hasn't got a chance of losing."[36]

For their part, top federal officials in the Public Health Service's Division of Air Pollution had no such doubts about automotive pollution outside California, affirming that it was a growing danger throughout the nation. Noting the eighty-five million vehicles then on the nation's roads, freely crossing state and local boundaries, division chief Vernon G. MacKenzie labeled the need for national controls "self-evident" and concluded: "The old plea for patience because our technical knowledge is inadequate to meet the rising demand for clean air is beginning to fall on deaf ears." Many in America agreed. At the end of June, the federal government named exhaust controls as one of a list of seventeen vehicle safety and other features it would seek in its purchases of fleet vehicles to help push the auto makers. Around the same time, New York State was seriously considering requiring exhaust controls, and as one auto industry official commented, "You get New York and California and we've had it," since if the two most populous states both demanded exhaust controls, it would be just as cheap for the industry to put them on all cars. Despite the industry's struggles, the federal government moved toward nationwide auto emissions control as President Johnson signed an amended version of the Muskie bill in October, 1965.[37]

Yet the federal government's promulgation the following year of emissions standards for 1968 model cars raised a question in the minds of California control officials: did the federal entry into regulation of auto exhaust preempt state standards? The new federal standards were based entirely on California's original emissions standards, including even the figures of 275 ppm of hydrocarbon and 1.5 percent by volume of carbon monoxide, but California officials had already determined that these levels of exhaust control would not restore Los Angeles' air to the level of purity in 1940—a long-standing goal from the early days of the APCD—or even prevent it from gradually getting worse, and they had therefore set even stiffer standards. In mid-1966, Louis J. Fuller of the Los Angeles APCD formally requested that the federal government clarify the impact of the new federal legislation on state policy. Thomas F. Williams, information director of the federal Division of Air Pollution, suggested that the issue might have to be settled by a legal test case, but that it should not matter much one way or the other, since the federal standards were based on California's and were similarly expected to be tightened in the future.[38]

The federal preemption issue came to a head during hearings on Muskie's proposed Air Quality Act of 1967, which was intended to promote interstate cooperation and a regional approach to emissions control throughout the nation. Although the original draft of the bill had not mentioned the issue, Los Angeles officials called for the establishment of federal minimum vehicle emissions standards that particular states could make stricter as necessary. The auto industry, which years earlier had spoken in favor of multiple standards, hoping this would actually mean no standards anywhere save California, now rejected this earlier disingenuous argument and called for one national standard applying everywhere. They claimed that multiple standards would cause confusion and chaos in their industry and would harm consumers, a position with which federal air pollution officials agreed, and the manufacturers even threatened legal action to prevent the states from setting myriad different vehicular emissions standards. Muskie also generally favored the idea of one uniform standard, so the bill the Senate would later pass was altered to establish formally a single national standard that applied automatically to every state save California. The bill also included an amendment authored by California Republican Senator George Murphy declaring that California alone could get an exemption from the federal standards by demonstrating a problem severe enough to require stricter controls.[39]

The motor industry struck back against this concession to California when the bill moved to the House of Representatives. John Dingell, a Democratic representative from Detroit who generally supported environmental causes, submitted an amendment authored by the Automobile Manufacturers Association deleting the earlier Murphy amendment, and the Dingell amendment passed in committee. As an auto industry representative told a federal official, "The tail is not going to wag the dog, and Los Angeles is the tail." Infuriated Californians put partisanship aside temporarily to rally together to defeat the Dingell amendment, as Republican and Democratic House members and other legislators joined to fight the provision. Even the APCD and MVPCB joined hands to fight for the restoration of the Murphy amendment on the floor of the house, providing expert advice to California legislators. In the end, with help from infuriated California citizens, the California representatives got their way and preserved their state's special right to set higher vehicular control standards than those of the federal government.[40]

California had another, more minor showdown with the auto makers the following year over Assembly Bill 357, the California Pure Air Act of 1968. Although California at this time had by far the most advanced control program in the nation or the world, it still had air pollution problems in its metropolitan areas, and some of its legislators and conservationists wondered whether more progress could be gained faster through different approaches to the problem. For instance, as part of this developing, more ambitious vision of air pollution control, some observers argued that the solution to vehicular exhaust lay not in trying to clean up the inherently polluting internal combustion engine but in replacing it altogether with newer power plants. There were numerous alternatives already under discussion and drawing attention both in California and in Washington, D.C., during the mid- to late 1960s, such as gas turbine engines, modern steam engines, and fuel cells; vehicles designed to run on compressed natural gas, propane, or even hydrogen; hybrid vehicles that could operate on gaseous fuels or electricity in the city but could switch to traditional gasoline in rural areas; and battery-powered electric cars. All of these technologies promised potentially lower emissions than the traditional reciprocating internal combustion engine could ever achieve, but although many existed in some form or another, most of these long-neglected propulsion systems were not then at a stage to compete with the long-perfected internal combustion engine in terms of range, reliability, cost, convenience, or ready power relative to weight. Meanwhile, there was more to be done to control existing vehicles as well.[41]

The Pure Air Act and other related bills set out to address such unfinished business. A.B. 357 would set state policy to purchase low-emission vehicles preferentially for state fleets. It also increased the power of the Air Resources Board to set and enforce stricter emissions standards; required the state to test the pollution performance of new model vehicles on the assembly line and in use rather than relying on the claims of manufacturers; demanded that auto makers control nitrous oxide emissions as well as hydrocarbons and carbon monoxide; and sought to raise the cost limit for used-car exhaust control devices from $65 to $150. To defeat the bill, the auto makers posted a full-time lobbyist to Sacramento for the first time, but bill proponents managed to outmaneuver Detroit. Although resistance from the auto industry, auto clubs, and rural counties watered down various provisions regarding nitrogen oxides and used-car emissions, the bill that was signed in July, 1968, remained a relatively strong, progressive measure that once again demonstrated the state's national and global lead in measures to control vehicular air pollution. Other bills proposed at that time looked even farther into the future. A.B. 356 sought to set emissions standards for state vehicles so low as to preclude the purchase of anything but electric or steam-powered vehicles for 25 percent of state fleet purchases, while another proposal would have required the California Highway Patrol to experiment with steam cars.[42]

Los Angeles County had one last great tussle with Detroit during the 1960s, and this time, it lost. On January 10, 1969, shortly before leaving office, the Johnson Administration charged the major American manufacturers of automobiles, trucks, and buses with violating federal antitrust laws by conspiring to prevent, obstruct, or delay the introduction of devices to control vehicular emissions from 1953 onward. The case grew out of the auto makers' 1954 research-pooling agreement on auto emissions controls, formalized as a cross-licensing agreement in 1955, and subsequent allegations by Los Angeles County officials that this agreement was designed to delay rather than promote automotive exhaust controls.[43]

After years of first politely then more forcefully trying to goad the auto makers to greater action, APCD director Griswold in 1964 publicly implied that the 1955 cross-licensing agreement had been nothing more than a pact to make sure that no single auto manufacturer would take the lead and put pressure on others by developing and making available exhaust control devices. This allegation inspired the Los Angeles County Board of Supervisors to pass a resolution on January 28, 1965, requesting that the United States Attorney General investigate "possible conspiracy and unfair competition between automobile manufacturers in the design, production and distribution of automobile smog control devices." In a letter to federal authorities, Supervisor Hahn pointed out how for all their previous disclaimers, the auto makers suddenly were able to produce devices when Los Angeles and California forced them to do so in the summer of 1964, so they presumably could have done so sooner had they really been trying.[44]

In July, 1967, a federal grand jury was convened to investigate the matter. After calling many witnesses and sitting in secret session for more than a year, the grand jury forwarded their findings to Attorney General Ramsey Clark in late 1967, but the Department of Justice sat on the matter until the executive branch was about to change hands. Finally, in January, 1969, the federal Justice Department formally charged the participants in the cross-licensing agreement with violating the Sherman Antitrust Act through an unlawful conspiracy not to compete in research and development or in the acquisition of outsiders' patents for automotive emissions control devices but rather to appraise jointly any patents submitted, to cross-license patents with one another, and to agree secretly that no member to the cross-licensing arrangement would produce a car equipped with new pollution control devices before mutually agreed-upon dates. The formal charges also stated that although crankcase ventilation devices could have been available for all 1962 model cars, the conspirators deliberately delayed the introduction until 1963; that the defendants knowingly lied to California officials when they declared it technologically impossible to introduce new exhaust controls on 1966 model cars; and that they unlawfully "agreed among themselves to attempt to delay the introduction" of such controls until 1967. As a penalty, the Justice Department sought an injunction against any future such cross-licensing agreements and sought to force the auto makers to issue licenses to those independent manufacturers who had requested these for their air pollution control patents during the period of the conspiracy from 1953 through early 1969. The Automobile Manufacturers Association protested their innocence and claimed that their arrangement had only hastened progress.[45]

Los Angeles County officials eagerly seized this opportunity to expose the industry with which they had been struggling for so long. They sought to join the federal action but were refused. LA County and California state officials also filed suit directly against the auto makers to try to recover the money spent since the early 1950s on fighting vehicular air pollution and on medical care for people injured by air pollution, though legal experts warned that such cases would be difficult to prove or win unless the federal government made available the transcripts and records of the earlier federal grand jury investigation. Supervisor Hahn took the lead in trying to mobilize support and joint action by other cities and states throughout California and the nation. When the new Nixon Administration proved reluctant to act on the issue during early 1969, California officials threatened to take action against them as well.[46]

Yet in September, 1969, rather than going to trial, the federal Justice Department announced its acceptance of a consent decree under which the auto makers promised not to conspire to obstruct development of emissions controls and also agreed to make royalty-free patent licenses and technical data on their air pollution control equipment available to all companies wishing to install it. In exchange for these concessions, the auto makers admitted no guilt in the proposed settlement. Federal officials claimed that this represented strong but reasonable governmental action, while critics in Congress as well as local and state governments charged that the Nixon administration had knuckled under to intensive lobbying by the automotive industry in letting the suit be "swept under the rug," and that the terms of the settlement implied that the original allegations were true. California state and county officials worked frantically and successfully to mobilize nationwide support for their demand that the settlement decree be withdrawn, while celebrated consumer advocate and auto industry foe Ralph Nader predicted that this settlement might become "the most widely contested decree in antitrust history," noting that the original grand jury had collected such damning evidence that it had wanted to indict the auto companies on criminal, not just civil, charges, a move blocked by the Johnson Administration. Yet on October 28, 1969, U.S. District Judge Jesse W. Curtis approved the consent decree, claiming that it effectively accomplished what the federal government sought, and on March 16, 1970, the U.S. Supreme Court upheld Curtis's decision.[47]

Thereafter, various state and local governments persisted in trying to accomplish what the federal government had declined to do—exposing and punishing the uncooperative, possibly deceitful American auto industry for its role in letting loose the plague of air pollution from which most cities by then already suffered. Ultimately, thirty-four states filed suits against the auto manufacturers in twenty-three separate federal district courts around the nation, showing how America's favorite consumer durable had become a favorite object of blame even outside California. Yet these suits all were dismissed by late 1973, partly because the crucial 1967 federal grand jury report and supporting evidence necessary to prove the cases remained sealed by court order.[48]

Notwithstanding this major legal defeat on the antitrust suit, Los Angeles and California had accomplished much to control automotive air pollution by the early 1970s. Primarily through the efforts of Angelenos, vehicular exhaust had been identified as a major source of pollution and had gradually been elevated to the level of a national issue subject to federal regulation. The State of California also spearheaded other new initiatives to tame auto emissions, including the first law to regulate evaporative emissions from fuel tanks and carburetors, in 1968; research into reducing pollution by controlling the composition and volatility of gasoline; and the nationwide push that developed during the early 1970s toward requiring new vehicles to use unleaded gasoline. An increasing number of experts, policy makers, and citizens continued calling for the banning and replacement of the gasoline-powered, internal-combustion-engined automobile by the 1980s or sooner. Some even thought about providing proper mass transportation.[49]

Yet, at a deeper level, most Angelenos had never fully faced their fundamental situation in their battle against automotive pollution. Although the automobile was recognized as the major source of the air pollution that so profoundly lowered the quality of life in Southern California over what it might have been, nearly all the pollution control initiatives advanced by Los Angeles and California presumed the automobile. Not just any automobile, moreover, but a modern, heavy, high-performance automobile capable of whisking its owner effortlessly over long distances in great comfort, insulated from the outside world. Californians, accustomed to such free, easy mobility, mostly never thought in terms of the minimum automotive means of transportation necessary to get them to and from places where they had to go. Rather, they took for granted their lavish chariots to carry them on any number of essential or frivolous trips, of any length at any time of the day or night, but wanted these to be emission-free and did not expect to have to sacrifice much money, range, performance, or convenience for this. Rather than considering possible alterations to their own behavior, Angelenos mostly railed unreflectively at others as the entire source of their atmospheric problems: politicians, regulatory agency officials, the auto industry.

In Southern California then and now, and more than in most places, popular wisdom holds that "you are what you drive," and even as they heaped blame on the car makers, Angelenos continued seeking to purchase glamour, success, and self-esteem by buying more car than they needed, just as the auto companies and their advertisers planned. As one observer resignedly concluded, "Whatever the public's distaste for air pollution, its attachment to the automobile is probably greater." Southern California's mostly unquestioning embrace of an opulent automobile culture sharply constrained the range of alternatives Angelenos would seriously consider.[50]

Not surprisingly, this consuming love affair with the automobile was reflected at the level of politics, policy, and urban planning. Throughout the 1950s and 1960s, when automotive air pollution was becoming such an emotional issue in Los Angeles, mass transit was still an idea struggling vainly to be born. Although local polls in 1957, 1968, and 1970 all found substantial majorities favoring mass transit, when asked to put their tax money where their mouths were, Southern Californians repeatedly rejected major mass transit initiatives. They also resisted carpooling. State policies influenced by powerful highway-users' lobbies further reinforced Californians' tendency to rely on cars more and to build more roads and highways rather than mass transit facilities. Moreover, however ecologically questionable they may have been, Southern California's sprawling development patterns and vehicular self-indulgence increasingly became the norm for various other urban areas throughout the United States, gradually making Los Angeles' environmental afflictions more typical than aberrational.[51]

After building their lives, cities, and culture wholeheartedly around the personal automobile, the citizens of Smog Town—and Americans generally—had limited leverage over the Motor City. They could demand changes to the automobile, that its emissions must be reduced or that its basic power plant must be replaced with an alternative design, but the auto industry usually could resist or delay changes, knowing that it was highly unlikely that auto-dependent Californians would ever put into effect their recurrent threats to ban the sale of automobiles. Even when the major auto makers underperformed or cheated a little on their emissions compliance during the late 1960s and early 1970s, Los Angeles and California (and later the federal government) could not do much more than cause bad publicity for the industry. During the first two thirds of the twentieth century, automobiles effectively had taken the city, state, and nation hostage, and this captivity could only have been broken through a massive and costly economic redevelopment and spatial and cultural reorganization that almost no one was willing to endure. Given this situation, probably the greatest leverage Californians had over Detroit was to purchase foreign cars—something done extensively, though more for reasons of economy and reliability than for emissions reduction—which ultimately hurt Detroit badly.[52]

Considering all this, it is still remarkable that Angelenos accomplished all that they did in gaining the support of higher levels of government to help extract lower vehicular emissions from an often uncooperative industry in a relatively short time.

No, smog clearly was not banished, and the air over the LA basin has never been restored to the purity of 1940. The achievement was in keeping the smog from getting even worse while a flood of new people, industries, and especially automobiles continued to pour into Southern California and Californians' lavish, mobile lifestyles remained mostly unchanged. Smog Town's ambitious air pollution control campaign was fundamentally conservative in nature, seeking to patch up existing socioeconomic and transportation systems rather than replacing them with something new, and in this, it accomplished its purpose. Southern California's overall ecological dilemma of how to sustain such rapid demographic and economic growth and high consumption levels in the long term never was solved, but the impending environmental problems have been held off, regulated, and controlled—for a while.[53]

# chapter 5

# Folklore

## *The Public Confronts Smog*

Southern California's severe air pollution problem, all the more shocking in an area once known for its stunning beauty and healthful climate and atmosphere, understandably forced many of the region's residents to become actively concerned about an environmental issue for the first time. Through the early postwar decades, however, such incipient environmental concern for most local citizens was not at all identical to the more radical, all-encompassing "environmentalism" that swept the nation at the very end of the 1960s. The early postwar environmental awareness was indeed moving in that direction, but the development of post-1970 environmental sensibilities did not happen overnight. Rather, the people of Southern California, and the nation as a whole, had to go through various stages and conceptual shifts in the gradual process of shedding some of their cavalier, environmentally exploitative traditional assumptions and taking on newer, more environmentally benign attitudes—a process still unfinished.

Although Angelenos were less willing than most other people elsewhere throughout history to give up dreams of cleaner air and to submit to living with a steadily deteriorating atmosphere, they initially showed many of the other traditional reactions to environmental problems in their demands for aerial cleanup: the assumption that it was somebody else's problem, not theirs; the corresponding resistance to any personal sacrifice to dispel the nuisance; the expectation of a technological panacea; and the conception of the problem in isolation, rather than as a part of a wider system in need of alteration. While such attitudes were stronger at the beginning of the postwar period than they were by 1970, they persisted even then.

As noted, Angelenos were already becoming aware of a smoke problem at the end of the nineteenth century, and sporadic public concern helped produce a spate of local smoke control ordinances during the early decades of the twentieth century.

However, sustained public alarm did not emerge until the early 1940s, when the new photochemical smog finally came to be recognized for what it was—not merely an infrequent aberration but an endemic problem during the summer and autumn months. During a severe siege in July, 1943, downtown workers "found the noxious fumes almost unbearable." An even worse attack during the first week of September that year brought a major public outcry, ironically only three weeks after Los Angeles mayor Fletcher C. Bowron had formed a committee to study the problem and had predicted the "entire elimination" of smog within four months. As a local newspaper later recounted, "Everywhere the smog went that day, it left behind it a group of irate citizens, each of whom demanded relief." In Los Angeles, there were reports that county employees and factory workers were starting to quit their jobs "rather than work in the gas-laden air," while a local court shut down and the County General Hospital struggled vainly to protect 350 tuberculosis patients from the choking fumes. About a dozen miles to the northeast, in the hard-hit Pasadena area, people begged for release from the stifling fumes that triggered chest congestion and asthmatic attacks and reawakened old pulmonary disorders that had caused some residents to relocate to Southern California in the first place. In the face of this firestorm of protest, worried local politicians promoted scientific studies of the problem and began their months-long campaign to clean up the smoky Aliso Street synthetic rubber plant, which angry citizens initially blamed for the smog attack.[1]

The county supervisors and other local government officials undoubtedly breathed a sigh of relief when the winter rains returned, bringing the 1943 smog season and public complaints to an end and allowing officials to return to political business as usual after some mostly cosmetic actions against smog. But the problem had not gone away. The late summer of 1944 brought even worse sieges, as the smog rose above 5,700 feet and even hampered the work of astronomers at the Mt. Wilson observatory above Pasadena. The new smog season also brought changes in the nature of public protest. While 1943 had seen disorganized complaints by unhappy individuals, in 1944, local citizens organized the first citizens' group to focus on the air pollution issue. During the summer of 1944, irate inhabitants of the little residential community of Altadena, in the foothills of the San Gabriel Mountains just north of Pasadena, revived the dormant Altadena Property Owners' League (APOL) and used it to prod county officials into action against smog. In July, the APOL addressed a resolution to the Los Angeles County Board of Supervisors blasting them for their failure to take action on the earlier report of a citizens' advisory committee and demanding immediate passage of an air pollution control ordinance and an end to further delays rationalized by the need for additional research. They sent another resolution to the county district attorney demanding the strict enforcement of existing nuisance legislation to control local air pollution. The APOL also sent copies of their resolutions and calls for community mobilization to an endless number of potential allies in the war against smog, including the Los Angeles City Council, local taxpayers' associations, real estate interests, insurance firms, even the United States Public Health Service and the Federal Bureau of Investigation.[2]

By August, the APOL's campaign began to bear fruit, as other citizens and organizations joined them in systematically pressuring the county supervisors to take real action against smog. Hundreds of local residents signed a petition demanding the control of two notorious industrial polluters in El Segundo. Going against the larger, more industry-influenced group in Los Angeles, the real estate-oriented Pasadena Chamber of Commerce publicly declared smog a menace to human health. A local savings and loans consortium adopted a resolution warning of the threat to the local real estate market from smog. Meanwhile, the APOL continued to lead the pack of citizen smog fighters, launching a petition drive for a grand jury investigation of the problem and requesting state intervention from Governor Earl Warren. When the county district attorney's office reportedly dismissed the Altadena group's persistent efforts as some sort of psychological problem—an early example of the long tradition of critics trying to marginalize environmentalists as unbalanced or irrational—the APOL's executive secretary, James C. Clark, invited them out to the foothills "to get their lungs full of psychology." Without a doubt, the angry Altadenans helped pressure the hesitant county supervisors to take more meaningful action on air pollution from late 1944 to early 1945.[3]

Unfortunately, though, Angelenos' early efforts at public mobilization against air pollution showed many of the characteristic hallmarks of immature environmental awareness. First, interest in the issue was cyclic rather than sustained, with general public concern flaring up only during the smoggiest months, then nearly vanishing during the smogless months of winter and spring. Public organization and mobilization was sporadic; even the militant, vociferous APOL promptly faded from view after its major push during the second half of 1944. Above all, the public tended to think air pollution was an easier and more straightforward problem to solve than it really was. Angry citizens tended to blame the whole problem on a few obvious major emissions sources, such as the butadiene factory or other petrochemical plants. As was typical of virtually the whole nation from the nineteenth century through the end of the 1950s, most ordinary people in Los Angeles long tended to focus their wrath on visible emissions and to see smoke (or steam that they confused with smoke) as the single direct cause of smog. While concerned citizens were right to be wary of the typical, often disingenuous and obstructionist arguments of business interests that endless further scientific research would have to be conducted before knowledge was sufficient to take action on air pollution, many members of the public long stubbornly refused to take into account the scientific complexity of the problem and the multiplicity of sources contributing to widespread, endemic atmospheric contamination. As was characteristic of Americans who had seen all the dazzling scientific discoveries and technological accomplishments of the war and postwar years, Angelenos often expected that there must always be some ready technological solution for any problem. Also, as in nineteenth-century England and America, ordinary citizens always tended to blame "the other guy," such as industries, oil refineries, and generating stations, while declining to accept responsibility for their own actions causing air pollution, such as the operation of backyard garbage incinerators and of automobiles. Such public attitudes would long remain a frustrating problem for regulatory agencies. Still, in Los Angeles as in the rest of the nation, public mobilization and concern was what directly or indirectly created and perpetuated air pollution control agencies at all.[4]

At the same time that some citizens were first organizing to demand air pollution control, others were mobilizing to limit the cost and scope of this effort. During the latter years of the Second World War, exactly like business interests had done in other American cities such as St. Louis, Pittsburgh, and Cincinnati, the Los Angeles Chamber of Commerce sought to discourage "precipitate action" against air pollution, action which it claimed would harm the war effort—a characteristic way to cloak economic interests in supposed patriotism. The dominant local chamber of commerce also formed its own Smoke and Fumes Committee to "assist" in the drafting of control legislation more favorable to business. A representative of this committee argued before the Los Angeles county supervisors that the proposed ordinance unfairly singled out industry and would put Los Angeles' industry at a competitive disadvantage during the postwar period—again echoing rhetoric from other cities with chronic atmospheric problems. Raising the specter of job loss, the business lobbyist argued: "We can't throw obstacles in the way of those groups in our community to whom we are looking for our payrolls"; he urged a redrafting to come up with a bill all parties could "get behind and be happy with." Such rhetoric had the desired effect on supervisors such as William Smith and Roger Jessup, ardent advocates of mostly unlimited free enterprise, and the drafting committee dutifully came out with a much tamer proposal. Meanwhile, the chamber of commerce, again in a manner typical of early postwar business proposals for air pollution control, sought to forestall government regulation by launching a "campaign for the voluntary abatement of smoke and fumes by commercial and industrial establishments," in the belief that "much of the present difficulty can be adjusted thru [sic] friendly cooperation"—long a favorite panacea of the nation's business community. In what would be a pattern for subsequent abatement efforts, the chamber also worked behind the scenes in Sacramento to derail state legislation that would have removed obstacles to controlling air pollution in the Los Angeles basin's uncooperative incorporated industrial cities.[5]

With citizen smog fighters only marginally successful at promoting air pollution control while business interests were enjoying greater success at limiting it, the problem only grew steadily worse during the early postwar years. This led to still greater public anger. Particularly in hard-hit Pasadena, citizens formed various short-lived anti-smog organizations and periodically marched on their city hall to demand relief during 1946. They, and their counterparts in cities throughout the area, were dissatisfied with the county's control effort and the conscientious, competent, but politically maladroit engineer then running it, Isador A. Deutch. Deutch particularly antagonized the public when in 1945, in an effort to help isolate all conceivable sources of pollution causing public discomfort, Deutch requested voluntary public cooperation in an experimental one-day limitation of garbage incineration, motor vehicle use, and the beating of dust out of rugs and mops. Local citizens were infuriated at what they saw as a trivializing of the local affliction, especially when most already felt certain that the problem was entirely caused by the major new industries that had come into the area during the preceding two decades. For his reasonable request, Deutch was lambasted in the press. As a result of the "dust mop" incident and similar missteps, Deutch lost the confidence of local citizens and the county board of supervisors and became one of the early casualties of the politics of smog in Los Angeles County.[6]

Sustained public agitation and the obvious failure of gentler approaches to air pollution control gradually pushed local officials toward sterner measures against smog, however. By the summer of 1946, county administrators recognized that for all the carping about local home rule, most incorporated cities would not stand in the way of a countywide control program, for "the entire citizenry was aroused to the point that it would not be patient with a jurisdictional fight as between county and city government." Public anger put increasing pressure on recalcitrant industrial polluters as 1946 wore on into 1947. Citizen smog fighters also gained immeasurably from their growing cooperation with certain segments of the business community. For instance, one of the area's foremost hoteliers, Stephen Royce, owner of the posh Huntington Hotel in the upscale neighborhood between Pasadena and San Marino, became a leader in the drive for meaningful air pollution control after regular customers announced their plans to vacation in Florida instead due to the LA smog. Similarly, some real estate interests finally gave up trying to deny the presence of smog to save the area's good name and instead began seeking a solution for the problem that was increasingly giving the region a bad name. They were helped toward this conclusion by the efforts of a new anti–air pollution group, the Citizens' Fumes Committee, led by Pasadena's Mrs. Fanchon Gary, who from May, 1945, onward wrote persistently to public officials and spoke before civic groups warning of the decline of the major local tourist industry if news continued to spread about "the present pall, as inexcusable and as unpleasant as that of any Eastern Industrial City."[7]

As in Pittsburgh and St. Louis, one of the key ingredients for Los Angeles' successful air pollution control campaign was the active support of the local press. In particular, the *Los Angeles Times* made banishing smog a pet issue during late 1946 and 1947 after Dorothy Chandler, wife of *Times* publisher Norman Chandler, was appalled by the vile atmosphere one day driving back into the region from cleaner areas to the east and marched into her husband's office to tell him, "Something has to be done." Chandler gave veteran columnist Ed Ainsworth the full-time job of smog editor to coordinate a campaign of publicity regarding the offensive conditions, and Ainsworth wrote numerous features on the daily work of the existing city and county control programs, along with editorials on the nature of the problem and what was needed to solve it. In addition, the *Times* invited area citizens to send in complaints regarding polluters, which would then be turned over to authorities, and the newspaper was soon inundated with such complaints, sometimes including photographs or license plate numbers. The *Times* also arranged and publicized Raymond R. Tucker's celebrated inspection tour of Los Angeles in late 1946, which many eager citizens believed would provide the answer to the area's affliction. About the same time, Chandler brought together a new citizens' committee, the *Los Angeles Times* Citizens Smog Advisory Committee, including the president of Cal Tech, leaders of the Rotary Club and the tourism-promoting All-Year Club, and other local businessmen. The committee set out to do "a great deal of missionary work" in convincing local business interests not to oppose a proposed new state law to allow creation of a countywide control district in Los Angeles. The respectably conservative *Times* also diplomatically helped to bridge the differences between angry citizens and business interests over how to abate pollution.[8]

The Tucker report and the passage in Sacramento of Assembly Bill 1, which rode over jurisdictional squabbles and industry opposition to create the countywide, county-supervised Air Pollution Control District, led to widespread hopes that smog would quickly be ended. Yet subsequent events soon dashed this false optimism, as local air pollution continued to prove much more complex than anyone expected. Many citizens also grew surly after control officials began to blame them for much of the problem and threatened to take away or control their backyard incinerators and automobiles. This all reached a head during 1953 and 1954, when public suspicion grew that by attacking incinerators and autos, APCD director Gordon P. Larson was trying to cover up for industrial interests. In early 1954, a frustrated Larson wrote a confidential memo to all the county supervisors complaining of the complexity of the automobile and incinerator emissions problems and the difficulty of explaining all this to the public. Although his agency was conducting "an ever increasing public relations program," with staff trained in the use of film, radio, and television along with the print media, still the message was not getting through. As he glumly concluded, "Experience clearly shows that a complex, scientific problem, such as ours, is difficult to present so as to receive popular understanding of accomplishments and the goals and will thus tax the abilities of the most experienced person." Within a year, for all his honest and conscientious efforts at fighting smog, Larson had been demoted as director, and he would leave the state soon thereafter.[9]

Though the public may have been clamoring for more pollution control faster, other powerful interests continued to pull in the opposite direction. Business and industry, fearing the introduction of new costs and new governmental interference in their operations, usually sought to slow down the campaign against dirty air and to keep control authorities as weak and tame as possible. Their typical arguments were that pollution control should be done voluntarily, not through governmental or legal compulsion, and that industries understood the problem best and would take care of the job themselves, after conducting thorough research. Since even most of the air pollution experts outside industry—control officials, scientists, and engineers—usually either came from or had close relations and extensive contacts with industry, they frequently shared this attitude, particularly in the early postwar decades. This traditional stance was clearly revealed in the report of a visiting expert from Chicago offering recommendations for the new Los Angeles County Air Pollution Control District. Noting that "the Control Office can do much to maintain a good relationship with the industries," he advised, "It is especially important to develop a feeling of cooperation and, whenever possible, to seek improvements through a spirit of civic pride, rather than by resorting to the law." The visiting scientist recommended an information service for businesses and regular discussions with industrial representatives, while in "difficult cases . . . when cooperation is apparent, variances from the requirements should be permitted to allow time for these to be worked out." All of these were sound, reasonable suggestions still appropriate for air pollution control agencies today. The problem lay in the recurrent tendency of control officials, like those back in the Progressive Era, to start placing good relations with industry above the main mission of reducing atmospheric contamination.[10]

Even when the Los Angeles County control program was in a relatively moderate phase, it was still too activist for local industry. In 1952, Franklin B. Cole of the Los Angeles Chamber of Commerce complained that the APCD's organization put too much power in the hands of a director who could stifle local business. Cole also disliked the permit system created under the 1947 California Air Pollution Control Act for dictating the appropriate equipment to be used; instead, he favored a system of merely setting standards, letting industry meet these any way they chose, and so allowing problems and inadequacies to be built in—the same argument Raymond R. Tucker had faced in St. Louis back in 1937. Cole further viewed as dangerous the APCD Hearing Board's power to collect fines from persistent violators. The business representative argued that no control action should precede comprehensive research, and that tough standards should only be applied against pollutants proven to be detrimental to health. He also claimed that the whole 1947 control law was unconstitutional but that nobody in the state had the money to prove it—presumably not even the wealthy, powerful local oil companies. Cole, in envisioning a mostly toothless lapdog of a control program, claimed to speak for the 11,023 small industries of Los Angeles County that could be hurt by precipitate action on emissions abatement. However, he did allow that some legislation and regulation was needed, even if industry did not wholly agree with it, and he noted that the county had lost few industries as a result of the 1947 law, since business owners all felt certain there was nowhere they could go in California that would not soon be subject to controls, too. Cole tendered some indirect support for the angry citizens who alleged that Gordon P. Larson was overly understanding of business when he offered his opinion that Larson was doing a "better and more reasonable enforcing job" than his predecessor, Louis C. McCabe, who had "cracked down too soon" on local industry. By 1955, Cole and his fellow industrialists would be longing for a return to the days of Larson or McCabe as they faced the much sterner, less "reasonable" enforcement approach of S. Smith Griswold and Louis J. Fuller, including expanded use of court action and penalties for violations.[11]

At the same time as public suspicion of the unlucky Larson was spreading, general public concern about the health dangers of polluted air continued to grow. Members of the medical community led the way in this, while state and local officials often tended to downplay the risks. An early official report from the mid-1940s on the Los Angeles situation soothingly declared: "Statistical analysis of vital records indicates that there is no cause and effect relationship between mortalities due to heart or pulmonary diseases and air pollution." Yet many local physicians were not so sanguine. In 1950, Dr. Francis M. Pottenger, Jr., who before the war had operated a respiratory sanitarium in later smog-plagued Monrovia, worried about what unknown effects the pollution might be having on people. Remembering how local citizens had initially viewed the smog sieges as a necessary sacrifice for the war effort, Pottenger, a top official of the Los Angeles County Medical Association, testified: "Those of us who have been interested in chest diseases and respiratory diseases saw a great change take place in our people." He also suggested a link between smog and asthma, skin irritations, and other afflictions and announced that the County Medical Association would dedicate itself to determining what risks local citizens faced. As he concluded, "We can go out in our gardens and . . . see dead plants and we know what happened. . . . But what does it mean for our boys and our girls—our children, and expectant mothers?"[12]

By 1953, local physicians had grown more certain that air pollution was a clear and present danger to health. That September, the Los Angeles County Medical Association published an open letter to the mayor of Los Angeles and other city and county officials, proclaiming that "smog is a deadly menace" and calling for greatly increased medical research on the public health effects of dirty air. Thereafter, local officials received a steady barrage of letters from local doctors warning of the "deleterious effects of smog," including respiratory complaints, eye irritation, and the sheer "irritability and frustration" from living under such conditions, which reduced "the joy of living in California" along with damaging civic pride. Dr. Robert B. Hope called for an all-out fight against the menace with "every bit of aid that is possible from every source available, Local, State and Federal." Dr. Boris Arnov felt smog was "undoubtedly . . . shortening the lives of many of our citizens," and he called for a "fearless forthright policy of eliminating the very sources of air pollution," conducted by a "committee with power and almost dictatorial rights" that would "padlock the establishments of the offenders," for, as Arnov concluded, "the greatest damage is undoubtedly emanating from the industrial plants." Other physicians warned of lung cancer, aggravation of allergies, and particular risk to children's health, as well as of the untimely death of Southern California's tourist trade.[13]

This doctors' crusade against air pollution caused yet another public relations problem for APCD director Larson, who defensively denied the local medical association's charges of dire health risks and inadequate health research by his agency. Sounding like a typical embattled bureaucrat in the face of criticism, Larson warned, "Dissension and confusion can only serve to delay the force which must be assembled to solve this smog problem." Yet the doctors were probably closer to the public's feelings on the issue than Larson was, and perhaps at least as close to the truth.[14]

The scientific debate over the dangers of air pollution continued on in a seesaw fashion for years, with some physicians and public health researchers producing studies finding relatively little risk to health—the position generally favored by most politicians and industrialists—while others revealed new grounds for worry. In November, 1953, worries about health risks of air pollution received further support when Dr. Paul Kotin of the University of Southern California's medical school, working under contract with the APCD, reported to the American Cancer Society that his studies of Los Angeles' smog had led him to conclude that urban air pollution was indeed a grave danger. As he reported, "We are creating a marked cancer hazard in the air over all our big cities by dumping all manner of fumes and gases into the atmosphere. The increasing frequency of lung cancer in cities as compared with rural areas all over the world indicates that the atmosphere may be the principal cause of this disease." Later, in 1955, he announced that in his experiments, smog had produced lung cancer in mice. In May of 1954, though, the state Department of Health began the first California Health Survey, which included the first statistical analysis of reported morbidity in Los Angeles County related to smog levels, and the agency found no significant increase in total morbidity during periods of high air pollution. Similarly, in a meeting of physicians, scientists, lawyers, and control officials brought together in late 1954 by Governor Goodwin J. Knight—who was running for reelection and so had a strong incentive to make polluted air not seem too dangerous—participants ultimately concluded that "smog is an annoying inconvenience but not physically harmful." This was followed by a later study of asthma patients in the Pasadena area, which found, contrary to the observations of many doctors and allergy sufferers, that there was at least no evidence of any *direct* relationship between smog and asthmatic attacks, since atmospheric oxidant readings and asthma attacks peaked at markedly different times of the day.[15]

Such reassuring negative findings did not settle the issue, however. In 1958, United States Surgeon General Dr. Leroy E. Burney warned of an apparent relationship between air pollution and lung cancer, while federal researchers noted accumulating evidence linking contaminated air to various forms of cancer and other diseases. As a result of such evidence, in December, 1958, the Los Angeles County Medical Association adopted another resolution, stating: "Air pollution represents a present deleterious influence on public health, as well as an imminent danger of catastrophic proportions." Later, in May, 1961, in what claimed to be "the most carefully controlled sampling of medical opinion undertaken to date," Los Angeles area doctors responded to a poll on air pollution and environmental health in greater than normal numbers, indicating a particularly "strong interest in air pollution." More than three quarters of the local physicians responding believed that "air pollution adversely affect[ed] the health of their patients," and two thirds named it as a "frequent culprit in cases of chronic respiratory disease." Slightly over a third of the doctors had advised at least one patient to leave the Los Angeles area altogether, usually because of air pollution, and the poll made a conservative estimate that more than ten thousand Angelenos had received this advice, "with at least 2,500 acting upon it." Moreover, "nearly one third of the physicians questioned [had] themselves considered moving from Los Angeles because of air pollution." This small exodus of people fearing for their health was in addition to the many more Angelenos who moved away from smog without doctors' orders during the 1950s and 1960s.[16]

Further health research and reports followed. For instance, to try to get more direct evidence of the impact of actual Los Angeles smog on experimental animal and human subjects, University of Southern California medical researchers during the early 1960s arranged to house a thousand animals in "smog shacks" between the east- and westbound lanes of the Hollywood Freeway for long periods, while also experimentally exposing one hundred emphysemic volunteers to air pollution in smog-filled rooms. Despite such efforts, it remained difficult to ascertain the health risks of long-term chronic exposure to subacute concentrations of air pollution—concentrations producing no obvious, direct result in terms of death or sickness—for which experimentation was endlessly complex. Most evidence suggested that there was a danger to health from such chronic exposure, but such evidence was never complete or conclusive. In March, 1966, at a major medical conference on air pollution in Los Angeles, medical researchers complained of the general lack of knowledge regarding contaminated air's health effects, the difficulty of differentiating the effects of smog from those of all the other health-impairing practices of modern humans, such as smoking, and the near impossibility of finding nonexposed people to serve as control groups in experiments. Dr. Frederick Sargent, a University of Illinois physiologist, further observed that the medical community could not assess the effect of air pollution on health because they still had no functioning objective standard of what a healthy person was! As Sargent argued, "The solution to the current environmental crisis simply cannot be solved on a long-term basis by short-term programs aimed at reducing morbidity and mortality." In other words, air pollution and similar pervasive environmental problems transcended the scope of ordinary public health research and policy planning, and confronting them effectively would require new ways of thinking. While other specialists discussed apparent links between polluted air and tiredness, respiratory infections, and lasting damage to cellular systems for metabolizing carbohydrates and proteins, such evidence remained suggestive rather than conclusive. So the question of the precise nature and degree of health risks from air pollution remained unresolved.[17]

Whatever the state of scientific understanding, many Southern Californians remained deeply concerned about air pollution as a possible health threat. In 1956, APCD official Dr. Leslie A. Chambers testified before visiting congressmen that when Angelenos complained about eye irritation from smog, beyond the mere fact of annoyance, they were "really concerned about it as indicating the possibility of more permanently damaging effects on health." Alarming predictions fueled public worry. In 1959, on the California Institute of Technology's weekly television program, Dr. Arie Haagen-Smit informed his fellow Angelenos that unless more stringent controls on atmospheric emissions were instituted soon, greater Los Angeles could become a death trap. Like a latter-day John Evelyn, he recommended the creation of more parks for breathing space and regional planning and zoning that would make further industrial expansion "practically impossible," and he concluded that even "relocation of industry, plus other unpleasant measures, may be necessary." Haagen-Smit was joined in these fears by many less famous figures.[18]

Together with the rising public fear and frustration over smog in the early 1950s came increased public mobilization, after faltering early efforts. Following earlier, short-lived citizens' organizations to fight smog, in 1949, Stephen Royce of the Huntington Hotel joined his friends James C. Clark, formerly active in the Altadena Property Owners' League, and Dr. Edwin P. Hubble, astronomer at the smog-plagued observatory on Mt. Wilson, in creating the Pure Air Council of Southern California. This group, intended to enlist thousands of volunteers in a mass campaign to promote public education and cooperation in the fight against air pollution in close cooperation with the APCD, actually gained only 343 members and lasted only through 1950—in part because of the disinterest of the technical experts running the APCD, who had no time for what they perceived as meddling laymen.[19]

Partly due to such high-handed treatment, some subsequent citizens' organizations that spontaneously arose to fight air pollution were hostile toward the APCD and harbored populist suspicions that it was merely a creature of the corporate polluters. For example, in 1952, more than one hundred residents of Arcadia, near Pasadena, formed the Pure Air Committee. A year later, at a meeting where Gordon Larson, an Arcadia resident, outlined and defended the APCD's program, Joseph Sternbach, vice chairman of the group, blasted the APCD for their "deliberate and snail-pace hunt" for solutions to the region's atmospheric emergency and charged that "politics also plays a big part, as those in high office hesitate to offend 'big business' by cracking down and enforcing air pollution laws." The group's leaders announced that they were "declaring war on smog, as too many lives are at stake," and they promised that with the cooperation of a "thoroughly aroused populace," they could bring enough pressure to bear on public officials to produce rapid results.[20]

By the following year, such groups' rhetoric grew even more strident. In October, 1954, during a terrible smog siege, six thousand angry Pasadenans thronged the local Civic Auditorium to demand that Larson and the APCD either "clean up or clean out." In emotional, populist tones, Francis H. Packard, co-chairman of the Citizens' Anti-Smog Action Committee, threatened local authorities with dire political consequences if they did not show more activity to end smog soon. Packard thundered to the crowd: "If you folks get behind this crusade and line up 100,000 members, we will get our supervisors clear out of bed. We say to the powers that be—get rid of smog or we will place someone in power that will." To Larson, Packard recommended that he "enforce the law without fear or favor," but if the job proved too big for him—and here committee members dramatically unfurled a large banner saying in all capital letters—"RESIGN!" Revealing the mounting public concern about the health effects of smog, Packard warned, "We're tired of talking. We want action—immediate action. We do not intend to be guinea pigs or sitting ducks for fifteen years." Packard concluded, and the audience enthusiastically agreed, that local authorities must promptly end the danger to public health without regard to cost. The same day, a group of Pasadena housewives calling themselves the "Smog-A-Tears" (in a spoof on Walt Disney's Mouseketeers) paraded through downtown Pasadena wearing gas masks and carrying signs. Even Gordon Larson, let alone elected officials worried about keeping their jobs, could hardly ignore this public uproar.[21]

Throughout the latter months of 1954, local newspaper headlines and articles screamed about smog almost without letup. In the enactment of local realtors' nightmares, in late October, the *Los Angeles Times* presented a photograph of Mrs. Walter S. Maben and her two children hammering in a sign in front of their San Gabriel home reading, "For Sale—Sacrifice—Leaving Because of Smog"; the Mabens hoped to escape from smog in Fresno, California, or somewhere in Idaho. In a further nightmare for the local tourist industry, the Elks' Club held its national convention in Los Angeles during a severe smog attack in July, leading many out-of-state visitors to seek first aid for nose and eye irritation and probably dissuading them from ever wanting to return. More shockingly for the whole community, on October 16, local papers carried the story of ten-year-old Betty Jean McMahan of Alhambra, who had a coughing spell at three o'clock in the morning and choked to death, after earlier complaining of a sore throat from the smog. McMahan's physician declared that smog "could have been a complicating factor." McMahan's sad story also received national attention.[22]

Given this state of affairs, and the public passions it aroused, it is hardly surprising that smog became a major political issue throughout the region during the elections of 1954. As early as March, 1953, Norris Poulson, a local Republican congressman who planned to challenge incumbent Los Angeles mayor Fletcher Bowron later, had already picked smog as one of his key issues, supporting air pollution control proposals in Congress and criticizing the sitting mayor's "laxity" on the issue, while Bowron's supporters responded defensively that smog was a county issue over which the mayor had little jurisdiction. This counterargument was to little avail; Poulson swept to victory over the four-term mayor in the spring of 1953, helped in large measure by the severe smog attacks of late 1952. As earlier noted, a worried Governor Goodwin J. Knight interrupted his reelection campaign in October, 1954, to fly down to Los Angeles with members of his staff and join in the investigation of the causes of smog, which was then creating a "state of near-emergency" in Southern California. Knight, who had previously dismissed smog as only a local problem, did an abrupt about-face, joined local officials in calling for federal aid, and promised to appropriate millions of dollars for studies at the University of California of causes of and cures for smog. This timely nod to public concerns in the most populous part of the state helped Knight hang onto his job for another term.[23]

Meanwhile, angry local citizens had demanded and won a grand jury inquiry regarding the absence of an adequate air pollution emergency warning system prior to the terrible siege of September and October, and Mayor Poulson was handily able to shift much of the blame onto the shoulders of already embattled APCD director Larson. Pasadena mayor Clarence Winder called for public prayer "to deliver us from this scourge," and members of the Los Angeles City Council put on gas masks to protest the fumes that were entering the council chambers. The following December, Francis H. Packard of Pasadena's Citizens' Anti-Smog Action Committee, which claimed to be affiliated with thirty-four other community groups throughout the area totaling thirty-one thousand members, issued a formal plea to the state's attorney general and later two-term governor, Edmund G. Brown, Sr., to take action against area polluters under the state's emergency and police powers. Pasadena's Office of Civil Defense commenced developing "disaster-evacuation plans to fit a possible atmospheric emergency." Residents in other states shook their heads in wonder at developments in what some were calling "Smog Town."[24]

Smog continued to be terrible in 1955, and it led to some rare soul-searching among Angelenos. September of 1955 brought what was called the worst smog siege ever, as ozone levels climbed to 0.9 parts per million in the atmosphere. This was well beyond the APCD's first-stage alert level of 0.5 ppm, at which outdoor burning was banned and unnecessary driving was supposed to be limited, and it was disturbingly close to the second-stage alert level of 1.0 ppm, the level at which a "dangerous health menace" was thought to exist. APCD inspectors were stationed at all oil refineries, ready to close them down if the ozone concentrations rose to the second alert level. As severe smog continued through the month of September, Los Angeles County sought to institute a new smog warning system to alert citizens to likely first-alert smog conditions the night before, though compliance was only voluntary. Area physicians worried that "smog spells could become bad enough to suffocate a lot of people outright, as happened in Donora, Pa., in October, 1948," while newspapers noted public concern that continued smog could "blight the city's development, jeopardize health and burgeon suddenly into a full-fledged disaster." Most tellingly, even the local business community began to confront the need for possible changes in behavior to battle the area's most notorious environmental problem. Regarding an air pollution conference held by the Los Angeles Chamber of Commerce in November, 1955, where the chamber resolved that "new industries to which smog-control devices cannot be applied . . . shall not be permitted to locate in the area," the *New York Times* observed, "Until recently, any such notion [of limits on industrial growth] bordered on civic heresy."[25]

This stab at facing up to the limits to growth was brief, however. After the public frenzy over air pollution in 1954 and 1955, Angelenos calmed down a little and gave the new leadership of the revamped, better-funded APCD some time to show what they could do. S. Smith Griswold and Louis J. Fuller were firmer and more visibly active in enforcement than Gordon Larson had been; they were also far more convinced of the importance of keeping the general public as well informed as possible. In addition to increasing the agency's public relations program from only one full-time employee to nearly forty, Chief Control Officer Griswold embarked on a long campaign to spread the message about local air pollution and the agency's efforts, problems, and achievements directly to the Los Angeles area public, often in person. Under Griswold, the total of 109 speeches delivered by agency staff to a total audience of 6,849 in 1954 rose to 471 talks to an estimated cumulative audience of nearly 30,000 in 1955, and this trend continued steadily upward. Griswold targeted civic organizations that were influential in forming public opinion, such as chambers of commerce, political organizations, and social or service groups such as the Kiwanis, Rotary, and Lions Clubs as well as various women's clubs. The district also sought to improve its relationship with the news media, with impressive results. As the agency bragged in mid-1955, in addition to the more than twenty-five hundred column inches of air pollution control news carried by daily and weekly newspapers in Los Angeles County every month, "radio and television news programs carry an Air Pollution Control story on almost every broadcast. Dozens of special programs, round table discussions, and TV interviews add to the airwave coverage." There were also thousands of copies of reports, pamphlets, and informational bulletins printed for distribution to the public, while information panels were set up on easels in public buildings, bank lobbies, and "other sites of heavy foot traffic." Griswold and Fuller, in this flurry of public relations activities, showed that they had learned well the lesson Larson had long resisted.[26]

Griswold and Fuller also won an early victory over a possible public relations problem: the Citizens' Anti-Smog Action Committee, the largest and most unruly of the anti–air pollution groups. After promulgation of "Regulation 7—Emergencies" in June, 1955, to set up an emergency warning system, the APCD was criticized by the California Manufacturers Association and the Long Beach Chamber of Commerce for causing bad publicity that might dampen tourism and damage the economy. The Citizens' Anti-Smog Action Committee further claimed that the smog-alert provisions placed an unfair burden on owners of backyard incinerators and automobiles, who were supposed to limit use of these in the event of high pollution levels, as against the oil refineries that the committee was convinced were the "real cause of smog." Thereafter, Francis H. Packard, head of the citizens' group, demanded access to oil refineries' daily operations records to prove the oil companies' culpability for all of Southern California's atmospheric woes. APCD Chief Enforcement Officer Fuller refused, noting that these records were given in confidence and could offer a competitive advantage to other companies if publicized. Packard then issued a press release alleging an APCD cover-up, and Fuller threatened to sue the committee for libel. Ultimately, Packard and his group backed down and formally apologized to Fuller. The organization never regained its former leading public advocacy role after reaching too far, and thereafter, like earlier organizations, it gradually faded from view. The APCD would not face such strident public criticism again for more than a decade.[27]

In addition to their new focus on public relations and information, the new APCD directors, following through on the research done under Larson, also brought results: for instance, the banning of garbage incineration in 1957 ended perhaps the single greatest source of public complaints about air pollution, while the continuing efforts against other stationary sources in the area brought some improvement. By the later 1950s, Fuller and Griswold also benefited from the increasing public acceptance of the automobile's primary role in causing the region's most notorious affliction, which allowed them to shift some of the blame and public anger away from the APCD and Los Angeles officials and toward Detroit. Moreover, as historian Marvin Brienes rightly notes, though the extension of Los Angeles' growing freeway system allowed Angelenos to "drive more cars more miles," it also eased traffic congestion near downtown and dispersed the greater total volume of automobile exhaust over a much wider area, so that it was less disturbing—for a while. Also, the federal government's somewhat hesitant entry into the business of controlling air pollution through the Air Pollution Control Act of 1955 raised hopes among LA residents that a solution might soon be found, though such faith in this feeble first federal law would prove misplaced. In tandem with the lessening of public protest over air pollution from the mid-1950s through the early 1960s, there is also statistical evidence indicating that total atmospheric oxidant levels improved for a time through the early 1960s before worsening again later in that decade due to the ever-expanding number of people and automobiles in the area.[28]

During this golden age of the Los Angeles Air Pollution Control District between the mid-1950s and the late 1960s, when the agency enjoyed overall public support and respect, new anti-smog groups emerged to work with the district rather than against it. Reflecting the shift of primary attention and resentment away from local stationary sources and toward automobiles and their distant makers, the major local citizens' organization of this period focused its information and mobilization efforts on that problem, although it remained concerned with more traditional emissions sources as well. This group was Stamp Out Smog (SOS), founded in 1958 by Margaret Levee, a housewife from the affluent neighborhood of Beverly Hills, along with nine other women. Levee and her group worked to contact and mobilize other organizations throughout California to fight against air pollution, including garden clubs, homeowners' groups, local chambers of commerce, labor unions, public health associations, and religious congregations. Levee quickly determined that success in swaying legislators went to those who could claim mastery of the facts and technical expertise, so she and her group set about learning the key data regarding automotive

pollution, to offer legislators seeking information an alternative to auto industry lobbyists. The group's seven-point plan of 1958 for dealing with vehicular emissions helped
influence the later creation of the California Motor Vehicle Pollution Control Board
(MVPCB) in 1960, and thereafter, the women of SOS proposed hundreds of other
pieces of state and local legislation and took every opportunity to testify on the issue
in both Los Angeles and Sacramento.[29]

Since Levee and her fellow founders of Stamp Out Smog were relatively wealthy
and well connected, they also had personal contacts with many influential state and
local politicians and used these to promote their message. In 1960, Levee was one of
the first appointees to the MVPCB. This gave increased legitimacy and a higher
profile to her organization and message, but as critics of the move pointed out, it
may also have tended to muzzle her somewhat, as she was constrained by the other,
more reticent members of the board. It may have made it more difficult, too, for her
organization to criticize the MVPCB, even when the board's public information
officer, Ray Kovitz, allegedly substituted cozy relations with the regulated industry
for a proper public information campaign and "considered smog-control movies
released by General Motors to serve the MVPCB's public relations needs." Nevertheless, whether muted or not, Levee still pressed for prompt emissions control while
on the board. At a 1964 meeting with auto industry executives where the president
of General Motors estimated that it would take at least another seven years before
exhaust controls could be installed, Levee exploded in anger and threatened the auto
makers with a "federal suit for colluding to prevent pollution control advances,"
showing that she was thinking this way long before the federal antitrust investigation of the industry. Stamp Out Smog, an organization with tough, committed leadership, was unusually successful and persistent, lasting from 1958 into the 1970s and
a new phase in citizens' relations with air pollution and control authorities. By the
late 1960s, the group claimed to be federated with and endorsed by more than five
hundred other organizations throughout the state, representing hundreds of thousands of members.[30]

While APCD officials and citizens' groups worked at lessening local aerial contamination, members of the public also tried to help in other ways. In a striking,
often comical expression of the characteristically American faith that there must be
some clever gadget to solve every conceivable human problem without fundamental
social or political change, a horde of inventors, from professional scientists and engineers to auto mechanics and backyard tinkerers, offered an endless array of devices
and suggestions as *the* solution for the LA smog. The four main approaches tried
were to somehow puncture the prevailing inversion layer to let the smog escape into
the upper atmosphere, to blow smog away from the area by mechanical means, to
change the physical or meteorological characteristics of the atmosphere so that smog
would not form, or to attach a wonder-device to automobile tailpipes and industrial
smokestacks. As was typical of the pre-environmentalist, pre-ecological mindset, many
proposals focused on removing the problem rather than preventing it.[31]

Once the role of the inversion layer in holding pollutants over the Los Angeles basin was discovered during the early 1940s, people immediately began dreaming up schemes to get around it. Some inventors proposed blowing the inversion away with huge horizontal fans mounted on towers, or with giant helicopters. Such plans were found impractical, requiring ten twelve-ton helicopters or a similar number of ground-based fans per square kilometer to churn up the atmosphere enough to accomplish anything, which would have been hugely expensive, noisy, polluting, and potentially dangerous besides. Other clever inventors suggested thermally puncturing the inversion layer. In one of the most elaborate proposals, an engineer suggested creating numerous black, conical mounds, perhaps one hundred feet high, with upward-pointing fans placed around this artificial hill in a spiral toward the pointed top. The great black mound was to absorb intense heat, while the fans would help send this heat up in a spiraling updraft that would tear open the inversion layer and set the city's smog free. Local officials, however, decided that it would take more than hot air to solve the area's smog problem.[32]

There were also numerous ambitious proposals to blow away accumulations of smog from the Los Angeles basin. Some recommended huge vertical fans. In particular, since the smog was hemmed in by the mountain ranges to the north and east, logical citizens suggested drilling vast tunnels in these mountains and fitting them with fans to blow the smog out over the Mojave Desert, where almost nobody lived and where Angelenos consequently felt free to dump their atmospheric garbage. However, there were problems with these proposals. Fans large enough and numerous enough to move the stagnant LA air to any significant degree, it was estimated, would have required "at least 12 Hoover Dams, more than twice the electricity produced in California, or at least one-sixth of all the electricity produced in the United States." Fan advocates then suggested using nuclear power instead, since that was supposed to produce unlimited energy too cheap to meter. As for the mountain tunnel project, one estimate held that to make any noticeable difference, there would have to be some "14,000 ventilation tunnels, each approximately 40 feet in diameter," and consuming another several Hoover Dams' worth of energy. Such schemes were seriously (or at least semi-seriously) considered for a time, though.[33]

Then there were dreams of atmospheric engineering and weather modification, which had long been sought as ways of bringing rain to the nation's dry western areas. Since sunlight helped to cause the photochemical reactions leading to smog, local citizens thought up various clever ways to reduce the sunlight over the area. One inventor called for "a gigantic parasol of white smoke laid by airplanes high over a city" to prevent the sunlight getting through, although it takes heavy clouds to block the ultraviolet rays that work upon smog precursors; it seems the inventor never stopped to think that the sunlight was probably the principal reason people had flooded into Southern California in the first place. Another proposal called for airplanes spreading a thin film of oil throughout the atmosphere on days that threatened smog, to reflect and screen out the sun's rays. An obvious and oft-suggested solution was rainmaking—seeding clouds to produce rain that would cleanse the air. Even if such efforts had been successful in producing rain, which they often are not, meteorologists pointed out that this would merely wash down particulates, which were not a problem; only strong winds could blow away the gases that produced smog. Finally, other would-be atmospheric engineers advocated chemically neutralizing the smog. However, no neutralizing agent was known, and those that were proposed were frequently objectionable in their own right.[34]

Probably most popular of all among smog-fighting inventors were exhaust control devices. There were myriad sketches of simple emissions control devices submitted to air pollution control officials in Los Angeles, Washington, D.C., and elsewhere, some by people with professional training, some by people with none, some by men, some by women, some for automobiles, some for other sources. Most were oriented toward the control of visible emissions, which long remained fixed in the public mind as the source of smog.

Some of these inventors were sincere in their efforts, while others were probably charlatans. One wonders at the claims of Henry C. Von Braun of Los Angeles, who offered a description with primitive charts of an automobile exhaust reactor that would eliminate smog while improving mileage. Said Von Braun, "My Carburation [sic] System is very simple. A carburater that splits the heavy droplets of gas into molecules cost not over 15 Dollars. An Intake Manifold that splits the heavy molecules into lighter once [sic] and splits the lighter molecules into Atomic Energy. Cost from 25 to 40 Dollars. . . . So if you are interested to save the Taxpayers Billions of dollars and have clean air, let me hear from you."[35]

Since many of the proposed solutions came from lower-middle- or working-class citizens, American inventors' campaign against smog was at least democratic, if not always reassuring. For instance, J. James of Los Angeles (later Pasadena) wrote in 1964 to try to get back the plans he had sent to the APCD regarding the "Matter of Smog Elimination of Air. Wind blowers etc." James complained to Los Angeles County Supervisor Kenneth F. Hahn:

In regaeds to my plans for Elimination of Smog, Whitch I submited to your office on about feb/64 The same your office refered to MR. Smith Griswold Air pollution control officer. However, I received A letter from MR smith Little or know explination, However I do want that office to send me Photo stat copy of my plans. to me. since this is my property, not the Air pollution. A few years ago I sumited plans to the city of Los Angeles, They claim the plans was lost. Will your office kindly request Mr smith to send me a photo stat copu of the Plans submited to your office.

Since the plans were lost, one cannot evaluate the technical merits of James's proposal, but it is hard to imagine that his presentation inspired confidence, particularly since it was apparently yet another scheme to blow smog away from the area.[36]

After the automobile was accepted as the prime cause of air pollution in Southern California, smog-fighting inventors naturally focused their efforts on it, and some of these efforts might have born fruit had the major auto makers been interested at the time. As early as 1953, Los Angeles area engineer Harry C. Barkelew was working on an exhaust-controlling muffler for cars, trucks, and buses, and he later offered more convincing technical reports and other evidence than most other individual inventors to indicate that his device might hold real promise of reducing motor vehicle emissions. However, like the other moderately successful first-generation catalytic and noncatalytic exhaust afterburners of the late 1950s and early 1960s, Barkelew's device was ignored by the American automotive industry, which wanted to produce devices of its own at its own leisurely pace. Subsequently, even after Barkelew's death, his wife, a member of the Alhambra Women's Club, conducted a one-woman campaign of writing angry letters to local, state, and federal officials to try to force the auto makers to use the products developed by inventors outside the auto industry on used as well as new cars. Other inventors also claimed, rightly or wrongly, to have been unjustly overlooked, helping to contribute to a suspicion in some quarters that the problem could have been solved but for the interests and/or authorities conspiring to bottle up ingenious technofixes. Thus could the great American faith in technology readily blend with populist paranoia when confronting a problem as complicated as air pollution.[37]

Members of the public were hardly alone in their faith in technofixes, though, for politicians, air pollution control officials, and other "experts" were all similarly focused on finding painless technological solutions. In 1954, Governor Knight, who was no scientist, forthrightly declared that "smog is a scientific and engineering problem and not a political or legal one," and other governmental figures generally agreed that the solution to smog would come through hiring ever more engineers and chemists rather than through changes in human behavior or economic or social policy. Authorities and the public agreed that smog would be banished without requiring any major changes in the notoriously self-indulgent, suburban consumer lifestyle that had come to dominate the region. Thus, even as the fight against smog was going on, the Los Angeles metropolitan area was frantically building more freeways, helping to expand an ecologically unsustainable Southern Californian way of life and to ensure a much greater total volume of vehicular emissions spread over a wider area. Although as early as the late 1950s there were some precocious proposals regarding the need for behavioral changes to fight the local atmospheric plague, including recommendations to promote carpooling, to revive or create a proper mass transit system, and to discourage private motor vehicle use, such suggestions were rare and mostly ignored until the end of the 1960s. As James E. Krier and Edmund Ursin insightfully observed, "The faith of the times was in mechanical, not human, engineering. The search for antismog devices was the preoccupying concern." Another favorite American way of sidestepping problems has always been to strike out for somewhere new in hopes of leaving all troubles behind, and smog along with cheaper housing was part of the reason thousands of Angelenos swarmed eastward to Riverside and San Bernardino counties during the 1950s and early 1960s, only to be surprised when the smog soon followed them and became worse than that over downtown Los Angeles. Southern Californians were aware of and concerned about an environmental problem, but most were not yet thinking ecologically.[38]

The general public could cause trouble for air pollution control officials in other ways than just inundating them with unlikely suggestions. As with other issues, people often favored environmental regulations until they were forced to change their behavior and put up some money themselves. For instance, the outcry about proposals to ban backyard incinerators lasted all the way from the initial suggestion in 1952 to the final prohibition in 1957. That outcry led to shrill, unfair charges against the APCD; helped fuel a brief circulation war among certain local papers, in which each vied to outdo the other on irresponsible, sensational reporting; and contributed to the downfall of Gordon Larson.[39]

Control officials' efforts to reduce automotive emissions would bring still more protest from a vocal minority of the public. This was particularly evident in the early 1960s, when the state of California, bending to the needs and desires of urban districts in the San Francisco Bay area and particularly the Los Angeles basin, sought to require the first functional emissions control devices and to institute a system of yearly vehicular inspections targeted at controlling pollution. As we have seen, during 1963, due to stiff resistance by automobile clubs and rural counties, legislation recommended by Los Angeles County to create a mandatory vehicle inspection program supervised by the California State Motor Vehicle Pollution Control Board (MVPCB) was watered down to a limp-wristed provision allowing counties to decide whether or not to implement such a program.[40]

Even more bitterly fought were the early state efforts to introduce crankcase emissions standards and controls for used cars. In 1960, as noted in the preceding chapter, the MVPCB first set acceptable levels for crankcase emissions. The first crankcase control devices certified by California authorities in 1961 for use on new cars proved defective, but various approved devices were available and required for new cars beginning with the 1964 model year, while the major U.S. auto makers had begun installing such devices voluntarily on all new cars sold in the United States starting with the 1963 models. Despite the trouble with the very first such device, this transition went off smoothly enough for new cars.[41]

All hell broke loose when the state began phasing in its program to require similar crankcase emissions control devices on used cars, however. After two certified crankcase devices for used cars became available during 1962, California passed a law in mid-1963 requiring that used vehicles that changed ownership after January 1, 1964, would have to be fitted with such equipment, and no used cars could be registered after December 31, 1964, without crankcase emission controls. Already in 1963, the creator of one of the used-car devices foresaw possible trouble, "a big psychological problem because of general resentment against anything mandatory," and urged a major public relations effort by the MVPCB to win over the public. He was right. Public protest began soon after the new law took effect. By January, 1964, public officials in Los Angeles were receiving angry letters about the new requirements. One "mechanically inclined" individual claimed to speak for all his friends and associates in declaring the new crankcase ventilating systems "over 99% useless" in controlling vehicular emissions, since, he incorrectly argued, 99 percent of vehicular emissions passed out of the tailpipe. Another angry citizen complained that he had put the new "SMOG Devise" on his 1953 Plymouth and his mileage had decreased by thirty-three miles per tank of gasoline. He affirmed his willingness to "try to cooperate in any way" to "help decrease this Killer Smog," but concluded that the new requirement was only "designed to help the oil company sell a lot more gas." Such criticism continued throughout the first year of the used-car regulation, as other Angelenos wrote in to deny that automobiles were causing any serious pollution. One professional chemist still insisted that industry and oil refineries were causing all the area's atmospheric troubles, while other citizens blamed the whole problem on impure motor fuel.[42]

For all the state authorities' efforts at public relations, the situation got much worse after January 1, 1965, when *all* used cars were to have crankcase controls. Citizens suspected that the devices caused trouble with their engines and were too hard to maintain. The automobile clubs opposed any sort of mandatory provision. Some mechanics feared the devices would make normal repairs more difficult, while used-car dealers disliked being stuck with the cost of fitting all the vehicles on their lots with the required devices to make them salable. Angry citizens again fired off a barrage of complaints at their elected officials, alleging greater repair costs and reduced performance. A frequent charge was that the new devices ruined spark plugs. William C. Van Meter spluttered that the equipment abated no smog but only ruined motors. In a remarkable but not atypical populist libertarian diatribe, he then declared that he would not register his 1954 Cadillac, bought used,

> until the thieving, racketeering, S.O.B.s in Sacramento are made to recind [*sic*] their smog abortion against the public. . . . I am certain that the state will lose many millions because of those bungling parasites in Sacramento. Boy . . . Am I going to vote for Ronald Reagan for Governor. . . . There is a real American. I say we will also defeat this Smog racket. They make no mention of putting smog devices on the oil refineries and the buses which gas you to death. . . . They are too big. . . . They pay off. They only pick on the poor unorganized public. What new conniving schemes are they cooking up next?

Others complained of the devices causing motors to misfire and to consume more fuel and oil. There were reports of motorists placing stacks of pennies in their crankcase control devices to prevent their operation.[43]

In the face of this welter of public protest, the state of California readily backed down. The Transportation Committees of the state senate and assembly both scheduled hearings on the matter in January, 1965, and in February, the legislature passed, and the governor immediately signed, a measure stipulating that failure to install the required device did "not constitute a crime," meaning that owners of used cars could register unequipped vehicles with impunity. This act was called "an emergency measure" necessitated by the "great concern and confusion" the new requirement had caused, and the legislature promised further legislation to "substantially eliminate this requirement," which was dutifully passed in June, 1965. Thereafter, only vehicles changing ownership in counties with functioning air pollution control districts would be subject to the crankcase control requirement. As the *New York Times* observed, "This was to have been Los Angeles' big year in its 17-year battle against smog, but now that it is up to individuals to do something about it, a sudden pall of inertia, confusion and resistance has all but supplanted the smog."[44]

Yet even during the crankcase controversy, not all voices were railing at the control authorities. One professional mechanic called the new devices "a definite step in the right direction" and urged that such efforts be continued "for our children's sakes and their children after them." Another correspondent, C. F. Petersen, blamed equipment malfunctions on lead additives in the fuel. Continuing in a populist-environmentalist vein strikingly radical for the times, Petersen declared:

> No bank in these parts, will finance rapid transit. They make too much from car financing. So we have no rapid transit, but we do have smog and tetra-ethyl-lead gasoline, and we have our busses (from GM, controller of Ethyl) AND I DON'T THINK THE LOS ANGELES BOARD OF SUPERVISORS WILL TAKE THE FINANCIAL EMPIRE OF GENERAL MOTORS APART. We may have to get Federal legislation IF we are going to whip the problem, and that can take a long, long time. In the mean-time we may look forward to Los Angeles gradually becoming the unhealthiest city in the United States as our population continues to grown [*sic*].[45]

By the mid-1960s, Petersen's warning was part of a growing trend in Los Angeles and urban California generally, as active public awareness and vocal concern about air pollution gradually began to reawaken after the relatively dormant period of the late 1950s. Statistics indicate that after a period of decline, total pollutant levels were again on the rise from about 1962 onward. At any rate, Los Angeles citizens again began writing their elected officials to express real alarm about smog. For example, in December of 1963, Hayes L. Pagel of Glendale informed his county supervisor, in words disheartening for real estate interests, "I am just about to leave my very fine home here in this area, which I like, and move somewhere else. I can't take it any longer. . . . I can stand the taxes but not the smog. Please concentrate on this if you don't do another blessed thing." Sandy Elster called on the county and all local utilities to purchase only Chrysler automobiles for their fleets to reward that company for developing engine modifications that supposedly reduced tailpipe emissions and to punish Ford and General Motors correspondingly; he declared, "I think hitting the automobile manufacturers on an economic level, is the only thing that is going to alert General Motors and Ford, to action." Later, reflecting growing concern over public health and mounting suspicion of corporations, both sentiments typical of the times, Mary Valleley fumed to her county supervisor, "Is *Big Business* or *Industry* that contribute to this poison gas so powerful that it is above our laws?"[46]

As part of this reawakening of local environmental anxiety by the mid- to late 1960s, professionals and experts in the Los Angeles area again started offering alarming predictions about what might happen were smog not controlled. In a widely reported story, Dr. Morris Neiburger, a University of California–Los Angeles meteorologist and long-time student of the LA smog, warned of impending disaster if humanity stayed on the same path toward quick and reckless industrialization, development, and consumerism that the United States and Western Europe had pioneered. "All civilization . . . will pass away, not from a sudden cataclysm like a nuclear war, but from gradual suffocation in its own wastes," he proclaimed apocalyptically. Focusing on the internal combustion engine and the personal automobile as the main culprit for the Earth's predicament, Neiburger predicted dire consequences if, for example, the eight hundred million residents of China all had cars. He explained: "Everybody in Los Angeles drives a car, and as the living standards of the rest of the world rise, all the Chinese, Indians and Africans will demand and get the same freedom of moving about that Americans and Europeans have." Later, in August, 1968, sixty faculty members of the UCLA School of Medicine signed a joint statement and distributed a formal warning to the public stating that those able to move away from the smoggiest areas of Los Angeles should do so immediately; the medical faculty claimed they felt a "collective responsibility" to warn the community of the serious health hazard from polluted air. In December that year, yet another medical expert forthrightly declared, "Within another five to ten years, it's possible to have an epidemic of lung cancer in a city like Los Angeles." This physician also blamed the situation on the personal automobile culture that had allowed public transportation to "wither and degenerate" to the point that cities were strangling in traffic and smog. Such statements naturally fanned the flames of public fear of air pollution.[47]

Not all observers were aware of the reawakening public concern about dirty air during the mid-1960s; some even wrung their hands over its absence. In early 1967, Ronald O. Loveridge, a political scientist at the University of California's Riverside campus, conducted a survey finding that Californians generally knew little and acted less regarding air pollution and glumly concluded that it was "sheer foolishness" to expect the public to take the lead in demanding environmental reform. Yet the level of overt public concern and anger over air pollution would soon skyrocket as the terrible 1967 smog season arrived. At the end of August, the Los Angeles basin suffered the worst smog in eleven years, as the ozone concentration nearly reached the second-stage smog alert level. September and October were also awful, with persistent smog tormenting Southern Californians at precisely the time when the debate over the Dingell Amendment to the federal Air Quality Act of 1967 was occurring in Washington, D.C. Representative John Dingell, the Michigan Democrat who had attempted to delete California's special permission to set vehicular emissions standards stricter than the federal government's, made the mistake of visiting Southern California in October, 1967. This helped draw even more public attention to the battle over the federal legislation and helped inspire local radio station KLAC to air a series of programs called *A Breath of Death*, which indignantly discussed the region's atmospheric situation, exposed the auto industry's role in the Dingell Amendment, and disparagingly referred to smog as "Dingell's dust." The bad smog season helped make Angelenos that much more receptive to the broadcasts, and with further encouragement from the APCD and Stamp Out Smog, they angrily followed the radio station's exhortation to write to their legislators about it, launching a blizzard of three to four hundred thousand pieces of mail at their local, state, and federal representatives—still possibly an all-time record for any single piece of legislation over a brief period. One state senator from Los Angeles got so many furious phone calls regarding his proposed pro-industry amendment to a state bill that he promptly altered his position. A California congressman observed that "air pollution is a bigger issue than Vietnam in California, and every Democrat and every Republican in the delegation will fight to the last ditch on this"—showing the distance many local residents had abruptly come from Loveridge's somber vision of obliviousness.[48]

After the Dingell debate, Angelenos' fury over smog only continued to mount along with the general expansion of local and national environmental awareness in the late 1960s. In 1967, Stamp Out Smog was joined by a new organization, the Group Against Smog Pollution, or GASP, founded by faculty at the Claremont Colleges in Riverside County, where people had once fled to escape the smog but which by then frequently had smog worse than in the downtown area. GASP ran ads in newspapers calling for citizens to boycott the auto makers and purchase no new cars until these were made smog-free; they also held panel discussions calling for stricter emission standards and development of mass transit and offered testimony before any local authorities contemplating tougher air quality regulations. By 1969, another new group had appeared, the Clean Air Council, composed largely of scientists, engineers, and other professionals hoping to stimulate a mass mobilization in support of stricter emissions standards, public transportation, and alternative vehicles; they, too, offered expert testimony in support of pollution control. During the same period, the older organization, Stamp Out Smog, reenergized and somewhat radicalized, suggested a consumer boycott of new automobiles to put pressure on Detroit. Later, around October, 1969, after the federal government had arranged its out-of-court settlement with the four major U.S. auto makers in the antitrust suit over their alleged conspiracy to obstruct and delay the introduction of motor vehicle exhaust control devices, Los Angeles area housewife Martha R.

Hemmer expressed the sentiments of many Angelenos when she angrily characterized the "President's decision in the Smog case" as "the biggest sell-out since Pontius Pilate did his hand-washing bit." By 1970, polls found that 70 percent of Californians felt air pollution was a "very serious" problem, and many saw it as the "most serious" social problem in the state.[49]

Unfortunately for the APCD, though, the new public uproar over air pollution of the late 1960s did not stay focused on autos or Detroit. The district had had something of a free ride in public relations since the late 1950s, with nearly all public anger over polluted air, the control program, or new control devices being directed either at the automobile industry or at the state Motor Vehicle Pollution Control Board. Indeed, the APCD had gleefully lent a hand in whipping up animosity toward those handling vehicular emissions. After 1967, though, newly reradicalized Angelenos grew increasingly critical of stationary-source emissions as well and began criticizing local control authorities, whom they charged with having grown lax, sloppy, and overly forgiving of polluters during the previous several years. They also sharply criticized the APCD for allowing gains in the control of individual stationary sources to be undone and canceled by the free proliferation of new stationary sources, just as was happening with automobiles.[50]

Ultimately, in a cutting blow to the APCD, even Stamp Out Smog, "the largest, most active, and most respected citizens' group," joined other groups and local politicians in withdrawing its support from the agency. In June, 1969, in a statement before the Los Angeles County Board of Supervisors, the president of SOS, Afton Slade, bluntly told the APCD to stop just harping on vehicular emissions and start cleaning its own house better. While accepting the major role of automobiles in causing the region's atmospheric affliction, she argued that the region was "going backward instead of forward" on "nonautomotive" emissions, declaring: "We have lost the energy and initiative that we had ten years ago. Smoke stacks are mushrooming, variances have been granted on an almost wholesale basis and companies have been given years, not weeks, or months, to clear up their pollution problems." She charged the district with putting off public complaints about stationary sources and called for them to forge ahead with proposed Rule 67, which would control nitrogen oxides and would effectively "eliminate future construction of large power plants in our basin, where, in truth, they no longer have a place." She questioned why a rule drafted fourteen months earlier was still not in force.[51]

In response, APCD director Louis Fuller quickly took action to push forward the new regulation. At a public hearing on the matter in December, 1969, Fuller forcefully rejected the claims of critics of the policy that the region would suffer power failures if further power plant construction and expansion were not allowed and noted warnings of health risks if local air pollution increased any further. The board of supervisors took this advice and approved Rule 67. However, this was Fuller's last professional triumph. In March, 1970, at the age of sixty-three, Fuller retired from the APCD.[52]

The battle for Rule 67 was also the swan song of the once-mighty Los Angeles County Air Pollution Control District. Thereafter, through the 1970s, the district and Fuller's replacement, Robert Chass, an engineer with extensive technical experience but little savvy in public relations, would face steadily mounting public complaint about air pollution in Southern California, not just within Los Angeles County but also increasingly from nearby counties that choked on their smoggy neighbor's emissions. In the summer of 1972, the mayor and city of Riverside even gained national attention by formally complaining to the state governor that air pollution in the South Coast Air Basin (including Los Angeles, Orange, Riverside, and San Bernardino counties) was "putting the citizens of this basin in a position of extreme peril and/or disaster," declaring a state of atmospheric emergency, and advocating immediate implementation of an emergency plan requiring conversion of all vehicles with four or more wheels to run on gaseous fuels. Ultimately, after a grand jury investigation and hearings before the California Air Resources Board, which respectfully but sharply reprimanded the APCD for laxity on enforcement, in the summer of 1976, the discredited Los Angeles County Air Pollution Control District was disbanded. It was replaced by a new, multicounty air pollution control agency mandated by state law, the South Coast Air Quality Management District, which still presides over the most difficult air pollution problem and the most complicated, comprehensive emissions control program in the nation.[53]

Angelenos' slightly ironic victory in the battle to replace what had formerly been the most successful, prestigious local air pollution control authority in the world represented the high point of another spin of the cycle of air pollution politics and public environmental activism in Southern California. During the preceding three decades, through times of greater activism or apathy, citizens of Los Angeles County and neighboring areas had grappled with a complex and costly environmental problem. With their active or tacit approval, as reflected in the county supervisors' steadfast support of the county APCD and in specific acts by individuals or groups, the issue of air pollution remained high on the local policy agenda throughout this whole period. This overall public support allowed the APCD to make major strides in controlling the world's most notorious air pollution problem. In periods of greater public activism and agitation, rather than merely leaving the matter entirely in the hands of professional proxies, citizens took a more forward role in the policy process and considered the possibility of requiring significant changes in the behavior of local citizens and industry.

If Angelenos' awareness and confrontation of air pollution represented a form of environmentalism, though, this was still an immature sort of environmentalism. Like people of earlier times who bemoaned smoke nuisances and pointed a finger at industry, the people of Los Angeles long had difficulty accepting the degree to which they and their ordinary daily activities were helping to cause the problem—as with the furor over the garbage incinerator ban, or the long, sometimes almost fanatical insistence that the oil refineries were the sole cause of smog and were conspiring with control authorities against the public, a claim that lived on into the 1960s. Even after most residents accepted the role of the automobile in causing the area's most hated attribute, the majority still assumed that they had an inalienable right not only to drive their automobiles but to commute longer and longer distances, usually alone, in their heavy, overpowered vehicles. In a manner reminiscent of the English men and women who would not give up their "pokeable, companionable fires" or the citizens of St. Louis who demanded a smokeless fuel as cheap as dirty soft coal, Los Angeles area residents felt that it had to be somebody else's job to make nonpolluting the cars they would not dream of giving up, and they viewed with self-righteous indignation the auto makers who did not produce technological magic bullets to solve their problem rapidly enough.

Angelenos, like people elsewhere, were capable of knee-jerk reactions when new policies placed a burden on them, as with the used-car crankcase control devices; they also long held a somewhat childlike faith in a technological easy way out that would allow them to resolve their immediate woes without changing the behavior that created the problem. Most Angelenos by 1970 still had a good way to go toward a more mature environmental or ecological awareness that would comprehend the interrelationships between discrete environmental and social problems, would accept both personal and communal responsibility for these in full measure, and would seek the fundamental systemic or behavioral changes that the situation demanded. Yet some, like the activists who began contemplating alternative vehicles, vehicle use restrictions, and the promotion of mass transit, had begun to start asking the sorts of questions that led down this path, while many—including industrialists as well as ordinary individuals—had come a long way from the 1940s in terms of accepting the need for environmental regulations. If the typical American faith in a painless technological quick fix remained, it had been shaken or strained a little by the lengthy, difficult battle against dirty air, while the understanding of environmental problems as deeper-rooted human problems requiring more comprehensive solutions had advanced.

Another point raised by the example of the sometimes heroic, sometimes neurotic, sometimes apathetic actions of the Los Angeles public in the local war on smog is that meaningful air pollution control programs did not start or continue without public activism and support, but the public needed to be handled with care. There is no doubting the truth of Gordon Larson's frustrated observation that it was hard to explain such a complex matter to uninformed, nonspecialist citizens; nevertheless, this effort had to be made in a methodical, persistent way. Technical experts and governmental officials may often have dreamed of being able to do their jobs without irrational public meddling, but an agency that becomes too remote from the public often tends to slide inexorably toward ineffectuality. With all its limitations, the public, like industry, is necessarily part of the policy process, and activist citizens help to form a necessary counterweight to the persistent and understandable desire of industry and other concerned interests to delay or erode burdensome regulations. Although even the most honest and thorough public information effort of course cannot guarantee an agency freedom from irrational attacks by members of the public, in the long run agencies ignore the public at their own peril. A somewhat complacent attitude toward the public that had developed in the APCD by the later 1960s helped to bring its downfall during the 1970s.[54]

Nevertheless, for all the shortcomings of the various participants, the story of air pollution control in Los Angeles was remarkable. By the later 1970s, James E. Krier and Edmund Ursin, chroniclers of the Los Angeles smog war, would be pointing out what they perceived as the failings and inefficiencies of the smog control efforts in Los Angeles and California, particularly the reliance on technology-based government regulations and the lack of market-based approaches offering economic incentives. While these criticisms are perceptive, and while there were unquestionably flaws in policy, the fact remains that in the historical context, using Los Angeles and California as a case study of the failings of air pollution control policy is somewhat like pulling the best student out of a class to serve as a model of juvenile delinquency. For whatever the shortcomings of the Los Angeles public, the county Air Pollution Control District, state authorities, and local industry, together they had accomplished far more than any other similar jurisdictions in the nation. The picture throughout the rest of the country was generally far less encouraging.

# II

# New York

# chapter 6

# Reinventing the Wheel

*Air Pollution Control Policy in New York City, 1945–70*

The story of air pollution control in New York City between 1945 and 1970 is more typical of the nation than is that in the city and county of Los Angeles. Rather than steady progress, there was usually a steady lack of progress. Unlike the great western metropolis, the great eastern metropolis long failed to come to grips with its serious atmospheric problems, despite major public protest. Like their counterparts in many other cities and states, New York City officials were generally content to nibble around the edges of the problem, showing enough activity to assuage public clamor for pollution abatement while avoiding actions that would upset local polluters. Ignoring obvious earlier lessons from cities such as St. Louis and Los Angeles, New York authorities long preferred education and persuasion to enforcement and penalties, and for many years, any brief efforts at stricter regulation invariably gave way to renewed laxity in a recurring cycle. The results of this approach to environmental cleanup were predictably modest.[1]

During World War II, air pollution in New York City was dormant as a political issue. There had been sporadic, usually abortive efforts to purify the city's atmosphere from the end of the nineteenth century through the 1930s, including neighborhood campaigns against smoky, coal-burning railroad locomotives and dustfall studies conducted by local, state, or federal officials. The onset of World War II brought an end to even these tentative efforts, and in 1942, a smoke control bill for the city was rejected by Mayor Fiorello LaGuardia and the municipal Board of Estimate as an obstacle to war production.[2]

For a long time, air pollution in the nation's largest metropolis was a lesser problem than other environmental problems such as garbage and water pollution, for which the city already had a besmirched reputation by the end of the 1800s. Throughout the nineteenth century, New York City had not suffered from the clouds of heavy coal smoke that were plaguing the great industrial centers of the nation's heartland. This was chiefly because unlike the industrial towns of the nation's interior, New York had relatively little of the most polluting heavy industry and, like other cities of the northeast, it had traditionally relied on harder, cleaner-burning anthracite rather than on cheaper, dirtier soft bituminous coal for its energy needs. With its breezy location on coastal islands, New York was naturally better ventilated than most other cities. All of these factors helped keep New York from developing a chronic air pollution problem for many years.

By the time of the Second World War, though, the Big Apple was developing a serious air pollution problem like that of America's other large cities. The first half of the twentieth century had brought more factories and industrial development to the greater New York City metropolitan area, much of it right across the Hudson River in New Jersey. Cheaper extraction techniques and improved transportation connections allowed bituminous coal to spread to the Northeast, gradually replacing anthracite and bringing worse smoke and soot as industrial and domestic fuel users alike shifted to the cheaper, dirtier coal. By the end of the 1930s, soft coal was firmly rooted in New York City, and various neighborhoods began to have chronic, noxious air pollution problems that even the city's strong breezes were unable to blow away.

As a result of this mounting problem, air pollution began to revive as a political issue in New York City soon after the end of the Second World War. In the face of public pressure, in January, 1946, New York City Council members reintroduced the earlier 1942 smoke abatement bill at the behest of Councilman Joseph T. Sharkey, a sponsor of the earlier bill who would long remain the city's single most steadfast legislative advocate of air pollution control. However, Sharkey's bill made little headway in a city preoccupied with more immediate concerns of demobilization and building a new postwar world.[3]

By the beginning of 1947, overt public frustration with the city's dirty air was spreading, leading the city government to make a show of action on the issue. On January 8, the New York City Health Department, then the agency with primary authority to abate air pollution, announced that it would be launching an anti-smoke campaign "later this month." The following day, Joseph Sharkey, by then Democratic majority leader in the city council, promised to propose a bill stiffening penalties for smoke violations and establishing a new, separate anti-smoke bureau within the Health Department. Soon thereafter, Health Commissioner Dr. Israel Weinstein, who insisted that he already had adequate legal authority but not sufficient funding, requested an additional $117,031 to hire new inspectors and staff for smoke control in the next year's budget, which Mayor William O'Dwyer promised would be forthcoming. Concerned New Yorkers cheered at the prospect of sustained, meaningful action on air pollution at last.[4]

Their optimism proved misplaced. In what would become a recurrent cycle for air pollution control in New York City, as public interest and immediate pressure waned, so did the city's control effort. By April, 1947, as the heavy heating and smoke season drew to a close, the control campaign abruptly ended as the mayor submitted to the Board of Estimate a budget message making no mention of additional funding for smoke control. As usual in New York, the money went to the many other more traditional claims on the city's finances, and Commissioner Weinstein was left lamely arguing that his seven harried smoke inspectors had been doing a fabulous job patrolling a city of eight million. Some citizens complained about this broken promise, but with the coming of spring, even most of them lost interest, and air pollution control got bumped off the agenda yet again.[5]

The next smoke season witnessed a similar cycle. After months of little activity, in October, 1947, Sharkey reintroduced the stricter smoke law he had offered in early 1946 to make the emission of dense smoke a statutory offense. Sharkey's move again spurred the city administration to some level of action. The same day Sharkey introduced his bill, the Health Department pointed out that Mayor O'Dwyer's committee on public cooperation for air pollution abatement was still functioning, while the department was making progress through persuasion. The following day, Weinstein gave a press conference detailing how the Health Department had persuaded owners of 101 local industrial concerns to improve their furnaces at a cost of $2,662,440; the agency also touted its smoke reduction classes for janitors and building superintendents and promoted self-policing industrial smoke abatement committees throughout the city. Weinstein again defensively asserted that Sharkey's bill was unnecessary and advocated "education in proper firing methods" and "persuasion to get smokeless equipment installed" in place of "legal compulsion." Like most state and local officials of the period between 1945 and 1970, Weinstein echoed earlier, Progressive-era counterparts in insisting that progress on air pollution control would come from cooperation, not coercion.[6]

The next day, Mayor O'Dwyer joined in the celebration of the status quo on air pollution policy, announcing plans of private industry to spend over eight million dollars on smoke improvements. In addition, the New York City Board of Transportation planned to reduce pollution from one of its subway system power stations, while the Long Island Railroad had agreed to replace coal-burning roundhouse locomotives with diesel engines at all yards in Brooklyn and Queens. In the face of this sudden and ostentatious show of activity, Sharkey backed down and held his bill in abeyance. He commended the Health Department's progress but predicted that his bill would eventually prove necessary because some interests "don't care about anybody and will not do anything unless they are forced to do it through public sentiment or laws enacted by the local legislative body." His words proved prophetic.[7]

As during the previous year, the city's anti-smoke campaign fizzled by springtime. Late in March, 1948, Alfred J. Phillips, a Republican city council member from Queens, introduced a local law to stiffen penalties on smoke offenders and to create a separate "smoke nuisance bureau" in the Health Department to help win compliance from industry, which Phillips found generally ignoring existing regulations. Despite demonstrated bipartisan interest in the issue, by early April, the city administration rejected the Health Department's modest request for additional funds to boost the smoke inspection force, hire other support staff, and make needed administrative changes. Air pollution was still a policy issue conceived but struggling to be born.[8]

In early June of 1948, Sharkey offered a bill to set up a single bureau within the Department of Housing and Buildings responsible for all air pollution control. Sharkey's bill would create a five-member Smoke Control Board including the city commissioners of health and of housing and buildings and an engineer, professional chemist, and nonspecialist, the last three to be appointed by the mayor. Sharkey's law would also create new permit requirements for combustion equipment, offer more precise definitions of offenses, and stiffen penalties. His bill was one of three pending proposals on the issue, again reflecting sustained, bipartisan public interest.[9]

Despite such interest and activity, the by then traditional air pollution control cycle—a flurry of public interest stifled by bureaucratic inertia—would probably have repeated itself in the 1948–49 smoke season but for a fluke. By October, 1948, Sharkey's bill had been stalled by stiff opposition from local business and industry, and as usual, city officials sided with the major economic interests. Sharkey publicly vowed to fight any efforts to return to the persuasion approach in regulating air pollution or to delay control through incessant further studies. Nevertheless, these longtime favorite tricks to postpone or prevent environmental regulation worked their customary magic. By the end of October, the bill appeared doomed to go the way of most previous initiatives.[10]

The unexpected circumstance that derailed the status quo was the "Killer Smog" of Donora, Pennsylvania, a small industrial town of about thirteen thousand not far from Pittsburgh. After an atmospheric inversion layer held industrial smoke and gases over the town for a week, the accumulation of contaminants reached a critical level at the end of October. Between October 30 and November 1, over a third of the local population sickened and twenty local residents died from this air pollution, an event that horrified newspaper readers in cities throughout the nation. More important, the Donora incident for the first time established a clear link between air pollution and disease or even death in America, for while earlier news reports and scientific studies had made hesitant suggestions of statistical links between dirty air and disease, after Donora, there was a smoking gun with air pollution's fingerprints all over it. Also, unlike the Meuse Valley incident in Belgium in 1930, the Donora episode happened within several hundred miles of most of America's population. Donora lent a whole new sense of urgency to the air pollution issue in the minds of many Americans, including many New Yorkers.[11]

More miraculous, the Donora incident stimulated action by the administration in New York City. By November 4, Mayor O'Dwyer was back actively participating in the development of the Sharkey air pollution control bill. On November 14, Joseph Sharkey predicted that his bill would be passed before the end of the year. By early December, even the New York State Chamber of Commerce was backing the war on air pollution in New York City, advocating centralized authority within one responsible city agency, the addition of trained technical experts, and adequate enforcement powers and penalties—in short, all the things the business community had earlier resisted. Donora was clearly the catalyst in this shift toward acceptance of slightly greater coercion by New York businessmen. However, New York's smoke fighters were able to capitalize on the incident precisely because they were already mobilized. For them, if not for the innocent victims of Donora, the disaster came at just the right time.[12]

Although Sharkey's latest bill did not pass before the end of 1948, it did remain on track and was not shunted aside at budget-making time as in the past. The bill did change before passage, however. On the recommendation of the new Health Commissioner, Dr. Harry S. Mustard, the proposal was altered in mid-January to set up a bureau in the Department of Housing and Buildings strictly for control of smoke and other byproducts of fuel combustion, while atmospheric radiation, toxic fumes, and other unforeseen new problems would remain the Health Department's responsibilities. On the first of February, 1949, the city council approved the new arrangement, creating a Smoke Control Board to govern the new Bureau of Smoke Control (BSC). The board would include the commissioners of health and housing along with a licensed professional engineer with smoke control and fuel combustion experience, a mechanical engineer, and a "stationary engineer" appointed by the mayor. O'Dwyer signed the measure on February 28.[13]

In late May, 1949, the city government chose William H. Byrne, a local engineer, to serve as temporary director of the BSC and to develop and test the new regulatory system before more permanent appointees were picked. On July 6, Byrne was sworn in, looking forward to visible improvements in about two years while admonishing New Yorkers that "none was without sin" regarding smoke pollution, though he optimistically noted the progress major polluters were already making. In what would become an oft-repeated criticism, though, smoke fighters warned of a grossly inadequate budget for the new agency.[14]

Byrne probably had no idea what a complex, politically charged situation he was getting himself into. Within two weeks of his swearing in, his fledgling agency with its ten inspectors was inundated by more than twelve hundred complaints about smoke. By August 6, he had received 2,485 complaints and had learned that "careless operation of apartment house incinerators and boilers" was perhaps the single most hated cause of smoke, while large utilities and industrial plants ran a close second, and locomotives, buses, cars, and even coffee roasters drew public ire. By mid-August, over three thousand complaints had come in, or three hundred per inspector, and the BSC was frantically trying to stay on top of enforcement while accumulating needed information on emissions sources. The new agency initially enjoyed general public goodwill and hoped for increased cooperation in the future. Nevertheless, New York City, with its eight million people burning twenty-seven million tons of coal and one and a half million gallons of fuel oil a year, as well as incinerating countless tons of garbage, obviously remained a huge problem for a small agency.[15]

As a result of this mismatch between assets and responsibilities, progress was naturally limited, and New Yorkers grew impatient. To reassure concerned citizens, in January, 1950, Byrne ordered a survey of trends in soot, fly ash, and dustfall in the city, which found substantial improvement during 1949. Local newspapers and smoke control advocates and even some United States Weather Bureau scientists questioned the validity of Byrne's figures, showing the degree of public suspicion that was already developing over the city's control program, while critics continued to demand major budget increases. Meanwhile, Byrne's temporary bureau had fallen farther behind, as complaints in 1949 mounted to 5,521 regarding 3,422 locations, of which 1,546 remained unchecked.[16]

Public pressure for greater results continued to mount until Byrne finally got off the hot seat in late April, when the mayor appointed permanent leadership for the Bureau of Smoke Control. The new chief, William G. Christy, had been a smoke abatement engineer in Hudson County, New Jersey, since 1930 and had organized the nation's first countywide Department of Smoke Regulation there during the 1930s. Christy had later made fuel and air pollution surveys for various municipalities, including St. Louis. The new deputy director was William S. Maxwell, a decorated veteran of two world wars who had started as a simple fireman in the U.S. Navy in 1916 and worked up to being a professional mechanical engineer and rear admiral before retiring in 1949. Most recently, Maxwell had headed the radiological section of the navy's Bureau of Ships, establishing a radiation laboratory on the West Coast and attempting to decontaminate naval vessels used in early atomic tests in the Pacific. Christy and Maxwell joined Housing Commissioner Bernard J. Gilroy, Health Commissioner John F. Mahoney, and a nondescript stationary engineer, Henry D. McEvoy, on the new, permanent Smoke Control Board. The directors also gained increased funding and office space.[17]

Christy started out talking tough, promising that the new smoke rules promulgated in early May, 1950, would be "strictly and uniformly applied." He offered hope that new technological improvements such as automatic coal stokers and improved apartment house garbage incinerators could virtually eliminate smoke and fly ash in New York City. He also made plans to introduce as a basis for enforcement the federal government's Ringelmann Chart for measuring smoke density.[18]

As was a common story with air pollution control programs in other cities and states, though, fairly soon after Christy took office, he started to back down and compromise on many issues. In early May, he supported a proposal by marine and harbor interests to form a voluntary association to police their own emissions, calling this "the right approach" and urging other industries to follow this example, since his agency depended on "public cooperation and voluntary effort" and could not possibly find and cite every smoke violator in the city. The marine and harbor interests' self-regulation would later be found to be lax at best. By late June, facing protests from industry, the Bureau of Smoke Control abandoned its plans to use the Ringelmann Chart and planned to adopt a more lenient system. It also dropped a requirement that furnaces and boilers burning soft coal should have approved mechanical firing devices. Still later, after proposing controls on the sulfur content of fuels in an effort to reduce sulfur dioxide emissions in the city—a recommendation from Raymond R. Tucker of St. Louis during a brief inspection tour to help city officials plan a stricter control policy—Christy promptly backpedaled in the face of industry complaints about higher fuel costs. In short, Christy proposed, then quickly abandoned, many of the proven smoke control techniques used earlier in St. Louis and Pittsburgh. In justification of Christy's backsliding, even some pollution control advocates agreed that the Ringelmann Chart might be problematic to use in New York City's skyscraper canyons and that certain proposed smoke regulations might be unenforceable. Still, Christy's hasty retreats boded ill for the creation of a stubborn, effective air pollution control program in the nation's largest city.[19]

Due to such perceived backsliding, Christy gradually came under fire from many directions. In mid-March, 1951, Joseph Sharkey urged stricter enforcement and harsher penalties, advising the BSC to "go after the big fellows" such as large industrial plants and the city's own facilities, along with smaller offenders such as apartment house owners. Sharkey pointed to the New York City Board of Transportation and Department of Sanitation as two "constant violators," proclaiming: "They have no more right to spew their smoke and soot upon the people of this city than any industrialist in town." At the same time, Christy was blasted by some industrialists for presuming to regulate them at all, particularly for requiring permits for combustion equipment prior to installation—a necessary feature of any successful air pollution control program but one to which many of the nation's independent-minded businessmen were not accustomed.[20]

There is no question that Christy and later control officials faced a difficult situation. Besides the innumerable small smoke producers in America's largest city, there were some intractable large polluters, each of which alone had smoke problems that would require millions of dollars to abate. Worst of all was the major electric utility in New York City, the Consolidated Edison Company (Con Ed), then the second largest such utility in the nation, which burned about half the soft coal in the whole metropolitan area and claimed legal protection in its quasi-official status as a chartered legal monopoly. Basically, any progress at Con Ed had to be purchased with politically unpopular rate increases. Next worst after Con Ed was the city's own Board of Transportation, with its huge generating stations to power the New York City subway system, along with thousands of dirty, smoky old buses. Various facilities of Con Ed and the Board of Transportation were hopelessly antiquated by 1950. Refurbishing these power stations would mean taking them out of commission for months or years even as demand for the power they produced went on increasing without limit. To build additional stations for reserve capacity while others were modernized would also take years and massive capital expenditures drawn either from rate hikes or equally hated tax increases.

As a result, Christy continued what would become a long tradition, begun under temporary director Byrne, of trying to coax and cajole major polluters to cleanliness through sheer force of rhetoric. Byrne, Christy, and later air pollution control commissioners sought to persuade the largest, most incorrigible polluters, against their will, to stay on a path toward gradual progress; at the same time, they tried to reassure the public that progress was indeed being made with the largest smoke violators, often viewed incorrectly as the sole cause of the city's dirty atmosphere. Already during Byrne's tenure, to soothe New Yorkers irritably waiting for visible signs of progress, the BSC had begun hopefully reporting the millions of dollars that Con Ed and the city departments had earmarked for air pollution abatement.[21]

While trying to squeeze these unwilling camels through the eye of the needle toward greater cleanliness, the BSC demonstrated official activity by slapping small fines on apartment house owners or managers. Apartment houses, with their often primitive, smoky furnaces, boilers, and garbage incinerators, were allegedly responsible for over half of all New York City's visible air pollution and were a frequent object of complaint, but many New Yorkers thought it unfair that a handful of apartment house owners should be dragged into court and fined while the Board of Transportation's notorious Fifty-Ninth Street generating plant or Con Ed's Ravenswood and Hell Gate stations were all uncontrollably, and seemingly unrepentantly, belching out more smoke than thousands of apartment houses put together. Multiple dwelling owners wanted their economic difficulties respected like those of the "big fellows," while militant smoke fighters wanted both big and little offenders cleaned up promptly and grew ever more vocal in their demands.[22]

The BSC had other headaches as well. In early May, 1951, in the wake of much-publicized U.S. Senate hearings on organized crime in New York City, it came out that there had been irregularities in the appointment of the stationary engineer on the Smoke Control Board, Henry D. McEvoy, back in July, 1949. Mayor O'Dwyer was legally supposed to wait for recommendations from professional engineering organizations before picking any of the three appointees on the board, but he did not wait on McEvoy, previously a boiler inspector for the Housing and Buildings Department who enjoyed no distinction save the "strong backing of the Queens Democratic Organization." Besides being a political hack with no professional qualifications, McEvoy faced allegations that he used his office to pressure smoke violators, under threat of court action, to purchase equipment from a Queens boiler company of which he was secretly a partner. There were also financial irregularities: officially, the stationary engineer on the board was to receive an annual salary of $5,850, but McEvoy was pulling in $8,500 a year. In mid-May, McEvoy quit; in early June he was indicted and charged with perjury. Underfunded and understaffed, discredited by laxity and humiliated by corruption, the BSC made an easy target, and criticism only grew sharper.[23]

In an effort to answer such criticism, in late June, Christy announced how "accumulated pressure . . . to prosecute smoke violators" necessitated a new BSC operating policy emphasizing the "summoning to court of as many chronic violators as possible." At the same time, Christy publicized the bureau's achievements of the past seventeen months: 11,467 complaints investigated, with 5,654 problems corrected and 2,062 hearings held. He also promised to add ten new inspectors by July 1, 1951, and announced that while violators would still be warned and given a chance to correct their smoke conditions, incorrigible offenders would be prosecuted "with renewed vigor." Christy insisted that there was a "definite reduction in the amount of smoke in New York now" compared to January, 1950, but admitted that "smoke elimination requires constant vigilance." New Yorkers remained suspicious, however, and public impatience continued to grow.[24]

In the face of such sustained public pressure for action, fault lines within the New York City air pollution control program began to open in late July, 1951. On July 19, Maxwell, temporarily serving as chief of the BSC in Christy's absence, announced a "Smoke Attack Set on 'Big Offenders,'" including Con Ed and divisions of the city government. Setting his sights on chronic major polluters such as the Department of Sanitation's big municipal incinerators and the Board of Transportation's most antiquated subway power stations, Maxwell declared: "A subpoena shall be issued to every person who has violated the regulations more than once." Suffering New Yorkers eagerly anticipated an end to air pollution control by persuasion alone for the big boys.[25]

Their hopes were premature. By July 23, when Christy was back in town, newspapers trumpeted the growing divide in the BSC leadership. In Christy's absence, Maxwell had slapped citations on Consolidated Edison, the two most incorrigible municipal departments, and even a local facility of the U.S. Army, and Christy was furious. Maxwell declared that city departments and other such big governmental offenders would "not get any special immunity" from him, concluding, "If we are not going to follow up on such violations, why have a Smoke Control Board?" Christy, however, doubted that the BSC had legal authority to force compliance from other city departments, let alone the federal government.[26]

The rift in the BSC continued to widen in the following days, spreading to the wider city government. Bernard J. Gilroy, head of the Housing and Buildings Department within which the BSC remained ensconced, proclaimed that the BSC would henceforth have "one point of view"—Christy's. Mayor Vincent R. Impellitteri supported Gilroy and Christy in this matter, though he tried to do so quietly for fear of losing political capital. On the city council, though, Joseph Sharkey endorsed Maxwell's firmer policy "without qualification," observing: "The only way this town can be cleaned up . . . is by pursuance of a 'get tough' policy, and when I say 'get tough' I mean just that."[27]

Maxwell himself continued to refuse to accept bureaucratic realities meekly, issuing further inflammatory pronouncements, such as advocating "using the big stick" to get results and affirming that the BSC had "tried the kid glove method now for a year and a half," and it had not worked. Maxwell also refused to pull back his "flying squad" of five inspectors specially selected for a drive against industrial polluters. The ex-admiral dramatically led his men on a sweep through Brooklyn and later created a second team so that the two flying squads could comb two different boroughs at the same time. Maxwell's first team went on serving violation notices requiring that problems be corrected or correction plans be submitted to the BSC within ten days; they hit Queens even as Gilroy was demanding harmony within the bureau, ordering Christy and Maxwell to make no statements to the press without clearing them with him first and insisting that there would be "no prosecution of Federal, state or city agencies for smoke violations." However, even Gilroy conceded to Maxwell that large private violators would no longer receive "special consideration." Through his tough stance on local violators, Maxwell became the first public hero of air pollution control in New York City. Christy, by contrast, became the subject of calls for resignation.[28]

Meanwhile, the embattled Christy had other headaches besides being outshone by his deputy. During the summer of 1951, when the Maxwell affair was going on, newspaper reporters uncovered a cozier relationship than some New Yorkers found appropriate between the BSC and the Commerce and Industry Association of New York, an organization with a reputation for working to delay any progress on air pollution control. On July 31, the *New York Times* reported how "an anonymous complaint" to the mayor's office had led to the disclosure that the business group had earlier released in the BSC's name an appeal for a "cooperative program with users of fuel" rather than a "crackdown" on polluters. A public relations representative for the association explained that he had derived the press release from a draft written by Christy and that the association's executive vice president made mimeographic and other services available for the BSC director. The association generously helped Christy after he visited their office to complain that he had been unable to "get our story before the public" due to inadequate public information facilities, which prevented his answering critics' charges that the smoke control office was "getting nowhere." The association spokesman emphasized that the stress on cooperation was all Christy's idea. This perceived collusion with major polluters brought further calls for Christy's resignation from anti–air pollution militants.[29]

Because of all the confusion and scandal within the BSC, a movement to replace it altogether gradually gained strength. In mid-August, 1951, Sharkey introduced a new bill to replace the BSC with an independent Department of Air Pollution Control, though he explained that he would not push for a new agency while current enforcement was fairly vigorous thanks to Maxwell's insubordination. The BSC ultimately got another year to prove itself, but the latter part of its existence was effectively probationary due to the constantly impending threat of replacement.[30]

In the meantime, during its final year, the Bureau of Smoke Control became entangled in two major enforcement battles that helped to keep New York's smoke control program largely on hold and that starkly pointed out the weaknesses of the city's early approach to air pollution control—weaknesses shared by many other early local control campaigns throughout the nation during the pre-1970 period. One of these battles was with the largest single polluter in the metropolitan area; the other was with numerous minor offenders who collectively added up to being one of the city's largest air pollution problems. Both of the BSC's adversaries proved stubborn and intractable.

The BSC's toughest single adversary was the Consolidated Edison Company. In addition to being the second largest utility in the nation, generating most of the electricity for the New York metropolitan area, and having a powerful voice in city government, Con Ed had long been the single largest generator of air pollution in the city. When Admiral Maxwell slapped violation notices on some of Con Ed's oldest and smokiest generating stations while Christy was away during July, 1951, he started a process that would embroil the BSC not only in a long drawn-out legal battle but in a life-and-death struggle for the agency.

When Maxwell first launched his attack on the big offenders and served violation notices to two Con Ed generating stations, he also requested that local magistrates cooperate in meting out stiff fines and even jail sentences for chronic offenders. The law provided for such penalties, but they had never been used, and despite Con Ed's repeated violations at a number of its facilities, it had never been fined even once and effectively remained above the law. Taking serious action against such a major offender required a total reorientation of policy, which the hesitant Christy resisted.[31]

Like the Board of Transportation, Con Ed was invited to explain its smoke abatement plans to the Smoke Control Board. Company representatives unveiled an eight-million-dollar, two- to four-year program to regulate its emissions to the point that they would cause no "public annoyance" or "detrimental effect on the public health." The Con Ed spokesmen also asserted that Con Ed was not the major air pollution problem and warned that its improvements would mean higher utility rates, noting that the company would be "going ahead with its program in the belief that its customers and investors were willing to pay the cost." They contested a violation notice served on the Hudson Avenue station in Brooklyn, arguing that no one could comply with the city's regulations on visible emissions if these were interpreted literally, since they would prevent *any* emissions or combustion at all most of the time. Maxwell, however, disclaimed such literalism, saying, "When I see chronic conditions, black smoke pouring out, that's when I step in."[32]

Evidently Maxwell's views prevailed within the Smoke Control Board this time, for on August 17, 1951, the BSC announced that Con Ed would face legal action for failing to comply with the agency's regulations or to make adequate progress at two of its smokiest stations. Con Ed's attorney tried to get the court summons thrown out as "inconclusive." This was only the first of many similar pettifogging ploys by the giant utility to obstruct justice, including calling for superfluous hearings, seeking frivolous changes of venue, questioning the jurisdiction of courts, and challenging the very constitutionality of the BSC itself. The BSC felt it had an airtight case due to Con Ed's long history of violations, and the agency pursued the recalcitrant corporation through every level of appeal. However, Maxwell alleged that other industrial concerns and building operators who had taken preliminary steps toward correcting their smoke problems had "slowed up their plans since the Edison Company's move," waiting to see whether the BSC might be declawed in time for them to avoid the costs of air pollution control equipment. Thanks to Con Ed's challenge to their authority, the BSC's whole regulatory program was partly hamstrung while the wheels of the legal machinery slowly turned and anxious New Yorkers awaited the outcome.[33]

Ultimately, the BSC prevailed despite all Con Ed's legalistic trickery. In late November, 1951, the Bronx Supreme Court found the BSC's enforcement procedure in the case to be legal and proper and dismissed Con Ed's latest delaying action. An elated Maxwell promised to forge ahead on twenty major violation cases that had been held up while the utility was conducting its legalistic maneuvers; he even anticipated reviving his flying squads to hunt down new violators. Con Ed's last trick also missed its mark, as the final appellate court in New York State confirmed the earlier decision and rejected Con Ed's writ petition in early March, 1952. On July 17, 1952, after nearly a year of legal obstruction, Con Ed was fined the maximum of one hundred dollars for creating a smoke nuisance at the Hell Gate power station, though company attorneys threatened more appeals and requested further adjournments. Despite the token fine for a mighty corporation, Maxwell hailed the fine as a great victory, showing that his agency pursued big violators as well as little ones. Yet though the underfunded, beleaguered BSC had successfully defended its authority against a challenge by one of New York's most powerful entities, Con Ed had managed to monopolize and obstruct the BSC for a year, tying up other enforcement activity, only to suffer a pinprick.[34]

The BSC's other major drawn-out battle was fought not with a mighty corporation on which the public could easily focus resentment but against a much more diffuse problem for which the public resisted taking responsibility—apartment house garbage incinerators. Though there had long been public complaints, the BSC first showed sustained interest in the domestic incinerator issue in January, 1952, when Maxwell, speaking to the Women's National Republican Club, branded apartment house incinerators cumulatively the worst air polluters in Manhattan, responsible for nearly 30 percent of the city's total atmospheric contamination, and called for a new city code to regulate new installations. Moreover, with their thick smoke, dust, and nauseating fumes, their frequent nighttime use when people were trying to sleep made them a particularly obnoxious nuisance. Maxwell labeled incinerators the "toughest problem in New York's air pollution picture," with each of the roughly thirty thousand of them poorly designed and "a potential violator." The deputy director predicted that a voluntary approach to solving the problem would be inadequate—compulsion would be necessary.[35]

Since the targeting of domestic incinerators might seem a backing away from Maxwell's customary policy of pursuing larger violators, one might think that this reflected a compromise at the divided BSC to return to going after the little guys. There may be some truth to this. However, as one of the largest cumulative sources of offensive air pollution and public complaint in the whole city, there is no question that incinerators were a proper object for enforcement. Yet, as Maxwell observed, incinerators were a thorny problem. Decades earlier, citizens and developers in New York and other American cities had found that, through the magic of combustion, phase changes, and the prevailing winds, they had a cheap, quick, easy solution to disposal of much of their sewage and solid waste. Consequently, during the 1930s and 1940s, cities like New York were largely built around the apartment house incinerator, such that by the 1950s, the incinerators were decidedly a built-in problem to which many citizens regularly contributed. While apartment dwellers might complain bitterly about the stench from other buildings' incinerators, they were stuck using them until either the costly repair or replacement of incinerators or the establishment of an expensive new system of sanitation for New York's millions of apartment dwellers, both of which would lead inevitably to tax or rent increases. Meanwhile, New York City's several thousand apartment house owners could decry persecution of the "little fellows" and point their fingers at the huge, belching smokestacks of the city departments and Con Ed. There was simply no easy, painless way out as regards either the big polluters or the smaller ones, and as such, in the absence of real civic determination, incinerator smoke, like Con Ed smoke, remained an ongoing, unresolved problem into the late 1960s. Unfortunately, front-end solutions to these atmospheric problems—such as using less electricity and generating less solid waste, or energy efficiency and garbage recycling—were never even considered; most Americans of the early postwar decades simply did not think that way.[36]

Despite the intractability of the incinerator problem, the BSC attempted to address it. Shortly after Maxwell's speech, Joseph Sharkey sponsored a bill to regulate the design and construction of new apartment house incinerators, as Maxwell requested, though this measure was soon stalled in the city council. In June, the BSC took action on its own and changed its regulations to require that future incinerators must have a dual-chamber design, with the second hot enough to consume any gases left unburned by the initial combustion chamber. The new incinerators were also to have fly ash settling chambers and automatic safety control devices. Noting the failure of voluntary compliance, Maxwell observed with disgust that even many new incinerators were still "no more than indoor bonfires."[37]

The incinerators were naturally still a problem with the onset of another summer season, when the incinerators' obnoxious smoke and fumes prevented urbanites from opening their windows to catch cooling evening breezes in the days before air conditioning was commonplace. As such, summer brought a whole new wave of complaints about incinerators from unhappy apartment dwellers. In late July, Maxwell noted that apartment house owners were making matters worse by using their faulty incinerators under cover of night to hide their smoke from BSC inspectors by day. Maxwell called on the city council to enact a year-old Sharkey bill to ban nighttime waste burning between 10:00 P.M. and 7:00 A.M.; the bill had been languishing unnoticed in the city council's General Welfare Committee since January. He also urged limits on daytime incinerator use.[38]

Progress on both pieces of legislation remained glacially slow, however, so the BSC again acted on its own, proposing in early September a new rule to limit operation of multiple dwellings' incinerators to between the hours of 7:00 A.M. and 7:00 P.M. In making this announcement, Director Christy noted that 70 percent of the summertime complaints to the BSC involved incinerators and that the bureau was swamped. The BSC also proposed a new rule that low chimneys within thirty feet of another building should be raised above the roof level of adjoining buildings, even if the neighboring structures were over five stories taller, since BSC inspectors in borrowed police helicopters had actually seen smoke from low chimneys pouring directly in through the windows of nearby office and apartment towers.[39]

The BSC's action helped force the city council's hand. On September 10, 1952, council members passed the Sharkey measure restricting incinerator use to between 7:00 A.M. and 9:00 P.M. Christy happily dropped the BSC's own proposal and predicted "Smoke Free Nights for All." Nevertheless, at the end of October, Maxwell and Christy still reported widespread "non-compliance" with the nighttime incineration ban and warned that inspectors might soon be assigned to night duty to catch violators. Despite all warnings, threats, and periodic crackdowns by the BSC and its successor agency, New York City made little progress on cleaning up incinerators in the face of stubborn opposition from apartment house owners who resisted corrective costs, taxpayers who feared having to pay for expensive garbage pickup, and the incinerator industry lobbyists, who religiously attended every public hearing on the issue.[40]

In the face of such seemingly hollow victories and limited progress by the troubled Bureau of Smoke Control, the movement to replace it with a new and improved agency grew steadily. On July 2, 1952, the New York City Council voted 20–2 in favor of a modified version of the 1951 Sharkey bill for a unified, comprehensive, independent air pollution control agency. On August 22, the Board of Estimate also voted 10–6 in favor of the measure to create a new Department of Air Pollution Control (DAPC) with carefully defined duties and authority to regulate "all forms of atmospheric pollution including smoke, fly ash, and chemical fumes emitted from apartment houses, industrial plants and other sources." Mayor Impellitteri signed the new law on September 24.[41]

The new commissioner of air pollution control was Dr. Leonard Greenburg, previously the executive director of the Division of Industrial Hygiene and Safety Standards of the New York State Department of Labor. Greenburg was eminently qualified for his post, with a degree in sanitary engineering from Columbia together with a Ph.D. in public health and a medical degree from Yale. He also made a promising early show of activity. Immediately after being sworn in on November 18, 1952, Greenburg promptly sought better office space, a proper laboratory, and a higher budget of around $450,000 for the next fiscal year—a long way from the token expenditures of 1947. He also laid plans for a comprehensive assault on air pollution and sought to put the city's control program on a sounder scientific footing, reflecting the characteristic belief of scientists and engineers that air pollution was basically a scientific and technological problem that would solve itself once scientific information was improved, expanded, and disseminated to the public. Many smoke control advocates eagerly anticipated "A Turning Point on Smoke."[42]

However, for all Greenburg's qualifications and commitment, and despite the best efforts of his staff, New York City's atmospheric woes did not go away after 1952. Rather, cycles established earlier merely kept repeating themselves. Ignoring the lessons of the BSC years, Greenburg, confident in the power of scientific research, returned to a policy of public education and persuasion. Meanwhile, the new DAPC still had to face the stubborn intransigence of large industrial concerns and apartment house owners with a budget and staff that were never adequate to face the problem, especially as this problem multiplied in costliness and complexity as understanding of it grew.

Like his predecessors, Greenburg directed much of his department's efforts toward quelling the smoke nuisance from apartment house incinerators. In June, 1953, Greenburg rediscovered what his predecessors had already learned—that apartment house incinerators in the city caused 30 to 40 percent of the total volume of air pollution. He later announced new rules requiring the installation of approved filters and limiting incinerator use to between the hours of seven in the morning and three in the afternoon. In May, 1954, the hours were extended to five o'clock and settling chambers were required for new installations. Two years later, the problem remained mostly unaltered, so Greenburg launched a drive against violators of the incinerator rules, which petered out after a few weeks. In April, 1959, Greenburg was still plaintively asking the city's apartment house superintendents to burn only between the appointed hours, a policy that continued to be widely ignored. In the face of such public uncooperativeness, the DAPC considered banning apartment house incinerators altogether. They backed away from this firm stand, but even modestly stiffened incinerator control regulations brought a collective scream from the city's builders and incinerator interests, who charged unfairness and hardship. Not surprisingly, Greenburg never got the garbage incinerators under control during his tenure in office, and science and exhortation alone did not help much.[43]

The story was the same for most other polluters that the new DAPC sought to rein in. Older sources such as the transportation authority's 59th Street subway system power plant remained out of control throughout the 1950s, with Greenburg throwing up his hands and declaring that further court action was useless as long as the city would not provide the money necessary to replace the facility, while the Con Ed station by the United Nations Headquarters still belched black smoke all over that and other buildings into the early 1960s, to the annoyance of foreign diplomats and the embarrassment of New York City officials. Like Christy long before, Greenburg in 1956 sought the authority to ban the use of high-sulfur industrial fuel oil, a proposal which was suffocated by the agitation of business interests and the overall disinterest of the city government. Despite sporadic enforcement drives, the overall air pollution problem only continued to expand.[44]

Part of the recurring air pollution control cycle naturally involved persistent requests for an adequate budget, requests mostly refused by the uninterested administration of Mayor Robert F. Wagner, Jr. Although in early 1954, Greenburg begged for $600,000 to "catch up on years of neglect," not until 1956 did the DAPC's budget reach that level or higher, and in 1958 it was cut in an austerity program that held it around $600,000 through 1962. In that year, Governor Nelson Rockefeller helped to enact a measure to use state funds to pay for half the cost of New York City's air pollution control expenditures up to a million dollars a year, and also during the 1960s, New York City became eligible for certain cost-sharing programs of the federal government. However, the city mainly took advantage of such incentives to maintain air pollution control budgets at about the same level while spending less of their own money. In 1965, the ultimately victorious New York City mayoral candidate John V. Lindsay pointed out how total New York City air pollution control expenditures rose above a million dollars for the first time only during the 1964–65 fiscal year, of which New York City itself was actually paying less than half, or less than they were spending in the late 1950s. A million dollars sounds like a lot compared to the 1940s, but Los Angeles, a less populous metropolitan area, had been spending over three times that much for many years.[45]

All the while, recurrent assurances of progress continued to collide with contrary evidence. In 1954, Greenburg claimed to see progress, like Byrne, Christy, and Maxwell before him. That same year, Mayor Wagner was more guardedly optimistic, cautioning that "complete elimination of air pollution is a gigantic task which will take years of relentless effort and laboratory experiments" but nevertheless predicting "substantial progress" within the next few years. Similarly, in November, 1955, the DAPC claimed to be gaining on air pollution. However, a 1956 study found average sootfall levels actually worsening, indicating that visible emissions were still out of control even as newer, invisible threats such as carbon monoxide were being revealed. This led Greenburg to change his tack in his budget request of 1957, noting ominously that "the shadow of danger from air pollution is looming ever larger over our city" and calling for preparations for air pollution emergencies.[46]

Later events would further suffocate the naive hopes of earlier years. A major blow against optimism came in March, 1960, in a report to the mayor from Acting City Administrator Lyle C. Fitch calling for a "Major Overhaul" of the Department of Air Pollution Control. Among other criticisms, the forty-page Fitch report found "little evidence" of any basic improvement in the city's air contamination situation since the department had come into being in 1952. Fitch concluded pessimistically:

"The best that can be said . . . is that the volume of pollutants would probably have been far greater if there had been no control program." Fitch found numerous specific faults with the management of the department. He even criticized the agency's program of laboratory construction and basic scientific research, one of Greenburg's proudest achievements, finding that these had never been adequately integrated with the department's total control program. However, the city administrator reserved most of his fire for the department's most basic functions, inspection and enforcement. The report branded the agency's inspection system "poorly conceived" for basing enforcement almost entirely on following up on citizen complaints rather than on routine patrol work—a practice the BSC critics had condemned prior to the initiation of Maxwell's dramatic sweeps by his flying squads. In recommending the establishment of a "motorized, radio-equipped, city-wide complaint squad," Fitch was only advocating the sorts of modernizing improvements already common in cities with better-funded, more effective control agencies, such as Los Angeles and Chicago. By 1960, the overworked twenty-eight inspectors of 1952 had increased only to thirty-six, still hopelessly low for a city of more than eight million, and due to disorganization, the city was not even getting its money's worth out of that inadequate number. Altogether, the report implied that New York City had spent seven years getting nowhere on air pollution.[47]

Above and beyond such administrative failings, Greenburg had stubbornly failed to learn the main lesson of the Christy/Maxwell days, a lesson control agencies throughout the nation took forever to learn: that scientific research and persuasion alone brought few if any results. After five years on the job, Greenburg himself finally confessed his disappointment with the results of his department's educational work and helpful advice in pursuit of voluntary compliance from polluters, and he admitted the need for stricter enforcement, a realization that critics noted came "a little slow." Even after this revelation, Greenburg persisted in slapping fines on petty violators while only holding discussions on control plans with major industrial polluters, opening his department up to charges of only taking action against the "little fellows." All in all, Greenburg showed no sign of having learned anything from the earlier, troubled experience of the Bureau of Smoke Control, and he repeated its mistakes for seven years.[48]

The Fitch report in 1960 did shake up the DAPC somewhat, as did the arrival of a new commissioner of air pollution control, Arthur J. Benline, that same year. If some institutional reforms followed in the wake of the Fitch report, though, the overwhelming fact remained that New York City's air pollution problem was much too big, and was expanding much too rapidly, for its control agency to handle. Upon taking office on March 8, 1960, just a few days after the critical report was released, Greenburg's successor Benline defended the past work of his agency and remarked that he'd happily fulfill Fitch's recommendations if the city government would raise his inspection force from the current thirty-six to about one hundred and boost his budget to $800,000 from $600,000. Although the budget gradually grew, the inspection staff barely did—Benline was still asking for an increase from thirty-two actual field inspectors to ninety in 1966.[49]

Throughout the 1950s and early 1960s, there were also other, darker intimations that the city was not getting on top of its atmospheric problems. In November, 1953, a week-long temperature inversion trapped pollutants over the whole northeastern United States from southern New England down to Virginia and west to Ohio. Authorities soothingly assured the public that the smog was irritating but not dangerous, though the *New York Times* reported in a front-page story how five telephone operators were overcome by the contaminated atmosphere and suffered difficulty breathing and twilight vision—symptoms of oxygen deprivation—while the city's birds were unusually agitated. Yet the experts still claimed that New York City would probably never experience lethal conditions like at Donora or in London, and public anxiety faded for a time.[50]

Later, the experts were not so reassuring. In September of 1963, the director of climatography of the U.S. Weather Bureau informed the International Biometeorological Congress at Paris that serious air pollution incidents would likely become more frequent and severe as population, industrialization, urbanization, and motor vehicle use increased, so that the East Coast of America could expect a major attack once every three years. As if to underscore this claim, in another front-page story in late October, 1963, Dr. Hollis S. Ingraham, New York State health commissioner, declared a "smog alert" over the whole southern third of New York State due to stagnant air conditions and asked citizens to curtail all unnecessary combustion as carbon monoxide levels rose dangerously high and people suffered from stinging eyes. These conditions blew away before much harm resulted, but the following May, Dr. Alvan L. Barach of the New York Academy of Medicine publicly testified that the city had escaped "near disaster" the previous fall "by chance alone."[51]

The 1963 episode brought renewed action from the New York City government as the cycle of public interest in air pollution swung back toward the agitation of the late 1940s and early 1950s after the relative quiet of the later 1950s. In 1964, City Councilman Robert A. Low, who replaced Joseph Sharkey as the city's leading legislative advocate for air pollution control after Sharkey got caught on the wrong side of a political squabble within the city's Democratic Party, introduced legislation to do what both Christy and Greenburg had sought to do long before—limit the sulfur content of fuels burned within the city limits to reduce excessive sulfur dioxide concentrations. Low held public hearings on his proposal. Later in the year, he became head of a committee appointed by the city council to study the problem. Unlike so many other such special committees, whose findings were ignored and soon forgotten, Low's committee in 1965 produced a hard-hitting report charging outright what many scientists and engineers still waffled on: that air pollution was a definite threat to human health causing increased death rates through respiratory disease and lung cancer. The report enumerated forty-four specific recommendations for clearing the city's air, including banning the combustion of more sulfurous coal and fuel oil in homes or factories by 1967 and requiring new control equipment for other users, such as Con Ed, that were allowed to continue burning soft coal—the exact same things St. Louis and Pittsburgh had done long before. Hearings before Low's special committee had reemphasized that automotive air pollution was a potentially serious additional threat to New Yorkers and their atmosphere, so the Low committee report also called for statewide requirements for automobile exhaust control devices and for new regulations regarding buses.[52]

This new activity on the city council was not immediately reflected in the Department of Air Pollution Control, however. Commissioner Benline admitted as much in a report released in late June, 1965, on the costs of dirty air. Noting that the nation lost some $11 billion a year, and New York City $520 million a year just from air pollution–caused property damage, Benline observed with resignation that the city's fight against atmospheric filth was presently merely "a holding action, hoping that we can keep things from getting worse." Sixteen years after the city took its first major steps to control aerial contamination, victory in the war on air pollution seemed just as far away as ever.[53]

As a result of reawakening public concern about the matter, air pollution control first became a major issue in a New York City mayoral election in 1965. Nearly all mayoral candidates from across the political spectrum that year chastised the departing Wagner administration for failing to get the problem under control during more than ten years in office and offered their own proposals for doing so. Public interest had revived enough that decrying dirty air was viewed for the first time as a sure way to win votes in a citywide election.[54]

The next few months brought reason for hope that New York City might finally get its air pollution problem in hand. The election in November, 1965, of Republican mayor John V. Lindsay brought to office a mayor with a greater personal interest in environmental issues generally, and air pollution in particular, than any of his predecessors had shown. Lindsay promised to bridge party lines and work with Councilman Low, a Democrat, to purify the city's air. Page-one news on May 4, 1966, was passage of a bill sponsored by Low, and praised by council Democrats and Republicans alike, to ban soft coal for heating, to forbid incinerators entirely in new buildings, and to require within three years 99 percent effective pollution controls on Con Ed and other users of bituminous coal. Mayor Lindsay agreed to sign the bill and announced that he would appoint a new air pollution control commissioner, Austin N. Heller, to replace Arthur Benline.[55]

On May 9, 1966, Mayor Lindsay's Special Task Force on Air Pollution, appointed right after the election and headed by the distinguished editor of the *Saturday Review*, Norman Cousins, released a report finding New York City's air the most polluted in any major U.S. city and offering numerous recommendations to revamp the current inadequate air pollution control program. Lindsay then vowed to revive the city's "dormant" DAPC. On May 20, Lindsay signed the Low bill, which its backers grandiloquently called the toughest city air pollution control law in the nation. In July, Air Pollution Control Commissioner Heller mapped out a new abatement strategy using more technology and modern management techniques made possible by the unprecedented budget of $1,649,063. In October, Mayor Lindsay announced a five-year plan to fight air pollution, which he called, truthfully, "the hardest-hitting, most aggressive campaign to clean up the air in New York City history." This plan would reorganize the Department of Air Pollution Control and nearly quadruple the field inspection staff from twenty-seven to ninety-four; would involve study of the fundamental nature of pollutants and closer monitoring through automatic sampling equipment, with release of daily records to the public; and would mobilize the business and scientific communities and the general public through expanded public education programs about air pollution. This was to be accomplished through a budget increase from $1.6 to $3 million, of which the federal government would pay $1.6 million and New York City and New York State would each spend $700,000. Austin Heller stated, "This will give you the kind of air you will be pleased with." The *New York Times* was more cautious: "The Lindsay administration has made a commendable start in strengthening a money-starved, understaffed program. But nothing is to be gained by promising a total victory that is nowhere in sight."[56]

As it turned out, the *Times* was right to be cautious in its optimism, for the DAPC continued to have trouble reorienting its inspection and enforcement program. In June, 1965, the city council's Special Committee for Investigating Air Pollution had found the city still "not gaining in the struggle," partly due to overemphasis on smoke and sootfall and relative neglect of the more recently discovered invisible gaseous pollutants, which were feared to be more deadly than the visible ones; this remained a problem in subsequent years. In May, 1967, the Department of Air Pollution Control rediscovered what the Fitch report had stated years earlier, that sending out inspectors to follow up on citizen complaints was wasteful of time and money. The agency vowed to rely more on the "Eyeball Surveillance" of trained inspectors and on technological aids such as helicopters and automatic movie cameras—the sort of enforcement Maxwell had sought to apply back in the days of the BSC. Con Ed continued to smoke while the thousands of existing apartment house incinerators remained almost totally uncontrolled, as entrenched interests linked to pollution, from soft coal sellers to fuel oil importers to apartment house owners and the incinerator lobby, successfully resisted regulation. As a group of air pollution policy experts from Los Angeles invited to New York City by Lindsay observed: "We have found it no mystery why New York has air pollution, or how to cure it. . . . The difficulty is that almost everyone is against air pollution and in favor of its control, until control invades some cherished personal interest." Moreover, the city government itself was still "the greatest single producer of air pollution" through its municipal incinerators, public buildings, transportation facilities, and government vehicles, as the Cousins task force reported in a follow-up report to the mayor in

May, 1967. Old patterns proved hard to break, and the same situation basically persisted through the end of the 1960s.[57]

Above all, the problem just kept getting worse. This was hammered home by the "Thanksgiving Weekend incident" of 1966, which brought the worst siege of air pollution New York City had ever experienced when a stagnant mass of air held pollutants down low over cities from Pittsburgh east to the coast and from Boston down to Washington, D.C., for five straight days. City authorities in New York urged those with cardiac or respiratory problems to stay indoors and pleaded for a voluntary ban on automobile and garbage incinerator use, while New York Governor Nelson Rockefeller later declared a first-stage air pollution emergency in the metropolitan area. Finally the trapped air and pollution blew away on its own. Ironically, the timing of the episode was indeed something for which to be thankful, since the incident might have caused thousands of deaths, like earlier air pollution sieges in London in 1952 and 1962, had it not by sheer chance happened during a holiday when many office buildings, factories, and automobiles were inactive. As was noted in a *New York Times* editorial, "On a regular work day the poisonous fog that held New York in Thanksgiving captivity could have been lethal." Later, in a study led by former Air Pollution Control Commissioner Dr. Leonard Greenburg, medical statisticians estimated that New York had suffered 168 excess deaths over normal statistical levels from this episode, but nothing like what might have happened had all the commuter vehicles, office towers, and factories been fired up.[58]

In the wake of this shocking incident, public awareness of air pollution surged throughout the New York City metropolitan area and throughout the nation. Support for more radical control measures also grew as warnings of doomsayers suddenly became more believable. In January, 1967, federal scientists announced that New York City had the worst overall sulfur dioxide levels in the nation, representing a potential ongoing threat to human health. The following August, the great city was declared to have the worst air in general, with dangerously high concentrations of carbon monoxide and particulates in addition to the sulfur oxide. In May, 1967, Norman Cousins announced that New York City could become "uninhabitable within a decade" unless the city, state, and federal governments joined in "an emergency action program of mammoth dimensions" for atmospheric cleanup, while Secretary of Health, Education, and Welfare John W. Gardner somberly warned in December of 1966 that Americans might have to live in domed cities or go around in gas masks. No one was laughing.[59]

Yet despite further encouraging rhetoric and bureaucratic reorganizations, New York City was never able even to come close to mastering its difficult air pollution problem by the end of the 1960s. In 1970, the city's roughly 135,000 smoky oil burners and 13,500 to 17,000 garbage incinerators were nearly all still operating in violation of city ordinances with impunity. Thirty of the city's eighty field inspectors had the impossible task of policing all these built-in problems that the city did not have the political will to confront in earnest. The May, 1968, deadline for the upgrading of all apartment house garbage incinerators, established under the earlier 1966 law, was repealed in 1967 when it became clear that there was no way the city could ever hope to meet it, while a new policy allowing apartment house owners to stop using their incinerators only exposed the inability of the city's Department of Sanitation to handle the additional unburned garbage. Ultimately, landlords were encouraged to start using their old, smoky incinerators again. A subsequent regulation to require incinerator upgrading by May, 1970, was challenged as unconstitutional by local real estate interests and in early 1969 was found constitutional but was enjoined from enforcement until a full trial could be held. This still had not been done by April, 1970, at which time only around two thousand of the city's incinerators were in compliance. Meanwhile, the city's other most incorrigible source of pollution, Consolidated Edison, was still a colossal problem, despite major expenditures on pollution control equipment that Con Ed claimed had reduced sulfur emissions by half and particulates by a third. The city's air remained perhaps the worst in the nation, and its vehicular emissions had grown to the point that they, too, posed a threat to health.[60]

The Thanksgiving Weekend episode and its aftermath starkly revealed that in more than twenty years of often half-hearted trying, New York City had been unable to get its air pollution under control. Such forward efforts as had been made were only barely able to keep in check a problem steadily expanding along with the growing total numbers of citizens, domestic and industrial furnaces and boilers, garbage incinerators, and motor vehicles. As in other areas of the country where similar approaches were taken, public education programs and attempts to gain voluntary compliance from businesses and individuals in the absence of any real threat of coercion had proved a failure—repeatedly. Scientific research similarly proved to be no substitute for enforcement or political will. Yet politicians and bureaucrats stubbornly persisted in reinventing the wheel over and over to avoid confronting the political, social, and economic difficulties involved. After two decades of various more established issues and interests invariably taking precedence over air pollution control, this issue was still struggling to be taken seriously in New York City.

# chapter 7

# A Fight to the Finish

## The Public's Crusade against Air Pollution

Long before there was anything that could be called an environmental movement, many ordinary New Yorkers were already agitating and organizing to fight air pollution, like citizens of other localities across America. This battle began soon after the end of the Second World War and continued through 1952 and the creation of the new Department of Air Pollution Control. After this point, public mobilization waned in a typical turn of the cycle as citizens hopefully but over-optimistically left aerial cleanup to the new agency, and the early postwar struggle for air pollution control in the nation's largest metropolis was gradually forgotten by the general public and historians alike. But public sentiment lay dormant for only a few years. By the early 1960s, agitation over the still-unsolved problem was building again. It remained strong throughout the rest of the decade, feeding directly into the wider swirl of concerns about humans' relationship with the natural world that engulfed the nation and that toward the end of the 1960s came to be known as environmentalism.

Air pollution, however, proved not to be a simple, straightforward issue, and the battle against it grew more complex than the earliest participants would ever have imagined. In addition to fighting the sheer physical presence of atmospheric contamination, New Yorkers were also engaged in a struggle as to how to conceive of and define air pollution, politically, legally, and economically as well as scientifically. Air pollution control radicals, who wanted to see the problem summarily banished by any means necessary, confronted business interests that resisted the economic costs of cleanup; apathetic citizens; and local politicians who reflected such traditional economism and inertia. Together, these forces represented a daunting obstacle.

Citizen advocates of pollution control also struggled to establish their legitimacy and right to speak in the face of a scientific establishment accustomed to being the only voice on the issue. Emissions control activists, together with some physicians and other professional experts, tended to construe air pollution as chiefly a threat to public health, demanding immediate action, while most of the rest of the scientific establishment, particularly corporate scientists and engineers, generally sided with the business community in minimizing the health risks and treating dirty air as a matter of technology and economics to be weighed against other such mundane matters—the same situation as during the prewar years. While citizen activists gradually won ground in their battle to be heard in the debate on air pollution, their struggle against bureaucratic inertia, political opposition, and exclusionary scientific professionalism was long and hard.

As in other cities making their early forays into environmental reform, local press support for pollution control was indispensable. New York's newspapers helped greatly to instigate action and facilitate public organization to pressure reluctant officials to confront the issue of air pollution. In particular the city's leading daily, the *New York Times,* offered regular, significant coverage and crusading editorials on the matter from the late 1940s through the 1970s as the issue grew from sporadic minor complaints about a smoke nuisance to dramatic front-page news. Such reporting frequently offers valuable insights into public actions and attitudes on polluted air, a crucial topic regarding which other sources are limited. Although the nation's largest metropolis, one of the media capitals of the world, is in many ways not an average American city, there are nevertheless enough common features to the story—including the cycles of public awakening and inertia and the typical political, economic, and scientific debates and arguments for or against control—for the public campaign to fight air pollution in New York City to offer a good early case study of the environmental mobilization of a community.

After spending years preoccupied with the Second World War, New Yorkers gradually began thinking about air pollution again in 1946. As already described, Councilman Joseph T. Sharkey's reintroduction of an earlier smoke control law indicated the revival of public interest in the issue. Concerned New Yorkers also could have learned of other developments on the air pollution front: scientists' discovery of an apparent link between urban air pollution and lung cancer, which was increasingly coming to be viewed as an epidemic; the mobilization of civic groups in Queens to study and combat air pollution; and an early report on the atmospheric woes of New York's young, fast-growing rival on the West Coast, Los Angeles. Already, letters to the *New York Times* excoriated New York City as the "dirtiest city of the world," noted that the local sootfall seemed worse than anywhere save Pittsburgh, warned of the "appalling" implications for the human respiratory system, and proclaimed to the city council and Board of Health, "Please take note that the New Yorkers want cleaner air." However, these reports, pleas, and admonitions were relatively rare.[1]

New York citizens' postwar fight against air pollution began on January 3, 1947. On that date, the *New York Times* published a letter from a "Constant Reader" together with a strident editorial on the smoke nuisance. "Constant Reader," who was more concerned with air pollution than with the spread of international Communism, beseeched the editors, "Can you possibly divert a portion of your valuable space from comments about the 'iron curtain' alleged to hang over Europe to the very real curtain that strangles, besmirches and mars the people and habitations of New York City?" This unhappy citizen blamed apartment house owners burning soft coal for most of the problem. Angrily noting how "appeals to the city Health Department bring neither relief nor even a reply," Constant Reader concluded, "When our City Administration is so palpably lost to its responsibilities and duties, a great circulating medium like The Times can do much to awaken office chairwarmers to some action. Perhaps they too have been smothered in smoke."[2]

In their corresponding editorial, the *Times* editors directed readers' attention to this letter and announced, "We share this reader's indignation and his belief too little is said and done to mitigate this nuisance." The newspaper's editors noted the estimated 40 percent increase in aerial soot, dust, and fly ash from 1936 to 1945, declared that "the most serious consideration is the effect on health," and listed some known or suspected links between dirty air and impaired health. The editorial then detailed some of the other known costs of air pollution, such as higher cleaning bills for clothing and buildings and extra lighting when the smoke shroud was keeping out 30 to 40 percent of the city's sunlight.[3]

Noting the complexity of the problem, the *Times* concluded by promising to discuss different aspects of the smoke nuisance in detail in the future, "in the hope that the City Government . . . and others with public welfare at heart will become interested in making a more energetic effort to relieve the city of this damaging nuisance." The *Times* published nine articles, seven editorials, and numerous letters about the smoke menace during the next two months alone. In this way, like the *St. Louis Post-Dispatch,* the *Pittsburgh Press,* and the *Los Angeles Times* before it, the *New York Times* helped to launch the postwar campaign to end air pollution in New York City, and it would remain in the vanguard of this crusade for more than twenty years.

In their editorials, the *Times* staff hammered away at what would become almost a litany, often repeated during the following two decades: the suspected health risks of air pollution, the proven economic costs, the sheer ugliness of dirty air, the need for proper combustion technology and for public education, awareness, and cooperation regarding the issue. The framework for understanding air pollution that would continue to shape the perceptions of experts, ordinary citizens, and journalists alike was still drawn from the earlier experiences of St. Louis and Pittsburgh and from even earlier smoke control efforts of the Progressive era.[4]

The key ingredient of the early postwar air pollution litany was the potential health risks involved, following turn-of-the-century pure air crusaders' warnings of danger to both physical and moral health from heavy urban smoke palls. By 1946, scientists had observed the statistical correlation between urban living and lung cancer and had already suggested air pollution as a factor. During its concerted anti-smoke drive of early 1947, the *New York Times* began by making relatively modest charges about potential health risks from contaminated air but emphasized these risks nonetheless. In addition to the warnings in the first editorial, the *Times* editors later also reminded New Yorkers of the conclusions reached in a New York Academy of Medicine report some years earlier, that New York City had an air pollution problem that was a "serious menace" to health. Editorials noted how smoke and fumes irritated the eyes, ears, nose, and throat, making them all more susceptible to infection, and pointed out the statistical links between air pollution and tuberculosis, the "white plague" that remained frightening and untamed into the 1940s. In a more holistic vein, the *Times* also alleged that "smoke diminishes the potential reserve, working capacity and well-being of the individual and increases fatigue and irritability," while blocking healthful sunlight.[5]

Such claims about health risks seem tame and tentative compared to later articles following subsequent scientific discoveries during the 1950s and 1960s, when the New York press reported frequently on the suspected relationship between air pollution and cancer along with proven links to emphysema and bronchitis, while *Times* editorials became ever more strident in tone. However, in the later years as in earlier ones, the health effects of polluted air remained tangled in scientific controversy, as they still do. After the triumph of germ theory over earlier miasmatic assumptions within the medical profession around the turn of the century, physicians and medical researchers devoted most of their attention to the identification of microorganisms as the direct causes of various diseases. Like other contaminants, air pollution often works in more indirect ways. Environmental health often does not neatly fit the microbial paradigm, and partly for this reason, it long suffered relative neglect from the medical profession.[6]

In addition, it is rare that air pollution can be fingered as the sole, proximate cause of any disease syndrome, since sources of stress such as harmful chemicals and noise attack the modern human in many forms. Air pollution is an endlessly complicated subject, and its character and reactions, which can vary by climate and geography, were barely understood before 1950, when there were hardly even any devices to measure air contaminants. Moreover, the effects of small amounts of air pollutants on living creatures may be subtle and cumulative over long periods of time; realistic laboratory experiments are difficult to set up, and opportunities for experimentation on live human subjects are obviously limited. Because of such difficulties with conceptualization, experimentation, and scientific proof, the medical and legal professions have had difficulty handling the health effects of air pollution.

To help get around this limitation, pure air crusaders often fell back on a second, more tangible part of the early postwar air pollution litany—the proven economic losses from filthy air. Already in the first editorial of the 1947 anti-smoke campaign, the *New York Times* reported federal officials' estimate that each U.S. citizen on average "suffered an annual loss of $10 to $30 from smoke, through damage to clothing, cleaning bills, disfigurement of buildings, damaged merchandise, injury to grass and plants, and loss of light," concluding that "the property [damage] is less serious [than probable health risks] but very costly." It had recently taken a hundred thousand dollars to scrub the soot-blackened New York Public Library building, while local department stores had lost millions of dollars in soot damage over the years, and the general public many millions or billions more.[7]

While the statistics and examples used to buttress this line of argument sometimes changed over the next twenty years, the basic argument remained largely unaltered and was frequently used to justify the war on air pollution in hard-nosed, dollars-and-cents terms. The *Times* was particularly fond of using the U.S. Public Health Service's rough estimate of annual damage per person in America to calculate New York City's total annual loss and then comparing this to the city's token appropriation for smoke control, promising in 1947 that "the city could save its taxpayers a great deal of money, besides removing a menace to health, by attacking this smoke and soot nuisance on a scale commensurate with its seriousness." In later years, the *Times* would draw frequent invidious comparisons between the limp-wristed efforts of New Yorkers and the relatively monumental achievements of citizens in Los Angeles and Pittsburgh, as measured by the money annually spent per person on air pollution control and the results gained. During the two decades following World War II, New York lagged far behind other major American cities, offering local journalists and gadflies ample statistics with which to support their economic justifications of air pollution control and their complaints of budgetary neglect.[8]

As with the health argument, though, the early economic arguments against air pollution had serious limitations in practice. Firstly, as with other unwanted residuals from production processes, damage from dirty air in a city like New York spread over a wide area and among millions of people, and damage spread so thin was relatively easy to ignore. As in many other American cities since the mid-1800s, most New Yorkers, particularly those in poorer neighborhoods, had learned to live with air pollution darkening their skies, houses, clothing, and faces, and only got worked up about it when it was noticeably heavier than usual. As we have seen, some of the most powerful interests in New York City did not wish to shoulder the costs of controlling their atmospheric filth and were not much moved by rhetoric about civic-mindedness or about economic damage suffered by others. Serious air pollution was by then built into the economic fabric of New York and other cities, and those accustomed to using the air as a cheap sewer system generally had no desire to change things.

To get around the first practical flaw in the economic argument—limited public concern—the *New York Times* and its allies sought to stimulate greater awareness of air pollution. In the first editorial of the anti-smoke crusade, the *Times* called for the creation of "an aggressive civic organization with smoke abatement as its sole purpose" to gather and publicize information, to agitate for greater control efforts, and to educate domestic and industrial fuel users about the issue throughout the metropolitan area. The paper resoundingly concluded, "A great field of public service beckons for an organization that will make a fight to the finish on smoke here." A few days later, the paper again hammered home public responsibility for air pollution control, as editors reminded citizens that "legislation alone will never solve the problem unless it is supported by an aroused public opinion." The *Times* would long continue drumming in the need for public organization and mobilization to confront air pollution.[9]

It also offered more detailed suggestions as to how this public mobilization might take place, presenting a more concrete vision of "an anti-smoke league," based on organizations earlier formed in St. Louis, Pittsburgh, and Cincinnati, which would combine various local civic groups to work with the city Health Department while remaining independent and duly critical of that agency. The proposed organization would be led by a "man trained in smoke control," possibly from Pittsburgh or St. Louis; it would conduct research into the sources, character, and quantity of New York's air pollution and the best way to fight this nuisance; and most of all, it would serve as an information clearing house for purposes of educating the public, issuing regular reports on the city's air pollution problem "in a manner to compel attention." The *Times* generously offered its services and those of its rivals to the yet-uncreated anti-smoke league: "We believe such findings would be published in the newspapers of the city, not only as news but as a public service." The newspaper also happily reported the interest of numerous civic organizations and the number of supportive letters received since the beginning of their anti-smoke drive, indicating that the hypothetical anti-smoke league would enjoy "enthusiastic support," monetary as well as emotional. Although the *Times* warned that the fight against smoke would be long and hard, it optimistically declared that there was no question that "a pugnacious civic organization backing its educational campaign with soundly developed facts would win such a battle and that beneficial effects could be expected soon."[10]

To try to get around the second problem with the economic argument—the intransigence of entrenched interests—the *Times* appealed to enlightened self-interest by reviving the Progressive-era argument that smoke represented inefficient combustion of fuel, which wasted money, noting how "the offender who creates the smoke . . . would save himself much money if he burned his coal efficiently with a minimum of smoke." Observing how soft, smoky, bituminous coal was replacing relatively smokeless hard anthracite, the *Times* acknowledged that the bituminous coal probably could not be gotten rid of—conceding away one of the principal techniques used in St. Louis and Pittsburgh—and so concluded: "The practicable approach seems rather to promote the use of modern equipment, education in proper firing and development of public conscience." This reliance on public conscience and education to secure voluntary public compliance remained close to the arguments of moderate turn-of-the-century smoke fighters, and it would prove equally naive.[11]

New Yorkers responded enthusiastically to the *Times*'s media campaign on air pollution, sending in numerous letters to the editor to vent frustration. On January 8, 1947, the paper published twelve letters responding favorably to the editorial of January 3, with writers ranging in power and prominence from Louise Vanderbilt to ordinary citizens complaining about matters from soft coal soot to incinerators to thoughtless beating of dust out of mops and rugs. One "Grateful Reader" from the Bronx lamented how her new curtains were blackened within a week while her lungs never enjoyed "good, clean air," and she concluded: "The soot is a menace to all of us. Please keep hammering away on this subject and if enough people become interested some good will come of your efforts." Emphasizing how the issue cut across class lines, another concerned reader noted: "With sufficient public education and prodding of officialdom, we should get results on a project that concerns all of us, whether we dwell in penthouse or tenement." Ms. Vanderbilt also congratulated the *Times* for its civic-mindedness and wondered hopefully: "Isn't it possible that something to prevent the smoke can be put on furnaces?" As if in answer to this characteristically American technological optimism, an unhappy Staten Island housewife pointed out the city's failure to enforce existing smoke laws and glumly concluded: "Maybe that cynical historian Henry Adams was not wrong in calling us the people most brilliant at invention in all history but least talented for administration."[12]

The public response to the *Times*'s media campaign offered hope that New Yorkers might indeed soon bring air pollution to heel. Already by January 10, the paper reported the creation of an anti-smoke committee within the politically influential Citizens Union, a local reform group, while an anti-smoke league had been proposed by several organizations, including the socially prominent Outdoor Cleanliness Association. The Bronx Chamber of Commerce, more conservative on the issue, saw no need for a new civic organization wholly focused on air pollution but nevertheless ordered a survey of emissions within their borough. About a week later, the *Times* printed a long letter from Arthur C. Stern, an engineer working on smoke and dust control for the New York State Department of Labor who would become an important leader in the city's fight against air pollution. Stern gave detailed recommendations for putting together an anti-smoke league and reluctantly offered his leadership—"So if no one else steps forward to try to organize this group, I will. Fools rush in where angels fear to tread." In early February, numerous civic groups participated in a meeting on air pollution at City Hall. All in all, the newspaper's mobilizing efforts appeared to be working. The *Times* had served as a midwife for the postwar rebirth of air pollution control efforts in New York City.[13]

However, developments proved frustrating. After making a show of effort early in the year, city officials quietly let the issue die as warmer weather brought the heavy heating season to a close. The complaints of local citizens brought no response. The next smoke season repeated the same cycle, despite mounting public frustration. In late October of 1947, when Joseph Sharkey reintroduced his stricter smoke law from early 1946 to make the emission of dense smoke a statutory offense, he explained, "For months I have received many complaints. Letters come in continually. Mothers buttonhole me on the street and say smoke in the air makes it risky for them to take their babies out in go-carts. Housewives complain of damage done to wash on the line.

Wherever I go to speak . . . people plead with me to do something now that the war is over. Many Councilmen get similar complaints and pleas. In my own section in Williamsburg the soot is sometimes an inch deep on window sills and sidewalks." Again, Sharkey's move provoked a brief show of activity from the city government, again the drive fizzled by springtime, and again no new funds for smoke control were forthcoming in the city's budget. The *New York Times* reflected the sentiments of the city's smoke fighters when it editorialized resentfully, "The budget omission is one we do not regret the less for the fact that we have come to expect it," reminding readers of Mayor O'Dwyer's broken promises of major funding increases for smoke control the year before. The paper again called for many more dollars and smoke inspectors, offering the successful example of Pittsburgh to blast New York's ineffectuality.[14]

In early June of 1948, Councilman Sharkey again answered the public's call for smoke control. In drafting his new bill, Sharkey sought the advice of Arthur C. Stern, by then chairman of the special committee on air pollution control of the Citizens Union, to produce a more comprehensive bill than two other pending proposals. Although the editors would have preferred an even more independent agency, the *Times* lent its support to the renewed campaign, drumming up public attention and concern, lauding Sharkey and Stern for their efforts, and reminding readers: "There is no record of any city making an effective campaign against smoke where it was not driven to do so by loud, organized public complaint and agitation." Evoking civic pride, the *Times* admonished local citizens: "Certainly if Pittsburgh can improve so much, New York, with a far less serious problem, can also clear its air to the benefit of our health and our tempers."[15]

Notwithstanding the valiant efforts of Sharkey, Stern, and many others, the new bill ran into stiff opposition from business interests fearing the costs of air pollution control. At a hearing on Sharkey's proposal in October, the bill was strongly supported by most civic organizations, parent-teacher associations, and homeowners' and tenants' groups but was opposed by big business interests such as the Association of Eastern Railroads, the Fifth Avenue Association, the Commerce and Industry Association of New York, the Real Estate Board of New York, and the New York Heating Oil Association. Anti-pollution advocates charged these business interests with shortsightedly placing their short-term expenses for pollution control equipment above the long-term health and well-being of the city; conversely, the Queens Chamber of Commerce argued that the Health Department was already doing "a good job" of controlling smoke, while the Commerce and Industry Association responded that they agreed there was a need for further air pollution control, but it would be better to improve the existing system than to introduce a new, complex, and costly bureaucracy. However sincere this very typical if hollow-sounding argument from the business community was, as of late October, 1948, it looked as though the Commerce and Industry Association and its allies would carry the day and again stifle meaningful control efforts in New York City.[16]

As noted previously, what kept the 1948 Sharkey bill on track was the killer smog of Donora, Pennsylvania, which left twenty dead and thousands ill in late October. On October 31, 1948, air pollution made the front page of the New York papers for the first time, with headlines blaring "Lethal Smog Sets Town Gasping" and "'Smog' Linked to 18 Deaths in Day and Hospital Jam in Donora, Pa.," while articles described the suffering of victims of the "poison gas." Polluted air had suddenly become a true media event in New York City as the alarming reports from Donora gave color to citizens' existing suspicion that air pollution was a threat to human health. Horrified New Yorkers read the vehement charge of a health official in Donora, Dr. William Rongaus, who called the incident "murder" and pointed his finger at local industry as the culprit.[17]

As a result of such exposure, the incident in Pennsylvania also had a particularly heavy impact on perceptions of air pollution in New York City and throughout the nation. On November 1, the governor of Ohio ordered an atmospheric survey of his state's industrial areas, prompted by recent events across the Pennsylvania border. In New York City, the Department of Health assured New Yorkers that there was "no danger of a deadly smog settling over this area" due to normally high winds and a lack of poisonous emissions sources. The Health Department's reassurance may have been over-optimistic, as later events would demonstrate, but their statement indicated that Donora had gotten New Yorkers wondering whether the same could happen to them. The Donora episode would continue to hover like a frightful specter over subsequent articles and editorials about air pollution in newspapers and magazines throughout the nation, while the *New York Times* in particular gave ongoing coverage of developments in Donora through the rest of 1948 and even into 1949 and 1950.[18]

As a result of the hugely expanded public concern and mobilization that Donora stimulated in New York City, the Sharkey bill creating the new Bureau of Smoke Control (BSC) was signed into law in February, 1949. Noting the elation of "many New Yorkers and especially housewives," the *New York Times* proudly congratulated itself along with Joseph Sharkey, Arthur Stern and the Citizens Union, the women of the Outdoor Cleanliness Association and other civic organizations, the other New York newspapers, and the city administration for their cooperation. It was a good start, said the *Times;* now they would all have to make sure that "the city provides adequate funds. Good law alone will not do the job."[19]

As it turned out, local control advocates would never be satisfied with the city's aerial decontamination effort up to 1970 and beyond. From early 1949 onward, the *Times* never ceased complaining about the inadequate funding of the city's air pollution control agency. Already in February, 1949, the newspaper's editors pointed out that salaries were simply not competitive with private industry for personnel with requisite skills. As for the overall smoke control budget, the paper worried that local authorities were "arming for a popgun war instead of the real article," since the budget of $186,300, while a major improvement, was still "inadequate," representing only two cents a year per capita while other cities and counties in America were spending anywhere from six to eighteen cents per capita to control air pollution. Returning to the economic efficiency argument, the *Times* reminded readers that a "decent budget" would be an "investment returning tremendous dividends."[20]

Smoke control crusaders had other topics for complaint besides the budget. Through the following years, they constantly demanded stricter enforcement and blasted local air pollution control officials for their weak efforts. After William H. Byrne was selected as the interim head of the BSC, citizen activists persistently demanded stiffer regulation and pressured the city government to find permanent leadership for the BSC. Later, after the new permanent director, William G. Christy, promptly backpedaled on stricter enforcement, the *Times* angrily asked, "Does the city mean business on smoke abatement?" Observing how the BSC's early policies were "half-hearted and left loopholes suggesting a fear of stepping on somebody's toes," the paper demanded that the agency "be prepared to step hard when the city's welfare requires it." With this uncompromising stance, the *Times* remained a megaphone for dissatisfied smoke control advocates.[21]

Among the most militant of the metropolitan smoke fighters were members of women's organizations. As noted, women had a long history of air pollution control activism, both in New York City and elsewhere. Even before the Second World War, New York women helped make the smoke nuisance a news topic of some frequency during the late 1920s and 1930s. Women and women's organizations also had a key role in agitating for smoke control during the postwar period. The New York City Federation of Women's Clubs and the Outdoor Cleanliness Association, a civic beautification society, were among the first groups to press for the creation of an anti-smoke league of civic organizations in early 1947. That March, Dorothy Frothingham Wagstaff, head of the Outdoor Cleanliness Association, and Mrs. John Weinstein, president of the local Federation of Women's Clubs, were among the ten leaders on the executive committee of a civic alliance to combat smoke. Months before the incident at Donora, Wagstaff and her successor, Mrs. C. Frank Reavis, were planning a public rally against air pollution for mid-November, 1948; the Donora deaths helped guarantee a larger and more attentive audience at this rally. Later that same month, the association screened the new film version of *Hamlet* to raise money for their fight against air pollution and other civic eyesores, and the organizers of this event included some of the wealthiest and most socially prominent women in the whole city: Mrs. Iphegenia Ochs Sulzberger, wife of the *New York Times* publisher; Edna Woolman Chase, editor of *Vogue* magazine; Mrs. Herbert H. Lehman, wife of a former state governor and future U.S. senator; and Mrs. William Randolph Hearst, Mrs. Louise Vanderbilt, and Mrs. William F. Morgan. The following March, Reavis awarded Joseph Sharkey a plaque "for his ten years of untiring efforts toward outlawing unhealthful soot and smoke in New York City."[22]

Not only social register types fought smoke, though. Other women wrote angry letters to the city's newspapers. One "Irate Housewife" complained bitterly of trying to clean up after all the incinerators belching "choking fumes" and "bits of burnt black particles." She wondered, "If . . . coal smoke can produce cancer of the skin, what can the effect of these fumes be upon our lungs?" and called for action from the city rather than more meetings and promises. Another local housewife, Genevieve Myers, excoriated the fumes of oil-burning water heaters that ruined otherwise pleasant summer nights and the "haze of blue smoke and strangling fumes from transportation buses" that suffocated local pedestrians and bus riders. As Myers concluded, in a perceptive commentary on quality of life and the inadequacies of traditional law and economics, "Until someone can discover a way to give discomfort a dollar value the situation seems doomed to go from bad to worse."[23]

Other women participated in other ways. For instance, they were prominent in the parent-teacher and neighborhood associations that were among the majority of civic organizations favoring air pollution control in the face of stubborn business opposition in October, 1948. A more dramatic illustration of the frustration and anger of ordinary housewives and mothers came at a public meeting on proposed new rules and regulations for the Bureau of Smoke Control in August, 1950. At this meeting, representatives of business and industrial interests whined that proposed limits on sulfur content of fuels would "interfere with the nation's rearmament" by draining "stocks of premium coal urgently needed by defense industries" gearing up for the Korean War—the old trick of wrapping oneself in the flag to avoid cleanup costs. Suddenly there was a commotion in the back of the room. Two Yorkville women, Mrs. Daniel Dolan and Mrs. Robert Wood of the Parent-Teacher Association of Public School 158, called out from the floor asking permission to be heard, then took the podium to demand action, complaining that their neighborhood had suffered long enough from the smoke nuisance. As the *New York Times* reported, "Given five minutes, the two angry women complained bitterly about the situation and graphically pulled soot-stained clothes out of a brown shopping bag, showing what happened to fresh laundry hung out to dry. Mrs. Wood observed that if they'd known they would be recognized, they could have brought hundreds of Yorkville housewives to confirm their story."[24]

In an editorial a few days later, the *Times* lauded these two bold women as "genuine representatives of the public" in stark contrast to industry, with its "little regard for the public weal" and its assumption that "tough smoke control was fine—for the other fellow." As the *Times* rhapsodized, "Through all this fog of technical verbiage and one-track reasoning, the unheralded presence of the two Yorkville housewives breezed like a breath of fresh air. In what they said they were non-technical, concise and to the point. . . . When the women sat down, there was a burst of applause from those at the hearing. We are certain that in what they said and what they did they represent millions of New Yorkers."[25]

Given this persistent, sometimes even heroic advocacy for air pollution control, perhaps it should come as no surprise that New York's most aggressive citizen smoke fighter during the early 1950s was a woman. At a time when female activists were typically genteel, upper-middle-class clubwomen who referred to themselves in the format "Mrs. John Q. Public," Elizabeth Robinson stands out as a stark and dramatic figure, even though the *Times* often transformed her into the slightly more domesticated "Mrs. Lyonel F. Robinson."

Robinson appeared like a whirlwind on the air pollution control scene in New York in late October, 1949, when she chaired the first New York commemoration of National Smoke Abatement Week, remembering Donora. In 1950, she chaired the second local observance of National Smoke Control Week and then headed the Committee for Smoke Control (CSC), a civic organization exclusively concerned with fighting air pollution and drawing members from other civic organizations, much like the anti-smoke league the *New York Times* had envisioned in 1947. She was also a leader of the New York City Federation of Women's Clubs and later served as vice president of that group. Robinson had personal reasons for her interest in smoke control. Although a naturalized American citizen and Queens housewife, she had been born and raised in a smoky suburb of Liverpool, England. As a child, she suffered from chronic bronchial troubles that sometimes kept her out of school and for which she blamed the filthy air of her hometown. Her physician father had spent his life researching respiratory illnesses. Having later married an American international exporter, she had gotten a chance to live in Canada and South America, "some of the most beautiful spots of the world," and had learned what truly clean, healthful air was like. Like many of the refugees to suburban Queens only more so, Robinson had seen better conditions and would not accept the steady worsening of New York's air without protest.[26]

Armed with the wisdom of her life's experience and an unshakable sense of mission, Robinson promptly set about turning up the heat on the city government and its Bureau of Smoke Control. In January, 1951, she sent an impatient letter to Mayor Vincent Impellitteri alleging that "not a single court summons has been issued to smoke violators since February, 1949, when the Sharkey anti-smoke law was enacted" and urging the mayor to act on the issue in 1951. In her letter, Robinson forcefully turned the tables on those Cold War–era industrialists who sought to avoid the expenses of air pollution control by hiding behind national security when she argued: "The dangers to health in polluted air are something more than a mere theory. . . . Today, with the nation mobilizing for defense and confronted by a shortage of physicians, the public health should be safeguarded as never before. Vigorous enforcement of smoke-control regulations should be one of the least costly means of achieving this objective." Robinson ambitiously directed her group to mobilize public opinion further, to coordinate the activities of like-minded groups, and generally to "serve as a public watchdog on all phases of air pollution," as she energetically put it. The *Times* editors praised Robinson and her allies for their efforts to compel action from the city government but warned that "the way will be long and hard."[27]

Undaunted, the redoubtable Mrs. Robinson spent the rest of 1951 rampaging like a bull in the staid china shop of New York City air pollution control policy. At the end of January, after a distinguished Harvard meteorologist had announced to the

American Meteorological Society that he had found no evidence of any significant decrease in sootfall in New York City over the preceding fifteen years—BSC claims of major progress notwithstanding—Robinson polled local women and publicized the results. Sixty-two percent of local housewives found air pollution worse than two years earlier, while a further third said it was no better. In April, responding to statements supporting BSC Director Christy's policy of cooperation and persuasion toward industrial polluters, Robinson forthrightly affirmed, "You do not know human nature if you don't realize the importance of adopting stern punitive measures. . . . One stiff fine will stop more smoke and soot than 1,000 paragraphs in any document of rules and regulations." And this was only the beginning, for she got steadily more strident in her demands for stricter enforcement.[28]

By the middle of the year, Robinson had become enough of a local celebrity that the *Times* published a substantial article focusing mainly on her alone. The occasion for this was the release by the Committee for Smoke Control of five thousand pamphlets blasting the BSC's leaders for preferring "education and cooperation" to enforcement and penalties and urging New Yorkers to monitor enforcement closely and to report violations of smoke laws. The public had received these pamphlets so enthusiastically that the CSC was ordering five thousand more. Robinson noted her full schedule of speaking engagements before women's groups in Manhattan, Brooklyn, and Queens and claimed that women, women's clubs, and parent-teacher associations were solidly behind her on air pollution control. As she explained, air pollution hit home "in the two most vulnerable places for a housewife, her children and her pocketbook. . . . Women are thinking of the health of their children, and where that is at stake, a woman will fight as primitively as any tiger in the jungle." She went on to argue that New York City housewives could save five millions dollars a year just on window and floor coverings if the smoke were cleaned up. Elizabeth Robinson's persistence, commitment, and total self-assurance made her an example for anti-smoke activists and a nagging headache for the city government.[29]

In the wake of the scandals that rocked the Bureau of Smoke Control in the summer of 1951, and particularly the revelation of Christy's covert dealings with the Commerce and Industry Association of New York, Robinson turned up the heat on the agency still further, launching a campaign to expose the reasons why progress on air pollution control had been so limited and to force Christy out of office. She retained Louis E. Yavner, former New York City commissioner of investigation, as special counsel for the Committee for Smoke Control to consider all possible legal steps that might help stimulate stricter enforcement of the city's smoke regulations. Toward that end, on August 1, Robinson, Yavner, and the CSC demanded permission to inspect the BSC's files regarding recent complaints to determine "how efficiently the bureau is functioning." This demand was grounded on the taxpayers' right to inspect public records under the New York City Charter, and Yavner announced he was ready to take the case to the New York State Supreme Court if necessary. Although Housing and Buildings Commissioner Gilroy later officially denied access to the bureau's records on the advice of the city's legal counsel, in mid-November, a local judge refuted the BSC's claim that the CSC had no "special interest" or legal right to inspect bureau records. The judge held that the public welfare *was* concerned and that there was "room for reasonable dispute" that the public interest was being adequately served by the BSC, so the inspection of documents represented a legitimate effort to allay "justifiably aroused" public concern. Like many environmental crusaders who followed her, Robinson successfully used the courts as well as public mobilization to put pressure on unresponsive bureaucrats.[30]

In the meantime, other New Yorkers were also growing more impatient with the BSC. In May, 1951, the *Times* editors irritably announced that the existing policy of cooperation and education had "failed, as any housewife mopping up soot in the apartment knows," while the BSC's budget remained "pathetically small." Proclaiming that it was Mayor Vincent R. Impellitteri's job to figure out why the BSC had "not lived up to its expectations," they demanded an end to the "mouse-like approach" and called for a "tough approach that means summonses to court and stiff fines and even jail sentences" as the "only way to get rid of smoke." Later, after bureau director Christy promised to stiffen enforcement, the *Times* and other unpersuaded BSC critics gave Christy and his staff an ultimatum: show real progress on smoke control during the next three months or face removal from office. Similarly, Magistrate Hyman Korn, the judge for yet another alleged smoke violation by Consolidated Edison, crossly asserted that "conditions are worse since the establishment of your control board than they were before" and urged the BSC either to "do a job or close up." The BSC was besieged by an increasingly vocal public impatient for signs of real progress on air pollution control.[31]

This sort of public agitation set the stage for BSC Deputy Director William S. Maxwell breaking ranks with his superiors and for the first time slapping violation notices on major offenders, including other city departments. By his stubborn refusal to back down in the face of pressure from Christy and Gilroy, and by keeping himself, the BSC, and air pollution prominently in the news for six straight days in late July and through much of August and September as well, Maxwell turned himself, as we have seen, into New York City's first real citywide air pollution control hero and the darling of the anti-smoke crusade. The public and the media loved his blunt statements about getting tough with smoke violators. For the *New York Times,* he was still invariably "Admiral Maxwell," a war hero turned peace hero. The tough-talking Maxwell got enthusiastic support from two other leading smoke foes, Joseph Sharkey and the equally tough-talking Elizabeth Robinson. When the Maxwell affair first surfaced in the news, Robinson sent a scalding telegram to the city administration, labeling Maxwell "of this moment the only member of the board who has the desire and the courage to enforce smoke regulations" and calling for him to be unleashed to show "what genuine enforcement can accomplish." Regarding Christy's hesitance to tackle major offenders, Robinson declared with customary candor, "The sooner he resigns the better." More surprisingly, even a business organization wrote a letter to the mayor supporting Maxwell. The New York Board of Trade noted that cleanup was costly but that air pollution was already costing local businesses about $150 million a year in damage. As they concluded: "The laissez-faire or take it easy attitude of Director Christy does not command our respect or approbation. . . . The Admiral seems to believe in law enforcement."[32]

Ordinary citizens cheered for Maxwell. For one example among many, Twyla K. Malone wrote to the *Times* commending Maxwell for pursuing big targets: "Thank God we have a public servant who carries out his duty and lets the chips fall where they may. I hope further pussyfooting will not nullify his courageous act. As a home-maker and business woman I have been experiencing over a period of years more and more difficulty and expense in keeping the home, the office and the person clean and respectable from day to day. I hope you will continue to give Smoke Control news a prominent place in your newspaper, so that the violators will have to get in line and stay in line." The *Times* fulfilled Malone's wish, for it, too, naturally loved the swashbuckling admiral, giving him extensive coverage and support for his pursuit of major polluters in articles and editorials throughout the summer of 1951, while excoriating unnamed "timid men who soft-pedal the danger of disease through pollution of the air."[33]

Christy tried pulling rank on Maxwell to get his outspoken deputy under control, but this effort backfired. After a public appearance at which Maxwell was polite and penitent, the *New York Times* revealed the reason for this change of style in a page-one, column-one story on August 29: "Maxwell 'Gagged' on Smoke Control." Christy had ordered his subordinate to make no statements to the press save statistics. The director argued that it was his responsibility to make public pronouncements about the BSC's activities and that the unconventional order was to prevent improper "leaks" of information. Gilroy and Mayor Impellitteri defended Christy, but the *Times,* along with many ordinary New Yorkers, felt otherwise. In an editorial titled "Admiral Maxwell Muzzled," the paper's editors noted how "Maxwell stepped on official toes" and stuck his neck out too far for smoke control, but they found him "guilty of a form of insubordination that we admire" in revealing "some unpleasant truths." Again, Maxwell won the battle for the hearts and minds of New Yorkers as citizen smoke fighters called for the replacement of Christy with Maxwell and the mayor avoided the political backlash from trying to remove him.[34]

Seemingly alone among the officials in the New York City government, Maxwell understood the depth of public sentiments about air pollution and the general dissatisfaction with a pollution control program that effectively allowed major polluters to select the pace at which to correct their problems. After his break with his superiors on questions of policy, he developed a flair for promoting himself and the cause of air pollution control through his flamboyant enforcement techniques. Almost everything he did caught the public eye and made good news copy. Besides his well-publicized quarrels with Christy and Gilroy, citizens and journalists enjoyed following the activities of Maxwell and his flying squad of investigators as they swept through New York neighborhoods, citing violators both large and small. In late July, 1951, as the bureaucratic battle between Christy and Maxwell over whether to tackle the biggest offenders was raging, Maxwell dramatically led his men in person "to prove that we mean business" and promised "no let-up in the industrial smoke campaign." His inspectors also issued citations to ferryboats of the New York City Department of Marine and Aviation, among other waterfront violators. New Yorkers, and the New York media, loved it. And Maxwell was just getting started.[35]

In a move that would have major implications for the air pollution fight in New York City, Maxwell made the front page of the *Times* again on August 2, 1951, by announcing that the BSC would soon begin using a helicopter to pursue smoke violators from above. The novelty of flight probably increased public excitement about this dramatic, innovative control initiative. Although the first flight in a borrowed police helicopter was canceled the following week due to severe weather, by mid-August, Maxwell joined his men in making the first aerial survey of New York City's air pollution problem.[36]

As astronauts looking back at the earth from space would later do, Maxwell and his inspectors gained from their aerial vantage point a new perspective on the problem they faced. In yet another page-one article, Maxwell described how the problem was "more serious than people in the city realize." He had seen plumes from hundreds of large chimneys, factory smokestacks, and ships' funnels blending with the emissions from thousands of cars, trucks, buses, locomotives, and other smaller sources to form one great, sooty blanket of filth stretching over the whole city. Maxwell announced that aerial photographs of the problem would be taken once weather conditions permitted; he later also arranged to have ordinary ground-based inspectors taken aloft to get this wider perspective on the problem.[37]

After the dramatic incidents of July and August, 1951, built up his public following, the bull-necked, media-savvy ex-admiral played to the hilt his role as local champion of smoke control during 1952. In his early January, 1952, report on the previous year's smoke control progress, again issued in the absence of Christy, Maxwell threatened further "crackdowns" on "willful and chronic" violators. Later, after many more months of the blunt, colorful statements and dramatic enforcement tactics the public had come to love, Maxwell's masterful campaign of promoting himself and his control program reached new heights in July, 1952. On July 7, the BSC invited reporters and photographers from the New York newspapers to ride along with inspectors as they pursued ships, boats, and other marine and waterfront violators in the *Marav*, a thirty-eight-foot launch borrowed from the city Department of Marine and Aviation. While inspectors were expected to be fair and reasonable in singling out likely targets, the journalists were also impressed to see the stealth and cunning of the BSC team in sneaking up and pouncing on chronic offenders who had been previously warned. The press and public enjoyed it even more when, on a later harbor sweep with reporters aboard, the BSC inspectors cited a ferryboat owned by the same city department that had loaned them the *Marav*. Maxwell denied that his agency was "biting the hand that feeds it"; he declared, "John Citizen will not trust us if we go after the little fellows but ignore obvious smoke violations by city departments. . . . We are trying to prove that we don't overlook anybody." The inspectors in the *Marav* also cited a local harbor sightseeing cruise boat and rescued two teenage swimmers caught in the East River tide—all of which raised the agency's profile and made further good publicity. Later, even the Statue of Liberty boat got a summons.[38]

Most eye-catching of all, though, were the combined air/sea/land operations of July and August, 1952, when Maxwell pulled out all the stops to put on a public display of enforcement. These operations reached new levels of sophistication as Maxwell arranged to "case" major smokers personally in a helicopter, then radio to a following seaplane to photograph the smoke sources. The photographic evidence thus obtained was then given to ground-based investigators who would cite the violators. Twenty of the twenty-eight BSC inspectors were involved in this operation, while others were simultaneously out in the *Marav*. Maxwell was elated with the results of this high-tech smoke fighting, which was not only sensational but effective; he noted that "one flight is worth a hundred ground-searchers" for catching violators, and he planned "surprise" aerial inspections to catch those violators who curtailed their smoke only when they knew the BSC inspectors were on the prowl.[39]

By the summer of 1952, however, despite Maxwell's enforcement display, time was running out for the Bureau of Smoke Control, as Councilman Sharkey's bill to create a new, independent control agency advanced. Not all citizens felt sure that this action would improve the city's air significantly, while many worried that the new law might remove their favorite smoke control official, William S. Maxwell, through stricter professional requirements for both the director and deputy director, such as advanced training in public health and sanitary engineering. Back in August, 1951, the three professional engineering societies that had originally recommended Christy and Maxwell reaffirmed their support, labeling the Sharkey bill a "deliberate attempt to remove the present members of the smoke board." Later, in 1952, when the BSC was fighting for its life shortly before the new law passed, this charge was frequently reiterated as Maxwell's fate became a hot political issue. In early July, 1952, the minority leader on the city council, Stanley M. Isaacs, a Manhattan Republican-Liberal fusionist, charged that Mayor Impellitteri did not want to fire the popular admiral but wanted him out. Making political hay from Maxwell's popularity, Isaacs alleged that Impellitteri had long sought to "starve out" the BSC through underfunding and that going after Con Ed was "enough to damn any official as far as this city administration is concerned." Sharkey vehemently denied that the act was only a subterfuge to get rid of Maxwell and Christy, declaring, "We are out of the experimental stage of smoke abatement now and it is time to launch a full-fledged attack on the smoke nuisance." To try to save Maxwell, many New Yorkers joined the *Times* editors in urging Sharkey to "reduce the qualifications for Commissioner and deputy to mechanical engineer in good standing," which ultimately was done for the deputy position.[40]

After the Board of Estimate voted for the Sharkey bill in late August, 1952, Maxwell finally accepted the inevitability of the new agency and humbly offered his services. On September 5, as Mayor Impellitteri was considering the legislation, various witnesses at a public hearing on the proposed new Department of Air Pollution Control all requested that the mayor keep Maxwell. Nonetheless, the retention of either Maxwell or Christy was "considered doubtful." The mayor merely promised to name "the best available experts to serve in the new department," affirming that "there will be no politics whatsoever in the new set-up." The *Times* urged the mayor in an editorial "to bow, this once, to public opinion," noting: "It is difficult for the public to recall an officeholder in this town who has shown such zeal, or who has had the public so solidly behind him." Other concerned citizens also wrote in to praise Maxwell, one calling him "a symbol of good and courageous administration" and adding, "If Mayor Impellitteri had more Maxwells in his administration he would now have the confidence of the people." Impellitteri did bow to public pressure when he finally signed the Sharkey bill in late September, and Maxwell ultimately was kept on as deputy director of the new DAPC. His retention represented a victory for citizen smoke fighters against the scientifically more sophisticated air pollution control professionals, who viewed the likes of Maxwell as mere amateurs.[41]

Despite their love for the crusty ex-admiral, New Yorkers were generally optimistic that their new, larger, more professional control agency would be able to finish the job the Bureau of Smoke Control had begun. There were similarly high hopes for Dr. Leonard Greenburg, Christy's replacement, who seemed the perfect man for the job with his advanced degrees in sanitary engineering and public health in addition to a medical degree. At a time of great, often unquestioning faith in science, many citizens assumed that Greenburg, a man of science far more than Christy or Maxwell, would be able to harness science and technology to banish air pollution from their city forever.

Assuming they had left the job in capable hands, New Yorkers generally turned their attention away from air pollution after the paroxysm of publicity surrounding the issue during the summer of 1952, to the extent that dirty air was barely even an issue in the mayoral election of 1953. As has so often happened in the history of environmental problems, after a time of intense public interest, the cycle receded and public interest gradually withered, as did citizens' organizations concerned with air pollution. Not that the issue died away completely: local papers still published stories, and local citizens still griped, about contaminated air. But air pollution lost much of its prominence as a newsworthy topic after 1952, even for the *New York Times*.

Part of the problem may have been that as air pollution science and control moved out of their pioneering phase, they became less accessible and comprehensible both for journalists and for the general public. Visible smoke seemed a relatively straightforward, traditional concept; air pollution, including newfound, invisible gaseous contaminants, was less so. The growing level of sophistication and professionalism in air pollution control can be seen in the figure of Dr. Leonard Greenburg, with his extensive training in engineering, medicine, and public health science. This trend toward greater sophistication and complexity can also be seen in the efforts that began while the BSC still existed to give air pollution more precise scientific and legal definitions. The initial attempts at redefining air pollution grew out of the complaints of business interests that the existing law, if interpreted literally, would effectively ban any combustion whatsoever within the city limits by its vague prohibition on "visible emissions" for more than two consecutive minutes out of any quarter hour. As such, in March, 1952, the New York City Council considered changing the legal definition of "smoke" from the vague "emission from a chimney, stack or open fire, or from heating fuel or refuse" to "small gas-borne or air-borne particles emitted . . . in sufficient number to be observable," among other proposed alterations.[42]

Not all New Yorkers were happy with the changes in air pollution policy. Elizabeth Robinson, for one, was disturbed at the proposed redefinition of smoke, suspecting that this was part of a businessmen's plot to weaken air pollution control. For her, smoke was perfectly straightforward: "Smoke is filthy, obnoxious, and wasteful stuff that comes out of chimneys—and that is what we do not want." She also demanded the disclosure of the names of those proposing new definitions.[43]

Later, at a meeting of a scientific panel brought together by the American Chemical Society and the Technical Societies' Councils of New York and New Jersey to discuss the nature of smoke as engineers, chemists, and physicians understood it and whether it was a health hazard or merely a nuisance, Robinson interrupted the discussion, declaring: "We know what smoke is. . . . It's that nasty, horrible, stinking, unhealthy stuff that comes out of chimneys and smokestacks. And you fellows better help shut it off." Robinson charged, with some justification, that endless scientific discussion about the nature of air pollution was only a delaying tactic: "We are fed up with this internal 'discussing' of the problem. . . . We women want the discharge of smoke into the atmosphere stopped. We don't care how, so long as it's done."[44]

Other panel members "squirmed or smiled weakly" as Robinson delivered her "blast." Then they set about undercutting her fundamentalist approach to the air pollution issue. Some panel members observed, reasonably if quietistically, that totally shutting off smoke would probably be impossible and that "a certain amount" was "part of the price of an urbanized, industrialized society." Dr. John J. Phair, professor of preventive medicine at the University of Cincinnati, said there was "no convincing evidence in any but the most exceptional circumstances that air pollution had any effect on health," and the only way to find out otherwise would be through a "costly and time-consuming morbidity survey." This claim, to the extent it was accurate at all, was really only a confession of present scientific ignorance, but this sort of argument has often been used successfully to justify further uncontrolled environmental abuses. Another panel member protested that "hysteria" like Robinson's only hindered the cause of air pollution control.[45]

This scientist's word choice, with its demeaning gender connotation, is revealing of the attitudes behind the new, professional order in air pollution control and its links to the prewar past, for what it implied, with smug condescension, was that while women could worry about beautification and other minor things, their input was not welcome in the masculine realm of the hard sciences where truly significant issues were handled. While that might seem to be pinning too much on one casual statement, it is unquestionably true that women and other citizen nonspecialists receded in importance on the air pollution control front when society delivered the issue into the hands of men of science. On the surface this may seem wholly reasonable—the women who had long carried the torch for air pollution control typically were not experts; the (monochromatically male) scientists claimed to be experts and were presumably better able to find the actual truth about air pollution. Yet, although their objectivity would seldom have been questioned back in the early 1950s, the scientists of the day, like their prewar predecessors, were often predisposed to agree with the industrial polluters' point of view. Of the professions represented at the meeting where Robinson delivered her fundamentalist "blast," the chemists and engineers were already close partners of industry, while many of America's medical practitioners were rapidly becoming enmeshed in the corporate-industrial culture as well. Because of this potential obstacle to objectivity, and due to the demonstrable limits of corporate America's capacity for self-criticism, self-regulation, or public-spiritedness, there perhaps remained a place for the sometimes simplistic but earnest oppositional attitudes of representatives of the general public such as Elizabeth Robinson to keep pressure on the regulators and keep the regulatory process more democratic and responsive. But such oppositional sensibilities were largely shunted aside by the new technocrats of air pollution control in New York City—at least for a time.[46]

By the mid-1950s, New York's pioneering reformers had won some skirmishes but lost control of the overall war on air pollution. For instance, after gaining such prominence for herself and the Committee for Smoke Control in 1951, Mrs. Robinson, evidently no longer head of a group seemingly torn apart by factionalism, dwindled in publicity during 1952 and basically vanished from the news by 1953. Another old-line, fundamentalist anti-smoke crusader, William S. Maxwell, briefly hung onto a job in the DAPC but similarly disappeared from the media and was gone from the agency within a few years. Of course, there was much room for improvement in understanding of the issue among early smoke reformers, and it was probably inevitable that earlier, simpler notions should be replaced by more sophisticated concepts. Irritated but ignorant citizens certainly could get in the way of effective air pollution control, as did those citizens of Los Angeles who long insisted that the automobile had no role in air pollution and who viewed this claim as part of a sinister conspiracy to get local oil refineries off the hook. Yet the gains in sophistication were balanced by loss of public enthusiasm and democratic input, while polluters gained new opportunities to hide behind scientific niceties and incomplete data.[47]

Along with the early amateur reformers went the naive optimism that the war against air pollution soon could be won. Despite warnings to the contrary, many New Yorkers, both experts and ordinary citizens, had presumed that the smoke nui-

sance soon could be beaten into submission or even eradicated by the clever application of science and technology of the sort that had unlocked the power of the atom,
won the Second World War, and would soon put humans in space. Such attitudes
persisted throughout the 1950s and into the 1960s, giving rise to the oft-repeated
question: if we can put a person in space (later, on the moon), why can't we lick the
air pollution problem? A good proportion of the articles on air pollution in the *New
York Times* during these years concerned inventions that were to help win the war
against dirty air, from the (rediscovered) electrostatic precipitator and various early
automotive emissions control devices to the upward-pointing propellers along the
New Jersey Turnpike that the designers expected, with Canute-like pretensions, to
generate updrafts sufficient to blow away hazardous accumulations of smoke and
fog. Like other Americans, New Yorkers assumed that technology would offer a quick,
easy solution to their problems without demanding changes in the way of life that
was generating the pollution.[48]

Yet, set against such naive hopes and technological dreams were the inexorable
scientific discoveries determining that visible smoke was, in fact, only the tip of the
air pollution iceberg, and that the greatest dangers lay out of sight. This unhappy
process of discovery began, at least as far as the American public was aware, with the
linking of the Donora deaths to sulfur oxide compounds. For some years after that
point, sulfur oxides were the only invisible air pollutants ordinary citizens and most
experts knew to worry about, and sulfur dioxide was the first contaminant other
than visible emissions such as smoke, fly ash, and other particulates to be monitored
by the new air sampling equipment that started becoming available during the late
1940s and early 1950s. A parade of other new causes for worry soon followed. As
described, Los Angeles research uncovered the role of unburned gasoline hydrocarbons and ozone in creating photochemical smog. Many experts cavalierly thought
this a problem confined exclusively to LA throughout the 1950s, until other cities,
from Washington, D.C., and Denver to Chicago and New York, began observing
photochemical smog locally by the early to mid-1960s. About the same time that
Haagen-Smit unraveled the LA air pollution mystery, carbon monoxide—a long-
recognized threat inside mines and buildings but assumed to be harmless in the open
atmosphere—was found in measurable concentrations in ambient urban air and
was recognized as a further potential health risk. Later, nitrogen oxides, largely invisible byproducts of high-temperature or high-compression combustion that in
sufficient concentrations produce a characteristic brown smudge on the horizon,
also joined the ranks of the unseen enemies of health and welfare. Along with particulates, these air contaminants remained the primary targets of air sampling and
monitoring programs through the 1970s and beyond.[49]

Concerned New Yorkers gradually learned of the discovery of these previously
unsuspected environmental dangers. Moreover, they would learn that whatever
progress might be claimed on controlling visible emissions, the invisible ones that
were thought to pose the greatest health risks were getting steadily worse. For in-

stance, in 1955, New Yorkers could pride themselves on having the second best air of that in five major cities according to particulate measurements; but by 1966, they learned to their dismay that New York City had the worst air of any American city due to its high concentrations of invisible emissions, particularly sulfur dioxide and carbon monoxide. Although as in earlier years, the evidence regarding health risks remained not wholly conclusive, New Yorkers could also read in the *Times* about various studies suggesting further links to respiratory diseases such as bronchitis, emphysema, and lung cancer. Such reports helped keep public concern about dirty air alive, as did the frightening, visible local air pollution incidents in 1953 and 1963 and newspaper accounts of fatal episodes in London, England, in 1952 and 1962.[50]

As a result of this gradually growing awareness of the danger of air pollution, it began to revive as a major public issue in New York City by the early 1960s, helping to give impetus to the Low Committee hearings from 1964 onward and making air pollution an issue during the mayoral election of 1965. Also during this period, a new citizens' organization concerned with air pollution, Citizens for Clean Air, led by another female English expatriate, Hazel Henderson, arose to take up the fight where Elizabeth Robinson's earlier Committee for Smoke Control had left off. Thus, New Yorkers were again becoming mobilized on the issue when another notorious disaster, this time the Thanksgiving Weekend episode of 1966, starkly revealed the seriousness of air pollution and that it was still very much out of control. In the wake of this shocking reminder, public agitation over air pollution in New York City continued to mount without interruption through the early 1970s. This, in turn, fed into the wider national awareness of environmental problems growing in America at the same time.[51]

All that had happened on the air pollution front in New York City between the end of the Second World War and the late 1960s happened because of ordinary citizens rallying and organizing to resist a threat to their health and welfare. Local citizens had won a considerable victory in breaking established patterns of bureaucratic inertia and ramrodding the issue onto the agenda of the city government through the late 1940s and early 1950s. Both the Bureau of Smoke Control and the Department of Air Pollution Control grew directly out of widespread early public agitation for an environmental cause.

Yet the early campaign against air pollution in New York City also revealed many of the characteristic weaknesses of early local environmental movements across the nation. The energy of the movement was difficult to sustain over the long run, leading to a decline in public interest during the mid- to late 1950s, such that when there was a resurgence of interest in air pollution during the mid-1960s, the experience of the earlier citizens' groups was already largely forgotten and whatever limited gains they had made were largely taken for granted. Consequently, New Yorkers had a hard time sustaining pressure on their city government sufficient to make officials treat air pollution with the seriousness many citizens felt it deserved, and air pollution long continued to be shortchanged in favor of more established policy issues.

Also, due to characteristically American, optimistic faith in science and technology coupled with corresponding technocratic assumptions, when New Yorkers did gain their new pollution control agencies, they merely turned them over to generally unresponsive control bureaucracies with supposed scientific expertise and professional pretensions that tended to marginalize citizen activists and ignore public input. Rather than facing the prospect of fundamental reorganization of living patterns to control air pollution, citizens waited for their scientists to produce technological fixes. While citizens were generally more aware of the need for compulsion to control air pollution than were most of the so-called experts, their proxies in the air pollution control agencies stubbornly refused to learn this basic lesson.

Most frustrating of all, every step made toward understanding and confronting air pollution served only to emphasize further the immensity of the problem and the inadequacy of existing efforts to control it. Over the first two decades of the postwar period, New Yorkers learned to their dismay that the smoke problem they had dreamed of banishing within a few years back in the late 1940s was a scientifically, economically, and politically complex air pollution problem firmly built into the very fabric of their society. The modern American way of life of the mid-twentieth century was, inescapably, a pollution-generating way of life, and New York's vast population made the cumulative pollution problem simply enormous. Environmentally progressive as they were relative to many other people in the nation and the world at the time, New York's early smoke fighters were not yet able to comprehend the full public complicity in the generation of air pollution, to recognize the extent to which it was entrenched in the social and economic fabric of American life, or to perceive air pollution not as a discrete phenomenon but as a single strand in a tightly interwoven knot of interrelated environmental problems demanding more comprehensive solutions. This understanding only gradually began to dawn during the mid-1960s.

Moreover, for other reasons besides sheer magnitude, the air pollution problems of New York City would prove beyond the means of local government or citizens to solve, and it is to this development that we now turn.

# chapter 8

# Jersey

## *The Interstate Dilemma*

If anything, New Yorkers' struggle with the interstate aspects of their air pollution problem was even more frustrating than their battle against strictly local emissions. In addition to spreading across various counties, the New York City metropolitan area also sprawled across state boundaries, into areas over which the New York City Bureau of Smoke Control and Department of Air Pollution Control had no authority, but which contributed to the general atmospheric pall smothering the whole region. Many such areas, particularly heavily industrialized counties in the state of New Jersey, long had little or no interest in bringing their aerial emissions under control.

This situation led to a protracted campaign to bring regional control covering the whole metropolitan area. However, while the obvious need for such a system was recognized early, implementation was much more difficult, as various jurisdictions jealously guarded their traditional local prerogatives and worked to delay and obstruct the establishment of a meaningful regional air pollution control authority. Most politicians and governmental officials from the local to the federal level always hoped the issue would be handled by the lowest possible level of government. Yet the many years of false starts and unfulfilled agreements on the road to regional control in the New York City area ultimately became a prime justification for increased federal intervention in interstate air pollution control, as did similar situations in metropolitan St. Louis, Chicago-Gary, and the Ohio River Valley between Ohio and West Virginia.

From the beginning, the long history of interstate friction over air pollution in the New York City metropolitan area was particularly focused on Staten Island, the traditionally more rural part of New York City closest to the dirty industries of northeastern New Jersey. Already between 1880 and 1883, a local physician undertook an investigation of interstate pollution in conjunction with the health departments of New York State, New Jersey, and Staten Island and concluded that northern Staten Island was indeed suffering from "injurious and unwholesome fumes." Further studies in 1908 and 1913 found elevated concentrations of sulfur oxide, arsenic, and lead in Staten Island vegetation, linked to New Jersey factories. In 1917, New York State even passed a law seeking to give state officials leverage against out-of-state polluters by authorizing the state health commissioner to take legal action against corporations emitting pollutants outside the state that "unreasonably" injured residents of New York State—action that could result in the stripping from New York-based corporations of their corporate charters or the revocation of out-of-state corporations' permission to do business within the state. However, like most other early air pollution control laws, this stern measure was never enforced.[1]

Increased industrial development in the region led to increased public outcry against the New Jersey fumes after World War I. This in turn led to the first systematic, comprehensive studies of the interstate movement of air pollution within the region. In 1931, the Stevens Institute of Technology in Hoboken, New Jersey, joined with Manhattan stations of the U.S. Weather Bureau to monitor the sources and quantities of airborne particulates over New York City, finding from careful studies of prevailing winds that much of the pollution in Manhattan came from the heavily industrialized area around Jersey City, as New Yorkers had long grumbled. In the wake of a formal complaint by the Richmond County (Staten Island) Grand Jury based on this discovery, New York Governor Franklin D. Roosevelt in 1931 ordered a special investigation of the situation by the state Health Department, which found heavy concentrations of atmospheric dust and sulfurous fumes on the part of the island nearest New Jersey and concluded in a 1932 report that New Jersey industries' "offensive, irritating, poisonous, objectionable and injurious" emissions were causing a "public nuisance and a menace to health . . . in Richmond County." Further studies found evidence of serious crop damage to Staten Island truck farmers, nursery operators, and commercial flower growers from Jersey's fumes.[2]

During this period, New Yorkers and New Jerseyites continued to blame each other for interstate air pollution. In 1935, David R. Morris, a meteorologist at New York's Central Park Observatory, blamed New Jersey for much of the city's aerial filth, while William G. Christy, then smoke abatement engineer for Hudson County, New Jersey, protested that Hudson County was no longer a problem due to engineering improvements that had reduced smoke pollution by 90 percent. Beyond this war of words, though, little action resulted as the increased general public awareness of air pollution during the early Depression years faded by the later 1930s.[3]

After the Second World War, Staten Islanders again complained to their local and state officials about New Jersey's atmospheric filth. Subsequent studies found that many of these local residents had "legitimate grievances" regarding detrimental health effects, including stomach disorders, irritated noses and eyes, and inability to sleep. Consequently, in 1949, the New York State health commissioner asked the

U.S. Public Health Service (PHS) to arrange a conference among concerned agencies in the region to address interstate air pollution. Representatives of the PHS, the health departments of New Jersey, New York State, and New York City, and the New York State Department of Labor (with experience of in-plant industrial pollution) held several meetings and undertook preliminary investigations, but "budgetary limitations" precluded a final report or further action by this first postwar interstate cooperative effort. In the early 1950s, a similarly constituted Subcommittee on Interstate Air Pollution Control followed through with an inconclusive report. Meanwhile, in 1950, the PHS called off its study of interstate pollution due to a scarcity of qualified technicians. Notwithstanding the valiant efforts of Staten Island's congressman, James J. Murphy, to get federal legislation calling for a nationwide federal study of the health effects of air pollution, the issue remained mostly dormant.[4]

In 1951, though, the matter was again forcefully brought to the public's attention in New York City. During mid-August, William S. Maxwell, deputy director of the New York City Bureau of Smoke Control (BSC), made his celebrated first aerial survey of air pollution in the metropolitan area and, as earlier described, saw the problem from a new perspective, with myriad individual sources all visibly contributing to one great pall of smoke hanging over the whole city. The most striking discovery from these early flights, however, was the graphic proof that the metropolitan area's atmospheric problem stretched far beyond the boundaries of New York City proper. For the first time, rather than merely trading unsubstantiated allegations, the aerial surveys produced conclusive photographic evidence of New Jersey industrial plumes blowing with the prevailing westerly winds to merge with the general New York City smog blanket. Maxwell himself reported how everything across the Hudson appeared to be smoking out of control. Maxwell's aerial expeditions provoked renewed outcries in New York City against interstate air pollution and helped to stimulate all subsequent trans-jurisdictional control efforts in the metropolitan region.[5]

By the late summer of 1951, New Yorkers besides those on Staten Island were growing more bitterly outspoken in their denunciation of New Jersey's aerial contamination. At the end of August, New York City's Department of Health requested that New Jersey health authorities abate obnoxious industrial odors wafting across the Hudson, citing "a large number of complaints." The city's assistant health commissioner himself called the stench "more nauseating than anyone can imagine." In early September, BSC director William G. Christy cited the ongoing harm to Staten Islanders and other New Yorkers from Jersey industrial emissions and declared, "Smoke is no respecter of state lines. . . . Our problem cannot be solved until we have some kind of regional regulation." He further announced that he would argue for regional control before the New York State Joint Legislative Committee on Inter-State Cooperation and would confront the chairman of New Jersey's ineffectual Commission on the Prevention and Abatement of Air Pollution with Maxwell's aerial photographs. New York State legislators accepted Christy's suggestion that regional interstate air pollution be put under the jurisdiction of the Interstate Sanitation Commission (ISC), an existing organization that monitored interstate water pollution between New York, New Jersey, and Connecticut. Meanwhile, city officials in metropolitan New Jersey predictably denied Christy's charges that they were doing "practically nothing" about air pollution, denied any contribution to New York City's aerial woes, and characteristically called for voluntary self-policing of emissions by polluting industries.[6]

The following year, New York City Mayor Vincent R. Impellitteri issued a formal request to the governors of New York, New Jersey, and Connecticut for legislation to give the ISC authority over regional air pollution control, noting the "interstate character" of a problem best handled by coordinated efforts among the three states. New York State Assemblyman Elisha T. Barrett, chairman of the Joint Legislative Committee on Interstate Cooperation, admitted that it was "pretty much agreed that some kind of interstate action is needed" for controlling air pollution in the metropolitan area. Thereafter, the New York State legislature passed bills to authorize and fund this new role for the ISC, while the agency prepared to study the feasibility of interstate regional control. Meanwhile, another congressional representative for New York City, John H. Ray, followed James Murphy in seeking legislation to start a full-fledged, nationwide federal investigation of air pollution to gain information for further legislation and control efforts. Ray alleged that when the wind blew from the New Jersey plants, "In many cases the entire crop of people who grow flowers for the market has been destroyed. . . . Vegetable growers have often lost much of their crops overnight." He found the problem far beyond the reach of local authorities and requiring federal intervention: "Local air pollution boards . . . cannot deal with the whole problem, because interstate features are involved. The States cannot cross State lines, and the cities cannot go outside their own jurisdiction." Ray's efforts to help his constituents were in vain, however, and the federal government would not pass its first weak legislation on air pollution until 1955.[7]

By the early 1950s, New York State and even New Jersey finally began to show some signs of life on the air pollution control front. Yet for a long time, the record of progress at the state level was weak at best. New York State's first major postwar action on the issue was Governor Thomas E. Dewey's 1948 veto of a bill to back New York City's new smoke abatement ordinances with state authority, with Dewey typically arguing that the purposes of the bill, which would have declared the emission of dense smoke in cities of more than one hundred thousand population to be a statutory nuisance, had already been accomplished by local law. Such neglect of the city's needs by the state government was a familiar pattern in New York, even though other cities, such as Buffalo, Troy, and Albany itself, were starting to develop significant air pollution problems. The state legislature did not resume even considering the issue before 1952, with the passage of the bill authorizing the ISC air pollution study. During 1954, the state Labor Department's Division of Industrial Hygiene conducted a survey to determine the need for air pollution control *outside* New York City, while in 1955 and 1956, the State Department of Health conducted more comprehensive surveys of the status of air pollution and control measures in the upstate areas. The resulting picture of a growing problem with little being done about it finally led to the passing of the 1957 New York State Air Pollution Control Act, creating a nine-member Air Pollution Control Board including five state department commissioners and four members representing local government, industry, and the medical and engineering professions.[8]

The picture was bleaker in New Jersey. After a progressive early step in 1931 with the establishment of the nation's first countywide Department of Smoke Regulation in Hudson County, little action was taken anywhere else in the state, and the Hudson County program itself was moribund by the late 1930s. After the war, public complaints both in state and out of state had grown to the point that the New Jersey Department of Health felt compelled to launch a minimal air pollution investigation program. In April, 1950, New Jersey created a special legislative commission to study the problem, and members were sent all around the state and to other cities and states with air pollution troubles to try to discover the full dimensions of the problem. In March, 1952, the commission issued a report tracing the state's pollution problems to its major oil, chemical, and metal-processing industries along with smoldering dumps, utility generating stations, and other more traditional sources. The report recommended action at the state level because local officials were found to be "powerless or unwilling to take proper action," while pollutants moved freely between local jurisdictions. Local chambers of commerce had formed active air pollution abatement committees, "hoping to avoid legislation," but while some companies were spending "large sums" to minimize their emissions, the report found, "there are also those who will do nothing unless forced." Some companies complained of pollution from neighboring companies ruining their products, while some major polluters "were not even willing to discuss the problem of air pollution." The commission also found that "the likelihood of atmospheric pollutants crossing political lines to create problems in neighboring states is real and indicates the desirability of New Jersey's participating in regional attempts at air pollution control." This report, together with the frightening 1953 air pollution episode over the New York City metropolitan area, created increased pressure for state control legislation.[9]

However, when a bill embodying the commission's recommendations was introduced before the New Jersey state legislature, it failed to pass due to the unanimous opposition of New Jersey's industry, which was happily accustomed to little or no regulation and could not accept a bill that might stiffen control policies beyond negligible existing local and county regulations. Instead, in 1954, the state created a nine-member Air Pollution Control Commission within the State Department of Health to make codes for controlling particular pollutants or industries. Four of the nine members were to be appointed directly by industry organizations, and the whole commission was hobbled by requirements calling for a lengthy process of "conference, conciliation and persuasion" prior to any enforcement action and mandating close industry participation in the drafting of any control codes. In a state largely controlled by polluting industries, this policy of seeking "co-operation and voluntary compliance" from the regulated and protecting them from "the possibility of administrative aggressiveness" guaranteed few meaningful results.[10]

In addition to obstructing control inside the state of New Jersey, major polluters in the "Garden State" were also effectively able to hamstring early interstate efforts to cleanse the atmosphere. The bill New York State had passed earlier in 1952 to authorize and fund an interstate air pollution study by the Interstate Sanitation Commission had died because New Jersey had failed to make a matching appropriation of thirty thousand dollars for the study. Again in March, 1954, state legislators from the New York City area introduced matching bills in Albany to authorize and fund an ISC study, which were soon passed and signed. In New Jersey, Assemblywoman Florence Dwyer, a Republican from the metro-area town of Elizabeth, introduced similar legislation, but she was opposed by the solidly Republican delegation from Bergen County and the solidly Democratic delegation from Hudson County, two heavily industrial districts in northeastern New Jersey that feared increased governmental regulation. In the end, obstructionism won the day, as the New Jersey state senate laid over the bill, already passed by the lower house, and concluded their legislative season. Despite New Jersey Governor Robert B. Meyner's condemnation of their "unstatesmanlike" behavior and his threats to call the state senators back into session to deliver the bill he had "promised" to New York's Governor Dewey, no further action was taken in 1954.[11]

Meanwhile, the regional atmosphere continued to deteriorate, suffering the emissions of more than thirteen hundred unregulated heavy industrial plants in northeastern New Jersey alone. A series of dramatic fumigations hit metropolitan Jersey towns during the early 1950s, with Newark "gassed" by sulfuric acid mists while Elizabeth and Linden faced "mysterious smog" linked to metal-refining plants. Such episodes led the New Jersey Health Department to send out one of the early mobile air pollution sampling laboratories to the area, although the technicians politely never monitored air pollution without announcing their visit first, giving industries a chance to clean up their act temporarily, and the laboratory only came after a request from a municipal health officer, offering another chance for industry interference. Across the Hudson, farmers and commercial flower growers on Staten Island produced evidence of heavy concentrations of characteristic industrial pollutants from Jersey's belching smokestacks, while citizens worried that if these dusts and gases were harmful to plant life, "what are their effects on human beings?" In 1954, local students at Wagner College began a program of sampling air pollution on Staten Island as part of a long-term PHS study program. In early 1955, a frustrated Dr. Leonard Greenburg, director of the New York City Department of Air Pollution Control (DAPC), noting that he was "just plain tired of waiting for New Jersey to take action," wrote a formal request to U.S. Surgeon General Dr. Leonard A. Scheele calling for an immediate, full-fledged federal investigation of interstate air pollution in the metropolitan area.[12]

Facing the threat of federal intervention, New Jersey's recalcitrant state senators finally gave in and passed the necessary legislation for the ISC study in 1955. Then it was Connecticut's turn to act up. Although at this time they did not feel their state was part of the regional air pollution problem, as either perpetrator or victim, Connecticut legislators wanted assurance that the state would not have to pay for any part of the ISC air pollution study and demanded that Congress approve the addition of air pollution to the ISC's traditional role in monitoring interstate water pollution. This federal authorization was gained during the summer of 1956, and finally, despite protests from New Jersey state legislators that the ISC was biased in favor of New York City, both New York and New Jersey were able to pass needed authorization and appropriation bills during the same year.[13]

At the end of 1956, the Interstate Sanitation Commission set out to discover the sources and extent of interstate air pollution in the metropolitan area, the types and degree of economic and health injuries caused, and the relevant meteorological and topographical factors in the region. The ISC would use staff and facilities of the U.S. Army Chemical Corps, the U.S. Department of Agriculture, the PHS, and the U.S. Weather Bureau in conducting its studies. Public hearings were held in New Jersey, Staten Island, and Manhattan, at which New Yorkers and New Jerseyites again traded blame for the problem. Frank Hauber, president of the Staten Island Air Pollution Control Association and a nurseryman, noted that his farm had been in his family for ninety years, but that only in the past eight years, since a major metal-processing plant on the banks of the Hudson River in New Jersey had doubled its stack height from two hundred to four hundred feet, had he faced chronic crop damage and recurrent ill health from the fumes, which no longer threatened residents of New Jersey but landed on Staten Island instead. As Hauber angrily recounted, "When the stuff comes over, it is sulphuric acid and anyone knows it doesn't do you any good when it gets in your lungs. We only want one thing here—clean air to breathe—and we think we are entitled to that." A professor of public health at Wagner College corroborated Hauber's account. Ultimately, the ISC's wind-direction and pollution-transportation studies using fluorescent tracer dust to track the interstate movement of particulates found that, mostly, New Yorkers suffered from Jersey pollution but that the wind did sometimes blow pollution the other way as well, leading to renewed ineffectual sniping between the two states.[14]

In January, 1958, the ISC produced a report on its investigations, finding interstate air pollution in the New York–New Jersey metropolitan area to be "a great and growing problem, affecting millions of people." The report noted "enormous" economic losses from air pollution and potential hazards to commercial flights from impaired visibility, along with clear evidence of injury to the health of individuals with respiratory troubles and further concerns that air pollution might contribute to increased incidence of heart disease and lung cancer. While the report identified countless sources of air pollution, New Jersey's heavy industries drew particular criticism. New Jersey's local approach to enforcement was also found feeble at best: "Not more than three or four of the cities can provide the technical staff necessary for reasonable control of pollution sources, and many of the communities are so small, and so dependent on their industry, that their efforts are ineffective." The ISC concluded that the regional, interstate air pollution problem could not be solved by the independent action of governmental agencies within existing jurisdictions and so recommended an interstate air pollution control district including New York City and the six dirtiest New Jersey counties, with a combined population of eleven and a half million inhabitants.[15]

However, when the ISC drew up legislation to give itself the necessary authority to combat the regional air pollution problem, the bill was rejected by the New York State Senate and "was never seriously considered" by either house of the New Jersey legislature. Existing control agencies in both states blasted the bill for setting up a "superbureaucracy" and for demanding unwarranted delegation of state authority. New Jersey, in particular, was frightened at the prospect of actually being forced to clean up. Thus, in March, 1959, officials from New York City, New York State, and New Jersey met to determine that the interstate air pollution problem could best be dealt with through the greater coordination and cooperation of existing state and local agencies in the region. To accomplish this, they formed the New York–New Jersey Cooperative Committee on Interstate Air Pollution, consisting of representatives from control agencies and health departments of the states and city. Instead of its original, more ambitious proposal, the ISC was given power only to study, to survey, and to oversee exchange of information among the other agencies. It was able to take action only through existing authorities and was required to keep secret the names of any polluters under investigation. Any minimal action the weak ISC might try to undertake could effectively be stopped by the failure of one state, usually New Jersey, to cooperate. This was still not ineffectual enough to satisfy the industrial lobbies and industry-dominated control authorities of both states, but the alternate legislation nevertheless became law later that year.[16]

While the ISC was being castrated, local officials in the metropolitan area were offering another quick fix for interstate air pollution in the form of the Metropolitan Regional Council (MRC), an organization started by New York City mayor Robert F. Wagner, Jr., which held discussions on various issues spanning the whole metropolitan area. In September, 1956, the MRC chose Karl Metzger, director of the county board of Middlesex County, New Jersey, as head of a new study committee on air pollution. In December, Metzger's committee called for an emergency air pollution control organization within the MRC with the power to order local health or air pollution agencies to shut down industries during atmospheric inversions, for while such conditions were rare in the New York City area, the committee found, "their effect upon the death rate and general health is real and serious." The MRC also approved the Metzger committee's recommendations for county air pollution con-

trol commissions with full funding and enforcement powers and for regional requirements for low-sulfur "smokeless" fuels and devices to eliminate automobile
exhaust fumes. Later, in May, 1957, the MRC called for the creation of a regional
warning system for air pollution to alert the public in the event of an emergency and
to clamp down on garbage incineration, factory operation, and vehicle use. New
York City officials argued that the MRC program offered the "best hope for area-
wide progress . . . at a reasonable cost."[17]

Despite these relatively ambitious early proposals, by 1961 the MRC was still merely
contemplating putting together a network of air pollution sampling stations to serve
an emergency warning function, and even this would only have the authority to
request that industries and individuals cut back their combustion voluntarily. Not
until 1965 was the system for a coordinated regional response to an air pollution
emergency actually being organized, and when it was needed during the infamous
Thanksgiving Weekend episode of 1966, it failed to work as planned. As was typical
of the interstate control effort in the New York City metropolitan area before 1970,
the MRC air pollution control program would never develop any real power.[18]

By the early 1960s, though, public concern about dirty air again began to grow as
the passing years revealed the inefficacy of the various state and local control programs. Even in New Jersey, citizens' groups from the metropolitan area went to Trenton to demand from their state legislators stricter regulations to control the "unbearable"
industrial fumes that forced them to keep their doors and windows closed all the time,
while the New Jersey Manufacturers' Association protested that the angry citizens'
definitions of air pollution were too broad and imprecise. Such renewed public agitation led to pressure for changes in air pollution policy at every level, local, state,
and interstate.[19]

There was still much room for improvement. Most unsatisfactory were the state
and local controls in New Jersey, where the industry-dominated Air Pollution Control Commission and State Department of Health conducted a feeble, underfunded
control program that basically offered polite suggestions and carefully refrained from
enforcing the law or prosecuting chronic polluters. Most state officials, including
the governor, were continually afraid that action to control air pollution would lead
to plant closings and state residents being thrown out of work, a prospect the industrial lobbies were only too happy to hold over their heads. By 1965, *Time* magazine
observed that although New Jersey's air and water was "hopelessly polluted," the
state still had only a "token anti-pollution campaign," while New Jersey officials
themselves admitted that funding was totally inadequate and the overall program
much too "sleepy."[20]

Yet New Jersey's powerful industrial lobbies still stubbornly fought any change.
In a statement typical of American industry's "technologism" with regard to air pollution control, and of the desire to keep a political issue "above politics," when some
more progressive state legislators answered public complaints and sought to replace
the wholly industry-dominated Air Pollution Control Commission with a new, more

responsive agency, the New Jersey Chamber of Commerce protested that this would "vitiate the entire concept of air pollution control in New Jersey" by substituting for the "existing technically oriented Commission one which would be emotionally oriented, and subject to the vagaries of a changing political scene." A representative of the state Manufacturers' Association further argued: "The present Control Commission includes a high level of technically trained people drawn from the major interests involved in the control of pollution. . . . The Commission must be free from politically inspired pressures." In words characteristic of business arguments against pollution control then and now, the Jersey business operators sought to hide behind supposedly objective science while failing to admit that they and their corporate scientists' actions regarding air pollution were of course largely subject to political and economic pressures, and they took their revolving-door relationship with the Air Pollution Control Commission very much for granted.

Meanwhile, the governor of New Jersey was left lamely arguing that New Jersey was doing all it could to control interstate pollution in the New York City area, since the state's major pharmaceutical, chemical, and petroleum-refining industries inevitably produced certain "obnoxious but otherwise harmless" odors, and that to do any more to control these might limit the state's industrial expansion. New Jersey still also hoped to take advantage of its lax pollution controls by luring industry away from New York City, as it long had done. As late as the summer of 1967, when New York's Senator Robert F. Kennedy toured the greater metropolitan area in a helicopter to review the interstate air pollution situation, he and his companions passed over a smoke-belching factory in New Jersey with a large sign offering an unsubtle reminder of the lax state of industrial regulation in the Garden State for New York industrialists considering relocation: "Hoboken Welcomes Industry, Come on Over." Not until 1967 would New Jersey get a new and improved state air pollution control agency, over the furious protests of the existing organization's chairman, who happened to come from the state's largest, most polluting electric utility.[21]

Renewed public concern over polluted air during the 1960s also brought at least the semblance of action on the interstate front. In order to circumvent New Jersey's continual obstructionism, New Yorkers ultimately threatened to bring in the federal government. City officials increasingly emphasized the interstate dimensions of regional air pollution stretching even beyond New Jersey, as when New York City Air Pollution Control Commissioner Arthur J. Benline angrily complained in October, 1962, that New York City was at the end of a "3,000-mile-long sewer from the California coast" due to the prevailing westerlies over North America, which brought atmospheric filth in from other states, particularly New Jersey. Mayor Wagner of New York City painted a similar picture for a visiting U.S. Senate subcommittee in February of 1964, emphasizing how the city could not solve its pollution problems alone. In February, 1965, New York City Councilman Robert A. Low further emphasized the interstate nature of the problem, observing how the fourteen hundred separate jurisdictions in the metropolitan area precluded any real progress through the independent actions of individual localities. Low also discussed a recent study that had found New Jersey contributing 20 to 30 percent of the pollution over the city. The final report of Low's special committee to investigate air pollution bluntly concluded:

> It is quite clear . . . that the interstate and state agencies have been unable or unwilling to establish standards or a plan for control, have been unable or unwilling to take vigorous enforcement action until standards or a plan are established, and have resisted the efforts of the Federal Government to undertake enforcement. Simply stated, the position of these agencies has been that they can do little without Federal aid, but they would prefer that the Federal Government merely give money and then go away without assurances of establishment of adequate standards or adequate enforcement. . . . Such an approach offers insufficient protection for citizens living under conditions conceded to be "virtually intolerable on some days."

In their statements, city officials were building a case for greater federal funding and other assistance under the new federal Clean Air Act of 1963, which greatly expanded the availability of federal grants for state and local control programs—though federal officials reminded them that even if there were no pollution wafting over from New Jersey, the quality of the air over Manhattan would still be inadequate as a result of purely local sources of contaminants.[22]

In 1965, New York State joined the metropolis in pushing for an increased federal role. Reversing his earlier opposition to federal intervention, Governor Nelson Rockefeller, noting contaminants from New Jersey believed to be "adversely affecting the health and welfare of persons residing in the State of New York," made a formal, public request to the secretary of health, education, and welfare, John W. Gardner, to call a conference of concerned officials in the region to develop uniform regional controls under the terms of the 1963 Clean Air Act. This action carried with it the potential threat of injunctions against the offending state and individual polluters within it, and it significantly ratcheted up the pressure on New York's noxious neighbor. Later, Rockefeller stationed mobile air pollution sampling laboratories along the state's border with New Jersey to gather data proving that New Jersey was contaminating the state of New York even outside New York City. Legislators in Albany also passed new legislation calling for strengthening the enforcement power of the ISC to control air pollution in the metropolitan region.[23]

These actions, together with the frightening 1966 Thanksgiving Weekend air pollution episode that blanketed the whole region, finally had the desired effect of bringing New Jersey to the negotiating table. To help deal with problems in the southern part of the state, where New Jersey was itself the victim of other states' pollution, as well as with the troubles of the New York City metropolitan area, during late 1966, Governor Richard J. Hughes began pushing for a four-state regional air pollution agreement. By December, 1966, Hughes, Rockefeller, Governor Charles L. Terry, Jr., of Delaware, and Governor Elect Raymond P. Shafer of Pennsylvania had all promised one another not to make their respective states "pollution havens" by neglecting air pollution regulations so as to attract industry, while state health officials were arranging joint planning and research on air pollution.[24]

Thereafter, the states tried to put together an enforceable, formal interstate compact on air pollution, which required approval by all state legislatures as well as the U.S. Congress. In early 1967, the states of New York, New Jersey, and Pennsylvania held a conference in Philadelphia where they announced plans for a Mid-Atlantic States Air Pollution Control Commission that would also include the federal government and possibly the states of Delaware and Connecticut. This first interstate air pollution control compact would set common air quality standards for all major pollution sources, from generating stations and factories to homes and automobiles, would create a regionwide common market for low-sulfur fuels and pollution control devices, and would establish a regional authority with power to investigate general causes and individual sources of air pollution and to mete out penalties of up to a thousand dollars per day for violations. The plan also would allow states and localities to set stricter controls if they wished, but no jurisdictions would be able to sink below the standards established in the interstate compact. New York's Governor Rockefeller, who had for years promoted proposals to strengthen the existing Interstate Sanitation Commission to address more effectively the problems of interstate air pollution in the New York City metropolitan region, was ultimately won over by New Jersey Governor Hughes's more ambitious regional plan and abandoned the idea of further tinkering with the ineffectual ISC. The legislatures of both New York and New Jersey promptly passed legislation authorizing entry into the new proposed arrangement, and a large bipartisan joint delegation of New York and New Jersey legislators introduced enabling legislation in both houses of Congress.[25]

There is no question that the sudden leap toward a more effective interstate control approach on the part of the states was motivated by a desire to forestall the impending possibility of more aggressive federal interference in the matter. In explaining his plan for an interstate compact, New Jersey Governor Hughes made this clear. He argued that there was a definite need for regional cooperation but hastily insisted that there was no need for a dominant federal role, proclaiming: "As for a super Federal agency, I don't buy that. . . . I don't think the time will ever come when we have to take air pollution control out of the hands of the states themselves." New Jersey, along with other states, was eager to see that ultimate control authority remained in state hands.[26]

Whatever the mid-Atlantic states might have wished, by the later 1960s, the federal government was inescapably in the picture due to the various earlier failed efforts at interstate air pollution control in the New York City metropolitan area. Following up Governor Rockefeller's 1965 request, by January, 1967, federal officials of the U.S. Department of Health, Education, and Welfare (HEW) held a conference on interstate air pollution in the metropolitan area. At the end of this conference, the federal government formally requested that both New York State and New Jersey pass stricter air pollution control laws and released a list of the nearly four hundred worst polluters in the area, based on earlier preliminary studies. Federal officials also called for the creation of a "formal interstate agency" with all necessary abatement authority and warned that federal control recommendations would "lay the legal ground for further Federal enforcement action if remedial action is not taken promptly." The mid-Atlantic states were thus on notice that their new proposed compact was their last chance to work out their mutual air pollution problems without further major federal intervention.[27]

Meanwhile, area residents grew steadily more exercised about the interstate pollution problem. This culminated in a report released in January, 1967, by Dr. Leonard Greenburg, a former New York City air pollution control commissioner who later worked as a public health researcher at Yeshiva University, finding that inhabitants of northern Staten Island, closest to New Jersey's industrial complex, showed markedly higher rates of lung and pulmonary cancer than their neighbors in the central and southern part of the island. While the northern part of the island was admittedly more crowded and had more pollution sources of its own, Staten Islanders angrily blamed New Jersey for their woes. Later that year, five hundred Staten Islanders formed a one-hundred-car motorcade to drive into New Jersey as a protest; there they were surprised and pleased to be joined by a further two hundred Jerseyites in fifty cars with anti-pollution banners pasted on their doors. William O'Connell, organizer of the Staten Island motorcade, charged that New Jersey had been "criminally lax in regulating industries that are prime offenders," while John Marsh, the mayor of Rahway, New Jersey, who joined the protest motorcade, noted that local governments were "reluctant to force industry to conform" to air pollution standards for fear that "they may lose them and thereby jeopardize the local economy." Such demonstrations put further pressure on state officials to come up with an effective, workable agreement.[28]

For all the public concern and the efforts of state and federal legislators, however, the Mid-Atlantic States Air Pollution Control Compact met numerous obstacles. Officials in various states worried that the agreement would tie their hands in undesirable ways, or that it did not adequately address the problems of greatest concern for their states. For instance, Governor Shafer of Pennsylvania wanted interstate agreements to cover only more specific regions, not entire states, since Pennsylvania shared a pollution problem with New Jersey and Delaware in the east but had to worry about pollution from Ohio and West Virginia in the west. He argued that air pollution problems were basically confined to particular "airsheds," such as river valleys or intermountain regions, a position seconded by the major federal pollution control advocate, Senator Edmund S. Muskie of Maine. Representative John M. Murphy of New York City/Staten Island, in turn, wondered whether a coal state like Pennsylvania or an oil state like Delaware could really be persuaded to accept strong measures limiting the sulfur content of coal and oil. He also pointed out, only partially accurately, that neither of these states to the south shared a common air mass with New York City or Connecticut, and wondered what was the point of bringing in additional states when this simply increased the possibility of one state obstructing the progress of the rest? Other critics fretted about the difficulty of getting all member states to enforce air pollution regulations equally strictly or even the frightful prospect of driving industry out of the controlled region altogether. At the level of actual practice, beyond mere rhetoric, the interstate compact could not solve the traditional problems of relative laxity between states in the absence of some higher authority. For all these reasons, the Mid-Atlantic States Compact was stalled through early 1968, with Congress and some potential member states hesitating to approve the measure.[29]

Meanwhile, the federal government was taking other action on national and interstate air pollution. In 1967, Congress passed the Air Quality Act, which ordered the HEW secretary to designate by May, 1969, air quality control regions in interstate areas experiencing significant air pollution. Thereafter, HEW was to issue air quality criteria, including recommendations for safe levels of various pollutants, "as soon as practicable"—within two years. After that, states would have a further fifteen months in which to establish air pollution standards and develop plans for control programs to meet these federal criteria. The law also sought to promote the creation of further interstate compacts to allow states to work out their mutual atmospheric problems with a minimum of direct intervention by the federal government. In a nod to industry lobbies and states' rights advocates, the Air Quality Act of 1967 refrained from establishing national emissions standards for polluting industries, instead authorizing a two-year study of the need for such standards.[30]

This legislation was the federal government's last, desperate attempt to continue the long-standing federalist policy of leaving pollution control in the hands of states and localities and keeping federal intervention to a minimum. As some predicted at the time, with all its deliberate time delays and expectations that states could effectively work together on shared problems they had often preferred to ignore, the Air Quality Act of 1967 was doomed to failure, in the New York City area as elsewhere. In 1969, a year after federal officials made the New York City metropolitan area the second official air quality control region in the nation, New York City and New Jersey were still ineffectually bickering about their unresolved interstate air pollution problem, New Jersey industries were still stubbornly resisting control of their emissions, Staten Island was still suffering major economic and health damage, and New York City mayoral candidate Norman Mailer flamboyantly declared that "the only way to end the smog is for citizens to get muskets, get on barges, go to Jersey and explode all the factories." The lengthy, time-consuming processes established under the Air Quality Act of 1967 could not show enough progress soon enough to satisfy the increasingly radical environmental sentiments of the American people as the nation moved toward 1970, the first Earth Day, and, not coincidentally, the Clean Air Act Amendments of 1970, which tossed aside the hopeful but unsuccessful interstate compact idea and finally gave the federal government ultimate authority in national air pollution control.[31]

By 1970, the states of New York and New Jersey and the various local jurisdictions in the metropolitan region had had more than twenty years in which to work out a voluntary, cooperative approach to their serious interstate air pollution problem, recognized by the late 1960s as perhaps the worst air pollution problem in the nation.[32] They had failed to do so, and good intentions and resounding rhetoric had repeatedly proven unable to overcome the political and economic pressures and interstate rivalries that contributed to the situation in the first place. The Mid-Atlantic States Air Pollution Control Compact, once touted as an example of "Creative Federalism" in action, ultimately proved no more workable or effective than the various feeble efforts that preceded it.[33] In the second half of the twentieth century as in the days of the Articles of Confederation, national problems transcending state and local jurisdictions inevitably required intervention by a higher authority not so directly subject to short-sighted, localized interests. Federal legislators certainly did not initially set out to supersede the states in air pollution control, not even in interstate air pollution control, where the federal government had clearer authority to do so. But the continual nonfeasance of squabbling states such as New Jersey and New York—or Missouri, Illinois, Indiana, Ohio, and West Virginia, among others—and the rising public outcry by frustrated citizens against dirty air gradually dragged the federal government, against its will, into regulating the issue.

# Florida

# chapter 9

# The Fickle Finger of Phosphate

*Industrial Air Pollution in Rural Central Florida*

Another place that developed a serious, notorious air pollution problem during the early postwar decades was central Florida—perhaps the last place anyone would have expected to find terribly contaminated air. Florida's greatest air pollution problem, like those in many other areas of the United States, was connected with a dominant regional industry: the agricultural phosphates industry, crucial in making chemical fertilizers, national and global use of which grew dramatically following the Second World War. Yet unlike most other areas plagued with chronic air pollution, which typically were urban, the Florida phosphate belt—located mainly in Polk County and neighboring counties midway along the peninsula east of Tampa—was a rural area devoted to ranching and agriculture. During the first two decades after the war, pollution from phosphate processing would wreak havoc on the traditional economic mainstays of this once-idyllic region, cattle and citrus growing. The battle against phosphate industry emissions in central Florida would drag on for nearly twenty years without resolution while an industry-hungry state government hesitated to demand cleanup and the federal government sat ineffectually on the sidelines. This pattern of inactivity, all too common throughout the United States during the 1950s and 1960s, vividly displayed the shortcomings of the pre-1970 policy of leaving air pollution control in the hands of states and localities.

The history of the phosphate industry was typical both of the pattern of resource-extraction industries in the state and of the perennial, unusually pronounced boom-and-bust cycle of Florida's frantic, unregulated economic development. Like so much of the former Confederate South, Florida, still largely a frontier area into the postbellum period, had almost no industry throughout the nineteenth century, and when industries did emerge, they were usually closely geared either to agriculture or to extraction of raw materials. After the Civil War, railroads were built all across the

"Sunshine State" to haul first cotton, then oranges, grapefruit, cattle, timber, and the many Yankee tourists already beginning to seek the sun and refuge from cold northern winters in Florida. After the appearance of the phosphate mining industry during the last decade of the nineteenth century, the twentieth century brought the wood-pulp paper industry, a foul-smelling, heavily polluting endeavor that came to dominate large areas of Florida's northern counties during the postwar era.[1]

Most of Florida's major industries went through wild swings of boom and bust during the period between the end of the Civil War and the beginning of World War II. A frantic boom in railroad construction and real estate, touched off by northerners' discovery of Florida's benign climate during the 1870s and 1880s, went spectacularly bust in the mid-1920s, though it did help suddenly increase the state's population nearly tenfold during that period. The same pattern applied to Florida's famous citrus cultivation, which expanded explosively throughout the state, even to more northerly counties, to satisfy new northern demand for oranges and grapefruit after the Civil War, until recurrent freezes and sagging prices drove this industry back south to the central part of the state, where it remained.[2]

The story was much the same for the phosphate industry. After tentative earlier efforts, the first great phosphate boom in Florida began in 1889 with the discovery of major hard-rock phosphate deposits in north-central Florida near Ocala and Dunellon. The resulting "phosphate fever" was much like an oil boom, as entrepreneurs flooded in to exploit nature's gift. The eighteen mining operations at work in 1891 burgeoned to 215 the following year, and more than four hundred by mid-decade. By 1894, Florida had already surpassed South Carolina to become the largest producer of phosphates in the Union. Predictably, this boom brought overproduction and bust. With the onset of recession during the 1890s, the industry faced bankruptcy and consolidation. As with oil, copper, and other mineral extractive enterprises, the Florida phosphate industry gradually came to be dominated by a few corporate giants, and only nineteen phosphate companies remained in business by the end of World War I. During the early twentieth century, phosphate mining also grew much more capital- and energy-intensive, gradually excluding small operators using mules, picks, and shovels as the industry moved away from the hard-rock phosphate deposits of northern Florida and introduced large quarrying draglines and flotation plants to harvest the softer, more scattered "land pebble" deposits of central Florida, mostly in Polk and Hillsborough counties to the east of Tampa.[3]

The Second World War brought back boom times to the Florida phosphate industry by disrupting production elsewhere in the world while hugely increasing total demand for phosphates in industry and agriculture. In 1941, United States phosphate production reached a new record high of more than four-and-a-half-million tons, and Florida accounted for 82 percent of the total. Production and profits continued to skyrocket during the postwar years. By 1960, the industry employed more than seven thousand workers, paid thirty-five million dollars in wages and salaries, and supplied 22 percent of the tax income of Polk County. By 1966, these statistics had risen to 10,550 workers receiving nearly seventy million dollars. Production doubled between 1950 and 1963, as a rapidly expanding global population created demand for ever more fertilizer, and by 1965, Florida's annual production of nearly sixty-five million long tons of phosphates represented 86 percent of the United States total and almost 30 percent of the world's total. During the 1960s, this extraction was increasingly performed by some of the largest and most powerful corporations in the nation, as the giant oil companies—among the few concerns with the massive capital and technological resources necessary to enter large-scale phosphate production, and attracted by seemingly limitless growth potential in the industry—gradually bought up most major phosphate companies, purchased millions of acres of additional phosphate lands and millions of dollars' worth of new plant and equipment, and tripled production capacity.[4]

The emergence of a major extractive industry monopolized by a handful of large, powerful corporations could not help but greatly alter the culture of a mostly rural area largely devoted to raising cattle and citrus fruit. Yet for many local citizens in Polk and Hillsborough counties, the most crucial change had begun, almost unnoticed, in the late 1940s and early 1950s, when phosphate mining companies branched out into chemical processing of fertilizer. Traditionally, the extraction, separation, and crushing of mineral phosphates had produced dust and spoil but not too much else, and the product was shipped elsewhere without further chemical refining. However, after 1948, when the Armour Agricultural Chemical Company opened the area's first chemical fertilizer plant and sulfuric acid plant for producing superphosphate and phosphoric acid, the phosphate industry's emissions to the air and water grew progressively more complex and damaging to local residents.[5]

Phosphorous is one of the three main elements necessary for soil fertility, the others being nitrogen and potassium. These three elements are consequently the main ingredients of commercial fertilizer, offered in varying proportions to suit different soils. Atoms of phosphorous, a common element in the earth's crust, are usually found chemically bonded to four oxygen atoms to form negatively charged phosphate ions, which in turn frequently combine with positive ions such as calcium or magnesium to form salts. The most common of these is calcium fluorophosphate, commonly called apatite, a combination of tricalcium phosphate with calcium fluoride having the chemical formula $3Ca_3(PO_4)_2 \times CaF_2$. This is the substance that created the Florida phosphate industry and later caused severe pollution of the air, land, and water in mining districts.[6]

Fluorine, one of the most electronegative and chemically active of all elements, renders the phosphates in apatite largely unusable by bonding tightly to the tricalcium phosphates, making them insoluble in water and hence unavailable to plants or animals. Plants growing in soils deprived of available phosphorous have difficulty developing and reproducing, while animals grazing on phosphorous-starved lands frequently will show stunted growth and malformed bones. Because of the tendency of fluorine to lock in phosphate ions, these symptoms of phosphorous deprivation can appear, ironically, on lands containing abundant but unavailable phosphates.[7]

In the 1840s, Western Europeans first began mining rock phosphates and crushing them into powder for direct application to the soil. Around the same time, in 1842, an English experimenter discovered that treating animal bones with sulfuric acid produced "superphosphate," a monocalcium phosphate compound with more available phosphorous than the tricalcium phosphate in bone or apatite. Borrowing this technique, early agricultural scientists soon began unlocking the phosphates in apatite by treating it with heat or sulfuric acid to produce more usable forms of the mineral, a process known as beneficiation. Since the acid method of beneficiation was cheaper, it soon came to dominate the manufacture of superphosphate. Since industrial pollution control techniques and recognition of the need for them were both minimal, at best, in the late 1940s, the introduction of large facilities in central Florida to produce sulfuric acid and process raw mineral phosphates with it would inevitably expose the surrounding environment to a host of new chemical byproducts, such as sulfuric acid and sulfur oxides escaping into the air or water.

A further level of beneficiation brought further threat of pollution, for in order to make triple superphosphate, which derived three more soluble, usable monocalcium phosphate molecules from any one molecule of tricalcium phosphate, the raw phosphates were treated with phosphoric acid, derived by mixing phosphate with sulfuric acid. Meanwhile, Florida's phosphate industry kept adding new, complicated chemical processes with new risks of pollution, such as the production of elemental phosphorous or the reacting of phosphates with ammonia to produce diammonium phosphate, a fertilizer base containing nitrogen as well as phosphorous. Each further stage of processing involved further complexities and further risk of pollution from the chemical plants.[8]

The various sulfur- and nitrogen-oxide byproducts of the chemical processing of phosphates were potentially harmful when released into the environment. However, of even more significance for local residents was the fluorine tightly bonded to the phosphate rock and accounting for around 4 percent of the total weight of phosphate mined. During treatment with sulfuric acid, fluorine compounds were separated from the phosphates along with various relatively harmless silicates and other impurities in the soil matrix where the phosphates were found. The resulting mixture of liberated impurities led to various new combinations, such as hydrofluoric acid and fluorosilicic acid, released during the processing, drying, and curing of phosphates. These fluorides—compounds containing fluorine—from the phosphate processing plants were emitted as dust or gas to blow freely through the surrounding countryside, with the chemically active fluorine atoms in them ready to react anew with whatever they contacted.[9]

Airborne fluorides were not only a potentially serious human health hazard if present in high enough concentrations; they also harmed various traditional supports of central Florida's economy, including cattle ranching, citrus growing, truck farming, and commercial gladiolus raising.[10] Flowers may not form properly or at all on gladiolus plants suffering fluorine exposure. Various other ornamental plants, fruit trees, and leafy vegetables are also sensitive to fluorine as well as sulfur dioxide, as truck farmers and commercial flower growers near major cities around the nation were to learn to their dismay when postwar industrial or urban air pollution problems around them worsened and spread.[11]

Hardest hit in central Florida's Polk and Hillsborough counties were the major livestock and citrus industries, although at first, few knew what was afflicting their livelihoods.[12] In 1949, Florida was one of the largest cattle-producing states in the nation, and Polk County had more cattle than any other county in the state. Florida was also the nation's biggest supplier of oranges, limes, lemons, and grapefruit, and Polk County was square in the center of the state's great citrus belt, producing a quarter of the state's citrus crop and 16 percent of the whole nation's citrus during the 1950s. However, by that time, Polk County was also the center of the nation's phosphate industry, most of which was squeezed into an area about twenty-five miles wide and thirty miles long, centered on the town of Bartow and including towns such as Lakeland and Mulberry in the western part of the county as well as slivers of Hillsborough County to the west and Manatee County to the southwest. Central Floridians would learn that their traditional economic mainstays and the newcomer were incompatible.[13]

After 1948, local farmers and ranchers saw cattle sicken and citrus yields fall markedly. As one observer reported, "Brahma cattle began dying of a mysterious epidemic. In the middle of plentiful grass and forage, they died of malnutrition, leaving their strangely-knotted bones alongside the barbed wire that wore out every few years. . . . In a county that by itself produces more citrus fruit than all of California, leaves were scorched, oranges grew the size of plums, and young plantings grew into middle-aged midgets." Citrus leaves that first yellowed around the edges, then turned brown and fell off prematurely, and the stunted growth of trees and fruit, were all symptoms of fluorosis. The stiffness of leg joints, the strange knobs developing on ribs and leg bones, the inexplicable starvation and prematurely rotted teeth that local cattle ranchers saw in their animals were similarly signs of fluorine poisoning. Local agriculturalists responded with confusion and frustration to what seemed to them almost a biblical plague.[14]

It is uncertain precisely when local citizens in the phosphate belt realized that they had an endemic problem and learned what the likely cause of it was. Dr. I. W. Wander, a doctor of soil science and horticulture at Florida State University's Citrus Experiment Station at nearby Lake Alfred, remembered first being asked to investigate a chlorotic leaf pattern in local citrus groves on April 19, 1950. The chlorosis—blotchy, uneven distribution of chlorophyll causing mottled, discolored, sickly leaves—resembled that symptomatic of a manganese deficiency in the soil; subsequent tests revealed that the cause was not a lack of manganese but an excess of fluorine. Later experiments at the Lake Alfred facility reproduced the observed leaf damage by fumigating citrus trees with low concentrations of hydrofluoric acid or other fluoride compounds. Thereafter, it was only a question of determining the source of the fluorine, and the obvious culprit was the phosphate industry.[15]

Around the same time, local cattle ranchers began noticing problems. Longtime Polk County rancher Donald S. McLean recalled observing changes in the once fine herds of the area by the early 1950s, as animals failed to fatten properly and lost their teeth. Ranchers also became aware that if the animals were moved out of the area—a change not practicable with citrus trees—their condition quickly improved. Local veterinarians suggested the possibility of fluorine poisoning, and animal nutrition and disease specialists from universities in Florida and Georgia soon confirmed this hypothesis, finding up to 7,000 parts per million of fluorine in the bones of a local animal when 200–800 ppm was normal. Again, the obvious cause was the phosphate industry.[16]

As understanding gradually dawned on them, local citizens mobilized to confront and abate the threat. A citizens' committee of afflicted citrus growers, ranchers, and other unhappy residents formed in Polk County during the early 1950s, documenting evidence of fluorine damage and presenting it to state and local authorities by 1954 at the latest. As a result of their agitation and a request by statewide citrus interests, in 1955 the state legislature created an interim committee to investigate the many public complaints about industrial air pollution in central Florida. This committee held numerous public hearings in Polk and Hillsborough counties between 1955 and 1957, then issued a report to the state legislature recommending that it establish a state air pollution control commission with authority to enact all needed control regulations. During the same period, the Florida State Board of Health began a limited program of sampling for atmospheric sulfur oxides and fluorides in central Florida. While in 1955 the Board of Health was also given statutory authority to adopt all rules and regulations necessary "to control pollution of the air—where created on private property, in public places, or in any place whatsoever," a later federal report remarked curtly, "No substantial action resulted from adopting this law." As we have seen, such an ineffectual statement of principle was typical of many states' and cities' first efforts at pollution control.[17]

Following the interim committee's report and the State Board of Health's inaction, the state legislature passed the Florida Air Pollution Control Law on June 18, 1957. This act created the Florida Air Pollution Control Commission (APCC), a panel of nine members representing government, industry, and the general public to hear and take action on complaints about air pollution in the state. The commission was given authority to establish air pollution control districts covering parts or the whole of one or more counties following a public hearing to determine the necessity of such an agency. Such a hearing, in turn, would be held if county commissioners requested it, or if 15 percent or more of the freeholders in an area petitioned for it, or on the commission's own initiative. The APCC was also given the power to promulgate rules and regulations regarding aerial contamination, which were to be enforced by the State Board of Health. Chronic polluters were to be warned and given a chance to correct their problems voluntarily; however, if such "conference, conciliation, and persuasion" did not work within a given time, the commission could give an incorrigible polluter a final ultimatum backed with the threat of a court injunction against all further violations. This enactment was similar to early anti–air pollution legislation in other states and cities, including the potentially problematic provision for authority to be divided between two or more state agencies. The original law was amended in 1959 to change the composition of the commission to ten members, including the state health officer, the state commissioner of agriculture, the director of the Florida Industrial Commission, one professional sanitary engineer, a representative of cattle interests, a representative of citrus interests, two representatives from industry, and "two discreet citizens of the state representing the general public."[18]

After Governor Leroy Collins appointed the first nine commission members to their four-year terms in September, 1957, unhappy residents of Polk County quickly called for a hearing to establish an air pollution control district in their area. Consequently, aerial contamination in the phosphate belt was the main item on the agenda of the first meeting of the APCC on December 12, 1957. Following a public hearing at Winter Haven on March 21, 1958, the commission created Florida's first air pollution control district, covering Polk County. In July, 1959, a similar district was created for Hillsborough County, and on June 10, 1960, both of these were merged to form the Polk-Hillsborough County Air Pollution Control District, which for several years was the only one in the state.[19]

Meanwhile, the APCC began setting standards for air pollution in its districts. Commissioners adopted regulations for acceptable levels of fluoride in pasture grass— forty parts per million—in Polk County in March, 1959, and the same standard was applied to Hillsborough County in October, 1959. In July, 1960, new regulations were established for fluorides in gladioli in the two-county area, while in December, 1960, a new rule was passed requiring review of engineering plans for new industrial construction or alteration. During the same period, the State Board of Health officials responsible for running the Polk-Hillsborough Air Pollution Control District surveyed ambient levels of fluorides and sulfur oxides in the air and fluorine concentrations in local vegetation. They also collected information from local industries on their emissions and started negotiations with company officials toward abating their pollution.[20]

There was widespread hope in central Florida that meaningful results would soon follow this initial flurry of activity. A joint federal-state report on air pollution control efforts in Florida optimistically promised that the APCC's diverse membership drawn from government and private interests guaranteed decisions "for the greatest general good." Admitting situations "where local authorities have been and will continue to be unable to secure abatement of troublesome air pollution" due to resistance from influential "special interest groups" causing pollution, the report claimed that the APCC would "speak for the people" and protect their air resources. Also, the agency could help handle the "complex technical matters in air pollution" that local officials could seldom handle on their own; to avoid sounding too activist, the report continued: "The Commission could restrict its activities . . . to situations where local control action is unfeasible." The APCC, with its longer reach and greater technical abilities, was expected to be a perfect complement to local authority in representing the public interest and gaining compliance from chronic polluters such as the phosphate industry, chief among the unnamed, uncooperative "special interests" opposing environmental cleanup. This policy, Floridians hoped, would allow local problems to be solved at the local level, with limited help from the state government and little or no interference by federal authorities.[21]

However, much of this optimistic faith proved to be misplaced. Events of these and subsequent years demonstrated that influential special interests were not so easily moved to clean up their destructive emissions by the voice of the people or their representatives. In central Florida, the road to clean air would be much bumpier and more winding than many naively hoped during the late 1950s.

# chapter 10

# Conference, Conciliation, and Persuasion

## Public Anger, Official Inaction

Rural dwellers in America have often been at a disadvantage in confronting pollution problems. Rural areas, by definition, generally are less likely to have the concentrations of either industry or population that produce major pollution sources, traditionally viewed as "big-city" problems. If a rural community is nevertheless afflicted with significant pollution, it is usually even harder than in more developed areas to mobilize enough local citizens to reach a critical mass of political influence to draw attention to the problem and ultimately to bring reform. The relative balance of power is often even more lopsided in rural settings, where one dominant, powerful industry with close connections to the local and state political establishment may face relatively few concerned citizens, often with less organizational expertise or access to the news media than citizens in urban areas. A job-scarce rural area with only a few major industrial employers is also more economically dependent upon those employers and less likely to criticize their misdeeds.

Despite these limitations, residents of Florida's rural phosphate belt did mobilize effectively and fought a long and successful battle to draw public and governmental attention to the pollution that plagued them. Much as some might have wished to, Florida state officials could not ignore the public outcry in Polk County. Central Floridians were fortunate to have some powerful friends, for phosphate industry emissions troubled the region's important citrus and livestock interests, and larger cattle or citrus growers in the phosphate belt sometimes were themselves wealthy and politically influential individuals. Together, the various individuals and interests opposed to air pollution from the phosphate industry—including major and minor citrus and livestock operators, commercial flower growers, the local tourism industry, some nonindustrial real estate interests, some chambers of commerce, along with scientists, public health crusaders, and angry citizens—kept the issue alive from the early 1950s through the end of the next decade. As such, the events in Polk County offer a striking example of how inhabitants of a rural area were turned into early environmental activists by an intractable local pollution problem.

Attention is different from action, however. Although central Floridians' pollution control campaign did provoke a response from state and local authorities, Florida officials long hesitated to demand changes from a major industry in a state still avidly seeking industrial development and jobs at almost any cost.[1] While the state officially recognized the problem and responded to citizens' complaints by establishing an air pollution control agency, state officials' persistent tendency to avoid confronting the phosphate industry while incessantly studying the pollution problem long made this government response a largely ineffectual, token gesture. Frustrated citizens then sought to appeal to a higher authority—the federal government—but these calls for help also went mostly unanswered, while the phosphate industry's air pollution problem remained unresolved into the late 1960s.

By late 1957, after persistent earlier agitation that helped to bring the first state air pollution laws of 1955 and 1957, local citizens in central Florida were well beyond the initial bewilderment that had existed in the early 1950s as to the nature and cause of their troubles. Their statements demanding the creation of an air pollution control district for Polk County revealed a strong sense of grievance against the source of their woes, the various phosphate plants in the area, which were proliferating and steadily increasing their output of both phosphates and pollutants. In October, 1957, Florida Citrus Mutual, a nine-thousand-member growers' cooperative based inside the phosphate belt in Lakeland, Florida, charged that fluoride and related emissions were causing "very serious damages" to citrus orchards, livestock, and other property and "adverse effect on the health of human beings" in southwestern Polk County. The organization passed a resolution claiming "serious impairment of health" from phosphate plant emissions, declared that the problem was not being abated, and resolved that the state Air Pollution Control Commission (APCC) set up a countywide pollution control district to "take at earliest possible date necessary actions" to abate conditions "disrupting the normal livelihood" of the organization's members. The unhappy citrus growers further claimed the support of many local organizations, including other agricultural groups such as the Polk County Farm Bureau and the county and state cattlemen's associations; business organizations such as the Federal Land Bank Association of Lakeland and the Lakeland Production Credit Association; local, regional, and national branches of the General Federation of Women's Clubs; other general citizens' organizations such as the Optimists Club and the Federated Garden Club of Lakeland; and local governments in such impacted cities as Mulberry, Bartow, and Fort Meade, along with the commissioners of Polk County. Many local citizens were already aroused and allied against the big new superphosphate operations, some nearly to the point of bringing legal charges against the phosphate companies.[2]

Similarly, in early 1957, at the behest of phosphate workers worried about general occupational health risks and particularly in-plant air pollution, the president of the International Chemical Workers' Union made a request to the surgeon general of the United States calling for a study of workplace health and safety conditions in the phosphate industry. The resulting study, conducted by the federal Occupational Health Program and the Florida State Board of Health with cooperation from local union officials, brought no conclusive results save that there was a lack of data on the subject; it failed to quiet workers' fears and suspicions.[3]

In the face of such mounting public concern, Florida officials conducted further investigations into the local pollution problem and how to confront it. In February, 1958, members of the Florida APCC, industry representatives, and two visiting federal air pollution experts conducted a day-long field trip around Polk County to inspect alleged fluoride damage and control measures taken by the phosphate industry. Local newspaper reporters accompanied the visiting experts during the whole day of field investigations, revealing local citizens' interest, concern, and anger about the matter of fluoride pollution. At an afflicted ranch only a few miles from nearby phosphate plants, Bartow veterinarian William Crum showed how local cattle were developing exostoses (large, abnormal bone growths) on their legs that "look like plaster," while teeth that developed during periods when the animals were grazing in surrounding pasture showed "extreme change" and deterioration or failed to develop at all. The unlucky rancher noted that he had been in the cattle business all his life but first discovered these conditions five years earlier when the new local phosphate plants began operation; he also claimed that during the previous two years, his bulls had become sterile. Regarding citrus damage, Dr. J. J. McBride of the state Citrus Experiment Station at nearby Lake Alfred observed that fluoride chlorosis in local orange trees first appeared "several months after the first triple phosphate plants began operation," and a longtime local citrus grower claimed that the production of his twenty-six acres of citrus groves had fallen by half during the previous five years. While the exact scientific details of the causes of allegedly declining citrus yields and deteriorating livestock health remained to be worked out, most indications—and all public suspicions—pointed to the phosphate industry as the root cause.[4]

Though the matter might have seemed obvious to local residents, scientists did face complications in trying to establish precisely how much damage local fluoride emissions were causing, if any. Citizens' claims that no such conditions existed before the big new superphosphate plants appeared were hard to test in the frequent absence of reliable records, since local inhabitants had not known they would be participating in an uncontrolled experiment. It was impossible to bring back preexisting conditions without closing down the fertilizer plants for some years, which was by then economically unfeasible. Then there were other factors possibly harming local cattle and citrus operations, such as soil mineral deficiencies, so that even if scientists determined that fluoride had indeed injured local producers, it might be hard to determine *exactly* what proportion of the damage could be attributed solely to fluoride at a time when general scientific knowledge of air pollution was limited. More central than scientific complexities were political difficulties, since by the postwar years, fertilizer and related products represented a substantial portion of Florida's economy, and the phosphate industry enjoyed major political clout in the state. Florida's air pollution laws and control program were designed to be very sensitive to the desires of industry, with three members of the ten-member APCC specifically representing industry and other members frequently selected for pro-industrial attitudes.

In short, the story of Polk County air pollution shared all the complexities of many environmental or public health debates, in which there are multiple variables that can seldom be fully controlled to allow unquestionable scientific certitude, and political complications may make authorities reluctant to pursue the truth. The difference was that in Polk County, the central focus of research was never the effects of airborne fluorides on human health. Throughout the battle over air pollution in central Florida, authorities basically ignored the possibility of human health risks and so reduced the immediacy of the issue, though local citizens continually complained of detrimental health impacts from industry emissions.

Whatever uncertainties existed regarding the exact mechanisms of fluoride poisoning and the exact extent of damage wrought, most scientists knew that there was a serious problem and what primarily was causing it. Writing to a friend in the federal Air Pollution Engineering Program, Harry E. Seifert, the first director of the Polk County APCD, observed: "Of course the fertilizer plants are the main source of pollution in this area." Whether or not control officials would be able to handle the political aspects of the local pollution problem was another question.[5]

Florida officials faced major obstacles in trying to get their new air pollution control program in Polk County up and running after March, 1958. Besides being rebuffed by the federal government in their pursuit of additional funding, the Polk County control program also suffered neglect from the state government. On August 5, 1958, Governor Leroy Collins rejected a request for an extra $31,959 to fund a study of Polk County air pollution. Collins felt that this air pollution situation was not an emergency, and Florida, under austerity measures typical of the 1950s, had formally restricted its ability to fund beyond legislative appropriations save in cases of emergency. Thus, Florida control officials were left trying to stretch the $45,000 remaining from their original funding of $65,000 to set up the laboratory facilities and hire the technical experts needed to investigate phosphate industry emissions thoroughly —a nearly hopeless task.[6]

Frustrated local residents sought to persuade their elected officials to increase funding beyond this totally inadequate sum. At a public hearing in the summer of 1958, Edwin N. Lightfoot, a retired engineer who by then had long been a vociferous local advocate of air pollution control and was chairman of the Polk County Citizens Committee on Air Pollution, argued that the situation developing in Polk County was indeed an emergency and urged the governor to give the control effort the funding necessary to be effective. Lightfoot was seconded by other members of the Citizens Committee and representatives of local livestock and agricultural interests. In response, APCC chairman Dr. Henry Lipscomb admitted that the original funding offered in 1957 had been totally inadequate, because no one had foreseen the difficulty and expense of investigating and controlling air pollution. Governor Collins replied that concerned citizens should approach the phosphate industry about sharing the cost of uncovering the facts about air pollution. Despite examples to the contrary, such naive faith in the conscientiousness and magnanimity of corporate America in voluntarily shouldering costs of environmental cleanup would often prove misplaced.[7]

Disappointed but not discouraged, Lightfoot, like other postwar citizen activists, promptly sought help from higher authorities. He wrote a letter to an acquaintance in the federal government, Assistant Surgeon General Mark D. Hollis, to seek help with Polk County's atmospheric problem. Noting how Hollis knew "something of our situation and the need for relief of the affliction that has been laid upon us," Lightfoot asked him to offer any support he could for Florida officials' request for assistance from the United States Public Health Service (PHS) in securing needed personnel for the county program. Lightfoot noted that he would be attending the upcoming first National Conference on Air Pollution in Washington, D.C., in November, 1958, and hoped to discuss with federal experts the local fluoride problem, "particularly in relation to toxicity of leafy vegetables"—a vain effort to persuade the human health–oriented PHS to pay more attention to the situation in central Florida. Yet the federal government mostly remained no more helpful than the parsimonious state of Florida.[8]

Meanwhile, public agitation in central Florida steadily increased. At the end of January, 1959, Julian C. Durrance, chairman of the Board of County Commissioners, wrote a letter to Dr. Wilson Sowder, Florida state health officer, noting a statement by a state health official at a recent APCC meeting in Jacksonville to the effect that phosphate industry emissions in Polk County were "not considered to be hazardous to human health." Durrance stated bluntly, "Many residents of the affected areas disagree with this viewpoint." Noting continual public inquiries on the matter, the Polk County official demanded to know exactly what public health authorities knew about fluoride's effects on human health, including the basis for the assumption of no significant impact, whether this conclusion was based on studies of Polk County, and what further investigations of the county were to be conducted.[9]

Sowder responded reassuringly and sympathetically. He offered a copy of the statement in question, noting that the state official had only said there was no *apparent* human health risk from *fluorides*, though not necessarily from other phosphate industry emissions. Sowder regretted his agency's inability to give a final answer on the dangers from Polk County's air pollution due to many still unresolved questions. Emphasizing the uncertainty and lack of research on the issue throughout the industrialized world, he quoted from a United States government publication stating how since "adequate methodology in determination of health effects and sufficient competent personnel were not yet available," experts concluded that a "crash program for definition of health effects was certainly not indicated at this time" and opted instead for "an orderly but aggressive" research program. However, Sowder promised that despite no present evidence of injury to human health, the State Board of Health planned to study the Polk County situation thoroughly in cooperation with local doctors and health officials, including a dental survey of school children around Fort Meade and Mulberry and an effort to detect and monitor any "subtle . . . damage to human health" from long-term, chronic low-level exposure to phosphate industry emissions. Legitimately pleading poverty, Sowder reminded Durrance of the agency's difficulty in securing necessary laboratory facilities and technicians with an inadequate budget. Sowder concluded by requesting that Durrance publicize his response to help quiet local residents' fears.[10]

Besides complaining to their elected officials en masse, Polk County citizens took other measures against industry emissions. In January, 1959, the journal of District Council No. 1 of the International Chemical Workers proclaimed that an "aroused membership" had called a "special open meeting" at their council headquarters in Mulberry, Florida, in February "to discuss the effects of air pollution on the health of the residents and employees in the local phosphate mining and producing area." Noting evidence of links between dirty air and disease, the union invited various health experts to discuss the topic, sent individual invitations to every worker in the industry, and encouraged the general public to attend. Contrary to more recent views of critics calling environmentalism a mere plaything of the rich, the independent actions of local chemical workers indicate that local concern about air pollution was widespread, extending beyond the middle-class members of the Polk County Farm Bureau and the local women's clubs to working people—even those who received their paychecks from the offending industry.[11]

Frustrated by the glacial pace of their state government's response to the phosphate industry's air pollution, central Floridians increasingly followed Edwin Lightfoot's example by seeking aid from the federal government. In April, 1959, Jane H. May of Plant City, Florida, a town in Hillsborough County near the Polk County border, wrote a desperate letter to federal officials complaining bitterly of the local pollution problem and pleading for help. May, a retired schoolteacher living on land her parents had homesteaded many decades earlier, reported serious and visible damage to her own small orange grove and neighboring farms and groves from the nearby phosphate plant, damage confirmed by private laboratory studies and local agricultural agents. She further alleged detrimental health effects from the local pollution, claiming: "Many residents of the area have been ill or even hospitalized from the amount of fluorine dust in the air. I, personally, have been under a doctor's care for many months due to dust allergies from same." May plaintively begged for "some action from your department concerning this serious . . . economic and health hazard." May's neighbor, A. B. Howell, who identified himself as a "truck farmer, living in a rich farm area on part of a parcel of land that was granted by a patent to my grandfather April 1, 1854, by President Franklin Pierce," similarly wrote to federal officials to protest. Observing how most of his relatives and neighbors were also small farmers and orange grove owners, Howell recounted how in the twelve years since a phosphate processing plant had been built in nearby Coronet, local residents had suffered serious injury to crops and citrus trees along with "many respiratory troubles." He urged federal officials to conduct an "impartial investigation" immediately.[12]

That May and Howell both turned to the federal government for an "impartial investigation" of the local situation suggests that they did not expect an impartial study or significant action from their state authorities. May and Howell also clearly were not the sort of professional busybodies that critics often presume environmentalists to be. Rather, they were apparently relatively ordinary citizens, native Floridians of modest means who took action to protect their community from pollution that many outside observers agreed was intolerable. May and Howell were probably among the large group of citizens living near Coronet, just outside Polk County, who in the late 1950s "had initiated action to have their area included in the Polk County Air Pollution Control District," as a visiting federal official reported.[13]

In 1960, other events on the air pollution front in Florida temporarily upstaged the drama unfolding in and around Polk County. Florida's Governor Leroy Collins, wishing to make a show of progress against aerial filth before his retirement the following year, requested the State Board of Health to conduct a study of air pollution throughout Florida. The agency would be joined by the University of Florida and the PHS. As was typical of early state air pollution surveys conducted with help from the PHS, the resulting *Report on Florida's Air Resources* mainly discussed the appearance in most of Florida's larger to midsize cities of noticeable quantities of the sort of ordinary air pollution from industry, vehicles, and solid waste burning that had become common throughout the United States. It mentioned the fluoride problem in the phosphate belt as one among many other problems, while the effort taken to conduct the statewide study diverted official attention away from Polk County and neighboring areas.[14]

However, the fluoride problem remained unresolved, and public complaints continued. In the summer of 1961, local citizens took another major step toward legal action against polluters when a grand jury in Polk County found sufficient evidence to prove conclusively that the phosphate industry was substantially harming local citizens and also cited the Board of Health for negligence in the handling of the air pollution problem. This decision was a clear vote of no confidence in state authorities' policy of research without results and interminable efforts at "conference, conciliation, and persuasion" mandated by Florida law, but it seemingly had little effect on state officials' approach to the problem.[15]

Floridians also continued trying to go over the heads of their own state officials to seek help. In August, 1962, Samuel P. Jones of Bartow, Florida, wrote an articulate, reasonable letter to Senator George Smathers, asking him to consider initiating federal action to control the air pollution problem in Polk County. Though Jones allowed that a problem affecting only about one-fifth of the residents of Polk County was not a "National issue," he explained to Smathers how toxic phosphate industry emissions were harming animal, plant, and human health in central Florida. Giving a respectful nod to the states'-rights attitudes that remained strong during the 1950s and early 1960s, he observed that the phosphate industry emissions should be a state and local issue, but state and local officials seemed "powerless to enforce proper control of such a giant industry that supplies vital products to all corners of the world." He requested a senatorial investigation to consider means of protecting local citizens. In effect, Jones was politely notifying Florida's senior U.S. senator that the traditional control approach rooted in meticulous federalism simply was not working in the fight against dirty air in Polk County.[16]

By 1963, growing numbers of central Floridians were threatening to bring lawsuits against the offending companies. Frustrated residents of eastern Hillsborough County joined together to sue a nearby phosphate mill for the damage it had caused them, although the court proceedings dragged on lengthily without resolution. As a result of such mounting public pressure, and probably in an effort to forestall a spate of private lawsuits, on May 10, 1963, the Florida State Board of Health took legal action against a phosphate plant for the first time by suing the Armour Agricultural Chemical Company for ignoring a permit requirement and failing to consult with the agency regarding the company's emissions control plans at its Fort Meade plant. This action temporarily encouraged local citizens that state control officials were finally stiffening their resolve, though the case ultimately would drag on for more than a year before state authorities agreed in September, 1964, to dismiss the suit if Armour limited its emissions to no more than 420 pounds of fluorides per day. To the disappointment of local environmental activists, the Board of Health carefully avoided any punitive measures for the company that had flouted their regulations, and the Armour case remained the sole example of Florida authorities moving beyond conference, conciliation, and persuasion during the 1960s.[17]

As frustration at the lack of conclusive action on air pollution increased among Polk County citizens, so too did political mobilization to confront the problem. In June, 1963, Donald S. McLean, president of the Polk County Cattlemen's Association, wrote to federal authorities to complain of the problem fellow stockmen and citizens faced and to call for help. McLean angrily charged that phosphate industry emissions were "highly poisonous or toxic" and "injurious to all forms of life," that they harmed "human health, comfort and well being" and caused "starvation, sickness and death of cattle" and damage to "citrus groves, commercial farms, . . . home gardens, ornamental plantings, native trees and vegetation." He branded the air pollution "an increasingly serious attack upon the economy of this area, upon the income and livelihood of its citizens as well as upon their health." The rancher offered photographic evidence of harm to livestock to support his claim that industrial fumes had caused the number of cattle in Polk County to drop from 130,000 to only 90,000 during the previous five years. Noting state officials' claim that the delay in controlling emissions was due to the "shortage of personnel and cost of monitoring equipment," McLean asked whether the PHS could loan the equipment necessary. He also requested the loan of any federal informational films that would shed light on the situation in central Florida and promised his organization's cooperation if the federal government ever sought to produce a film concerning the local atmospheric woes. Apparently many local ranchers were mobilized and sought to educate themselves and their neighbors regarding their environmental troubles.[18]

Public agitation against polluted air in the phosphate belt persisted through the rest of 1963 and into 1964. During this period, like female activists elsewhere, Harriett Lightfoot, wife of Edwin Lightfoot and chair of the Community Improvement and Air Pollution Committees of the Women's Club of Lakeland, emerged as one of the leading local activists in the fight against aerial contamination in central Florida. In August, 1963, Mrs. Lightfoot wrote to the Florida APCC to complain of the unusually severe air pollution during the summer. She recounted how on numerous occasions, her eyes and skin had been so badly burned by the chemicals in the local air as to cause excruciating pain and sleep loss and to require medical attention. Repeating a refrain common nationwide—"Gentlemen, if this air contamination can do this to one's skin, what does it do to one's lungs?"—Lightfoot observed that she had come to the area nineteen years earlier "to enjoy the fresh air and sunshine," but had been forced to stay indoors for the past eight years whenever the wind blew from the phosphate mills. Noting the number of times she had requested help from state authorities before, she argued that this time, things were so bad as to demand immediate action "to stop this evil which descends upon our unsuspecting citizens from the phosphate processing plants."[19]

When this letter brought no meaningful result, Lightfoot wrote to Governor Farris Bryant to urge him to request help from the federal government. Noting the uncontrolled growth of local phosphate plants and sulfuric acid plants in size, number, and production capacity, the angry citizen activist alleged that by lifting injunction warnings on various phosphate plants, allegedly without due process or public notice, the Florida APCC was still treating the phosphate industry gently while disregarding the rights and needs of local citizens. She also reported that the past summer had been "the worst in our history as far as air pollution is concerned. . . . Plants, flowers and trees were killed. People were coughing and sneezing and suffering head pains and sore throats. . . . A health officer told me that one doctor alone treated eleven patients for nose bleed and spitting up blood in a day, yet nothing seems to be done to relieve this situation." Noting that her physicians had told her "not to go out unless I was completely covered from head to foot when the wind blows from the Phosphate plants," she asked angrily, "Is this what we cam [sic] to Florida for, to be steamed to death in the hot summer, with temperature ninety-five in the shade and no shade, robed in mummy fashion to keep from getting burned by Sulfuric Acid?" Sounding a note of warning for a state still heavily economically reliant on tourism and emigration from colder northern climes, Lightfoot continued, "Friends have written me that they were coming to see us and I had to write and tell them that this was no place to visit until the Air Pollution was controled [*sic*]." She concluded by requesting an immediate response and prompt action from the governor.[20]

After receiving a hollow note of reassurance from state control authorities that everything was well in hand and that industry was cooperating fully, Lightfoot again wrote to the governor. Referring to the letter from the Department of Health, she pointed out that while the phosphate industry was "spending millions in correcting this pollution," they were "spending many more millions in constructing new and larger plants and the emissions as a whole are far greater than they were." Contrary to state officials' claim that bringing fluorides under control had revealed previously unsuspected trouble from sulfur oxides, Lightfoot denied that the fluoride emissions were under control and further declared, "The State Board of Health has been aware of the sulfur oxides for years." In her frustration, she charged that even the Soviet Union more adequately shielded its citizens from sulfur oxides and alleged that the phosphate industry was receiving special protection from local and state authorities: "Whenever we complain to the Board of Health their answers are, we do not have enough money to get control of the industry's air pollution. There have been whisperings that the industry is of such magnitude that it would hurt this part of Florida to push them too hard." Bemoaning the death of the local cattle industry and the impending similar fate of citrus growers and local tourism, she concluded by arguing that since the federal government was offering financial help and the State Board of Health was always pleading poverty as an excuse for their inactivity, there was every reason to request federal intervention.[21]

Lightfoot also made her best effort to go directly to the federal government herself by writing a letter to U.S. Attorney General Robert F. Kennedy. Having heard Kennedy on television discussing how the United States Constitution guaranteed all citizens equal protection under the law, probably in relation to civil rights issues, the Florida clubwoman perceptively decided that this federal protection against the taking of life or property without due process must also apply to citizens facing air pollution from the phosphate plants. Telling of years of minimal action in Florida following the initial studies and legislation, she pleaded for federal intervention on behalf of the air pollution victims who were not receiving equal protection under the law and whose property was being devalued and ruined without due process, including "poor people, made poor by the vicious fumes from the processing plants," "helpless widows," and "old people whose life's savings are in their land, which now is worthless." Like many other early environmental activists who sought to go over the heads of their state and local officials, Lightfoot received a polite letter from a federal bureaucrat explaining that under the Clean Air Act of 1963, the federal government had authority to intervene in intrastate air pollution matters only if a state governor specifically requested it or if a mayor asked for it and the state government concurred. Needless to say, this had not happened in Florida.[22]

Other citizens and representatives of local organizations also sought help from higher authorities. In December, 1963, Paul B. Huff, director of the Polk County

Farm Bureau and chairman of its Air Pollution Committee, wrote to Senator Smathers seeking advice as to how local citizens could get help from the new Clean Air Act of 1963, which significantly expanded federal funding and efforts to control air pollution. Citing a by now familiar litany of complaints, Huff wrote of how central Florida's atmospheric problem was "considered by many to be one of the worst pollution problems in the United States," how millions of dollars' worth of damage were occurring each year, forcing cattle and citrus operators to relocate, and how local residents had "gotten very little sympathy and action out of local and state authorities," while phosphate operations only continued to expand production and emissions. Putting Smathers more directly on the spot—"We would like to know how you stand with us"—Huff requested his personal help in securing federal assistance.[23]

This and earlier letters led Smathers to seek information from the federal Division of Air Pollution and probably helped lead to the single event that did most to focus national attention on central Florida's atmospheric woes. As a result of the sustained pressure from concerned citizens such as Huff, Lightfoot, and McLean during 1963, the air pollution problem in the Florida phosphate belt got onto the agenda for field hearings on air pollution throughout the United States planned by Maine's Senator Edmund S. Muskie and his Senate Subcommittee on Air and Water Pollution for early 1964. These hearings would for the first time turn the spotlight of national attention on Florida's greatest atmospheric problem; they would also graphically display some of the reasons why the existing approach to air pollution control in the United States was not working.[24]

On the morning of February 20, 1964, the Florida leg of the Muskie subcommittee's field hearings convened in the federal courthouse in Tampa, Florida. The central Florida situation was the only air pollution problem on the schedule of the senate subcommittee not linked to a major city, and Muskie noted how "striking" and "incongruous" it seemed in a state "famous for its magnificent climate, its hundreds of miles of beaches, its resort areas." Yet Muskie proclaimed hopefully that Florida could find a successful balance between its environment and industrial growth. He was followed by Florida state officials, who explained the workings of the state's control program; noted the need for further research; painted a picture of generally harmonious cooperation between representatives of industry, government, and the public within the state's Air Pollution Control Commission; emphasized the overall cleanliness of Florida's air; and worried over the possibility of federal encroachments, under the Clean Air Act of 1963, on state and local prerogatives. Hearing this standard testimony, the visiting senators never would have known there was a serious problem. This soon changed, though, as various subsequent witnesses told tales of suffering at the hands of the phosphate industry, questioned the efficacy of state control efforts, and clashed angrily with the industry's representative.[25]

Edwin N. Lightfoot, chairman of the Citizens Committee on Air Pollution, briefly reviewed the history of citizen mobilization and legislative action on polluted air in central Florida from the mid-1950s through 1963. He characterized fluorine as "toxic to all human, plant, and animal life when released in quantities" and warned of newer sulfur oxide emissions that grew "more noticeable each day" and threatened to "become a more deadly peril than fluorine" if left uncontrolled. The retired engineer then leveled harsh criticism at the existing air pollution control program in Polk and Hillsborough counties. He pointed out administrative weaknesses, including divided authority whereby the APCC could adopt rules and regulations, as it had done, but depended on the State Board of Health to enforce these. Quoting from the state control law, Lightfoot blasted the requirement that the Board of Health seek to end air pollution through conference, conciliation and persuasion, since this process could "go on indefinitely" without bringing results. Although the board was supposed to submit any matter to the APCC "for a hearing and action" any time conference, conciliation, and persuasion failed, and although in 1963, the Florida state legislature had changed the law to require enforcement action if such methods did not bring significant results within sixty days, Lightfoot continued in exasperation, "As far as we know since the passage of the act in 1957, the board has never submitted a case to the commission for a hearing." The "idealistic policy of conference, conciliation, and persuasion," Lightfoot insisted, was a proven failure.[26]

Noting a 1958 report by the APCC declaring that "satisfactory air pollution control equipment" was "available to control practically all emissions," Lightfoot then attacked the common argument from governmental and industrial representatives that such equipment was prohibitively expensive. He first pointed out how such expense was only relative and should be computed "in relation to the value of the product being turned out," by which measure it would likely prove only "a very small part of the total." One way or another, he argued that the cost was far from being the primary consideration: "We do not think that the element of cost should be considered an important factor when it is clear that on one hand a right is violated and on the other a wrong committed." Continuing with this line of reasoning, and reflecting the growing liberationist ideology and rhetoric of the 1960s, Lightfoot offered a forceful argument for strict pollution control as a basic human right:

No industry . . . however important, has the right by its operations to release noxious gases day by day into the atmosphere which . . . may become injurious to human, plant, or animal life. Under such circumstances the duty to desist is absolute. . . .

In this period of the world, when the right of every human being to live in comfort has become a universally accepted principle in American life, we do not believe that any community of citizens should be required to suffer perpetual discomfort or injury resulting from the discharge of noxious gases into the atmosphere.

Demanding firmer enforcement from state officials, Lightfoot hoped that the visiting senators would agree that "the noble experiment of conference, conciliation and persuasion" had failed to clean the atmosphere sufficiently and would "find ways and means to help us from a Federal level."[27]

Responding to West Virginia senator Jennings Randolph's question as to whether he felt that state health officials had been negligent, as the earlier 1961 grand jury report had alleged, Lightfoot replied that "it seems to us there is either something wrong in the machinery set up to do the job, or lacking in the will to use that machinery." He approvingly mentioned former Surgeon General Leroy Burney's suggestion in 1958 that air pollution was serious enough that control action would have to precede exact scientific certitude, as control authorities in Los Angeles County had decided long before. Lightfoot wondered why such action could not have been taken "from a general knowledge of facts" in central Florida years earlier, before so many cattle and citrus operations were ruined and so many residents were forced to leave to preserve their health. In defending his control program, State Health Officer Dr. Wilson T. Sowder responded that it was "wishful thinking" to imagine that any judge would convict a polluter without very solid "technical and scientific evidence," especially when an underfunded state agency was up "against a multimillion-dollar industry with a great deal of skilled legal talent." In raising these legitimate points about the difficulties of taking novel environmental complaints before traditional courts of law, Sowder described a situation that would frustrate control officials and angry citizens throughout the nation during the early postwar decades.[28]

Subsequent testimony concerned economic damage to farming and ranching operations in the region. In a prepared statement presented by a subordinate, Florida Agriculture Commissioner Doyle Conner offered his department's statistics on what phosphate industry emissions had done to cattle and citrus production: 25,000 acres of citrus plantings significantly damaged and 150,000 acres of grazing lands abandoned, in addition to further known damage to local flower growers and truck farmers. Conner noted the serious toxicity to livestock of inhaled or otherwise ingested fluoride, offering graphic details on the destruction of teeth, deformation of bone, and reproductive harm. Such circumstances caused the number of cattle in Polk County to drop from 120,000 head in 1955 to less than 95,000 in 1960, and those that remained were in the eastern areas of the large county where phosphate plant emissions seldom reached. Hardly any range cattle remained near the phosphate mills, even though much of the affected area was once "prime grazing land." Conner also discussed the history and nature of air pollution damage to the local citrus and gladiolus crops at length, although equally detailed statistics on production declines were not available.[29]

The lengthiest and most powerful testimony of the day came from Donald S. McLean, former president of the Polk County Cattlemen's Association. Above all other witnesses, McLean exemplified the anger, confusion, and frustration of ordinary people whose lives were being overturned and ruined by a scientifically complex new air pollution problem over which neither industry nor government officials seemed to be taking sufficient action. McLean graphically described how the problem first manifested itself like a biblical plague for unlucky local ranchers, deforming cattle and reducing the value of herds until most ranchers were driven out of the afflicted area. Notably, even sick cattle usually improved once out of range of the phosphate industry's emissions. McLean further recounted how after local veterinarians had diagnosed a "massive outbreak of fluorosis," suffering ranchers had called in university experts on fluoride poisoning in livestock who confirmed the earlier findings, warned that "unless this situation were altered, the entire cattle industry in this area would be wiped out," and advised the local cattlemen to seek legal help. The witness further alleged that the pollution similarly produced illness in humans, which also cleared up once sufferers moved far from the phosphate plants. Observing how any improvements in the emission levels of individual phosphate plants were negated by the increase in the total number of plants and total emissions, the angry rancher exploded in exasperation, "Can nothing be done about the 7 tons of poison gas being sent down upon us every day by these plants?" McLean further offered laboratory reports showing that permissible fluoride levels still were being regularly exceeded; he asked rhetorically, "Is not 14 years long enough for arbitration, conciliation, and persuasion?" Complaining of being driven from his home by fumes and warning that without stern governmental action, the declining cattle population would only sink further, the cattlemen's leader pleaded for federal help.[30]

Referring to the various photographs of deformed cattle and fluoride-contaminated bones the frustrated rancher had brought as exhibits, Muskie then asked McLean whether such evidence had been presented to either state or industry officials and whether they disputed the evidence. McLean responded that such evidence had been made available, but to no avail; he noted that "they have been kind enough not to deny it; they haven't done a thing about it." When the Maine Democrat then asked what reason the authorities gave for taking no action, McLean answered with another frustrated outburst, responding to state officials' earlier testimony about the difficulty of taking corporate polluters to court: "Sir, that is the reason I wrote you and asked you for Government help. Now, then, after all these years, if you will look at this, and when any State agency makes a statement that there is not enough evidence here to bring suit, what do you need to bring suit, sir?" When Muskie again asked how industry officials responded to all the accumulated evidence, McLean alleged that they were naturally happy to feign ignorance of the consequences of their acts—"Well, I imagine, not knowing, they were pretty well content, wouldn't you be?" Professing no desire to drive the phosphate industry out of business but only to make them clean up their operations, the unhappy rancher then warned the federal visitors that the pollution would be temporarily reduced for the senators' inspection tours of phosphate plants the following day: "You come down here and see us sometime when they don't know you are here. It will be a different picture." He also offered to procure further public testimony on the local scourge, promising, "If you will, sir, send us a man that you can trust, fully trust, and let him come down here; I will get you hundreds of people that will tell you lots of things that you need to know. They will not come in here. There are a lot in here that won't say anything." Through his sometimes ungrammatical but always earnest testimony, McLean in particular put a human face on the complex, politically and legally involved atmospheric problem of central Florida and implied that much of the problem still remained concealed or unaddressed due to pressure from a powerful industry.[31]

Further witnesses testified on damage to Florida citrus groves and saw no sign of improvement in total pollution damage. Despite all the definite evidence of damage cited, one witness asked only for further research on the problem. This puzzled Muskie, who observed that if all this damage had happened and was known to be from industrial emissions, then what was the need for so much further research?— "Why aren't we getting down to some action?" Another local anti-pollution firebrand, Paul B. Huff, chairman of the Air Pollution Committee of the Polk County Farm Bureau, stridently agreed that there was no need for more interminable research while citrus growers were being put out of business and were sustaining millions of dollars in cumulative damage from ruined groves and stunted trees; Huff called for an end to the "pussyfooting around."[32]

Like previous witnesses, Huff recounted how corporate irresponsibility had brought serious economic and health damage to local citizens. He further accused the phosphate industry of using economic blackmail to avoid the costs of environmental cleanup, long a favorite tactic of polluting industries. As Huff said angrily, "What is the attitude of the phosphate industry? One leader in May 1961 was reported in the Lakeland Ledger to have said that agriculture and the phosphate industries were incompatible and that the people of Polk County should decide which they want. Many industry leaders in speeches to the public have reminded workers and patrons that if the phosphate industry costs are forced up that they will have to close down or move to the Carolinas." Huff also charged state health officials with complicity in the ongoing crime and cover-up: "They [the industry] have the attitude that because they provide jobs and spend money in a community that they should be left alone. The State board of health has sided with the industry on these views. When we tell the State board of health personnel that the phosphate industry must clean up, they ask, 'Do you want us to close them down, put them out of business?'" Huff described how the industry and its allies in the state government had long sought to placate local citizens with soothing statements that the fluoride problem was already taken care of, but recent studies of local fluoride concentrations had found the highest levels yet. Meanwhile, the advisory committee for suggesting new air pollution regulations on which Huff served had not been called to order by the chairman in over two years, and while state control officials complained about inadequate funding, Huff charged, they made no real use of the funds they had. Huff concluded that local citizens found it "very difficult . . . to see how anyone could get anywhere with conference, conciliation, and persuasion if there seemed to be no intent to do any more than persuade."[33]

Next to testify after the fiery Huff was Floyd Bowen of the Florida Phosphate Council. Like any good industry representative, Bowen sought to come across as a calm voice of reason and moderation, offering arguments typical of industrial polluters nationwide. He did not deny the possibility of some problem due to industrial emissions but insisted that there was insufficient research data to determine anything one way or another. Bowen adopted the lawyer's trick of emphasizing all other conceivable variables to draw attention away from the most likely suspect, arguing that not everyone had trouble raising citrus or cattle near the phosphate mills, that much of the damage to local citrus groves was from winter freezes such as in 1962, and that cattle were leaving that area of Polk County only because the land value was rising and hence ranchers were selling their land for high sums to make room for more housing subdivisions and citrus groves. Bowen claimed that the industry had spent sixteen million dollars on research into their air pollution problems and that they of course would not have done this had there been no need for additional scientific data. Regardless of the alleged lingering uncertainties, Bowen further claimed that his industry had made great progress in cutting its total emissions even as production levels and the number of plants increased.[34]

Given all this alleged progress, Bowen suggested that lingering public dissatisfaction resulted from a disingenuous conspiracy to defame the phosphate industry. In a classic early depiction of environmental activists as merely a meddling minority, he suggested that "a great deal of this so-called problem is not what it seems, but is the result of vigorous agitation by a highly vocal minority well skilled in ways and means of obtaining headlines." Rather than branding these environmental activists as Communists, though, as some anti-environmental critics later would, Bowen instead accused them of capitalist speculation, suggesting that they sought to force the phosphate industry to purchase their land at many times its value.[35]

The industry representative was careful to emphasize that the phosphate industry and state and local government could take care of their own problems without federal help. Bowen assured the visiting senators that "the phosphate industry which I represent welcomes any intelligent, impartial, and unemotional survey of air pollution." "However," he pointed out, "this is a local condition since the area is 200 miles from the nearest State border. Furthermore, the situation is being handled by Florida industry and Florida agencies." Thus, the phosphate industry reminded the federal visitors that this was an intrastate matter and hence really none of their business under 1950s interpretations of federalism.[36]

Bowen also restated for the visiting senators the economic blackmail argument, more politely and subtly than the version local residents had heard—that if the phosphate plants were forced to clean up, they might be forced to leave instead. He noted that despite the industry's best efforts, "it should be understood that technology has not yet advanced to the point where all traces of fluorine can be eliminated." He went on to discuss the economics of the situation in the same terms used by many other polluting industries, describing truthfully how pollution control was costly and became more so as it reached higher levels of perfection. He warned ominously: "It is conceivable that pollution control costs ultimately could reach a point where the companies would no longer be able to compete in the world phosphate markets. . . . When and if this should happen, the economic welfare of some 7,500 phosphate employees and their families—plus thousands of workers in allied fields—would be at stake." This implicit reminder that an industry held thousands of voters as economic hostages usually had a powerful effect on state officials and legislators. Bowen also adopted another favorite tactic of polluting industries: pleading poverty. At a time when the phosphate industry was riding high and production and profits were rising, the industry representative was careful to point out how the Florida producers had to face competition from all around the world while the price for agricultural phosphates had risen little in twenty years. Thus, Bowen presented further cleanup as an economic hardship for the industry and a potential economic disaster for the region.[37]

During the hearing, the volatile Huff tussled verbally with the unflappable Bowen. When the industry representative spoke of improvements in total emissions, the farmers' spokesman offered laboratory reports showing a very different picture. When Bowen glowingly described the abundant citrus crops his company, International Minerals, had harvested from groves they owned near the phosphate plants, Huff accused Bowen of lying outright. He produced photographs showing a sickly grove able to produce no fruit and recounted how after he had told a former employee of International Minerals that he had gotten these pictures, the company bulldozed this afflicted grove to the ground, presumably to hide the evidence. Bowen, however, insisted that the grove in question had merely been badly damaged by freezes in 1958 and 1962. Muskie and Randolph tactfully kept the opposing witnesses from each other's throats.[38]

The following day, after Muskie had to leave early, Senator Randolph and his staff went to observe "the actual conditions in the field"—or so they thought. A few days later, Paul Huff sent a note with photographs to Senator Muskie observing that when Randolph saw the triple superphosphate plant at Bonnie, Floyd Bowen could claim that it "was clean as a candy factory," but within twenty-four hours after the federal visitors left, "the phosphate plants were back to full operation and operating as usual." Huff sent photographs as examples of "what normal operation looks like here." He also provided a report on an analysis of children's baby teeth collected by a local dentist, who found elevated levels of fluoride. As Huff observed, "Dr. Gore collected these teeth for the State board of health department, but they refused to pick them up." It was not easy to get the straight facts on a politically charged, legally involved air pollution problem such as that in central Florida.[39]

The visit of the Muskie subcommittee led to a surge in hope among local citizens that finally their stories of woe would be heard by higher authorities and that meaningful action might result. Some local citizens began sending pleas for help directly to Senator Muskie. Jane H. May and A. B. Howell, who had sent earlier SOS signals to the executive branch of the federal government, wrote to Muskie to tell how his committee's visit to the area "was followed with the greatest interest and heartfelt hope that at last we had a chance to be heard about the devastating blight to homes that have been our homes for generations." In the earnest, simple language of ordinary folks, they described the plight they and their neighbors faced at the hands of an unresponsive industry and an ineffectual state bureaucracy: their crops ruined, their lands devalued and impossible to sell, their homes and fences damaged and corroded, their health impaired by industry emissions. Both noted how repeated pleas for action from state officials had brought no response.[40]

Howell followed McLean and Huff in claiming that the local uproar over air pollution would be even more pronounced but for direct or indirect intimidation by the powerful phosphate industry. He described the difficulty of getting action in a company town—or, for that matter, a company state, observing: "A high percentage of my neighbors . . . are employed by the phosphate industry. Either from lack of concern, unawareness of the seriousness of this situation, fear or loyalty to their employer's [sic], they do not publically complain of these conditions." Robert H. Taylor, a twenty-year veteran of the United States Coast Guard, also wrote to express his horror at the deteriorated state of the region where he was born and where he had hoped to retire. He claimed that the air pollution in central Florida was worse than the infamous variety in Los Angeles, where he had been stationed, and he blasted the industry that had done this to his home. Taylor was blunter than Howell, charging overt intimidation of critics by the phosphate industry. He wrote: "THE EMPLOYES WILL PROBABLY BE THE LAST TO COMPLAIN FOR FEAR OF LOSING THEIR JOBS. . . . SOME PEOPLE HAVE BEEN SCARED FROM MEETINGS ABOUT THIS BEFORE, BY EITHER COMPANIE'S OFFICIALS OR LAWYERS. . . . [A local committee to discuss air pollution abatement] WAS FORMED AND DID NOT HOLD TOGETHER LONG FOR THIS VERY REASON, OF BEING SCARED AWAY FOR FEAR OF LOSE OF JOBS. YOU JUST DONT BITE THE HAND THAT FEEDS YOU." He then concluded his letter with the earnest call of an ordinary American for true patriotism, interestingly equating love of country with environmentalism and protection for traditional rural ways of life: "I HAVE SPENT THE LAST 20 YEARS OF MY LIFE SERVING MY COUNTRY AND AM PROUD TO BE ABLE TO DO THIS. NOW I WOULD LIKE TO RETURN TO THE FARM AND PEACE AND QUIET. IF CONDITIONS LIKE THIS ARE ALLOWED TO CONTINUE EXISTING IN FLORIDA OR ANYWHERE IN THE USA THEN MYSELF AND MANY OTHER PEOPLE SEEM TO HAVE FAILED MISERABLY." Phosphate industry emissions were turning ordinary Floridians such as May, Howell, and Taylor into environmentalists long before that term gained its current usage.[41]

Yet despite public hopes for help from higher authorities after the Muskie subcommittee came and went, the wheels of bureaucracy continued to grind slowly in central Florida, producing limited results. On May 29, 1964, a public hearing was held at Lakeland to air citizens' opinions about proposed new regulations that would limit total fluorine emissions in the Polk-Hillsborough County area to four thousand pounds a day. Phosphate industry representatives branded the regulations as arbitrary, implying a court test were they to be instituted; Lawton Chiles, then a state representative from Polk County, similarly questioned the legality of the proposed rules. Dr. Louis C. McCabe, a former government official who had become a private consultant and founder of a company offering research for hire to polluting industries, warned state air pollution control authorities not to "freeze a great American industry which is extremely important in world power" by imposing the new emissions restrictions on the phosphate industry.[42]

Thereafter, Florida control authorities took a few encouraging actions. In January, 1965, after over a year of "rather discreet discussion" among state health and air pollution control officials, the APCC instituted a new permit system to prevent total daily fluoride emissions from increasing and gradually to bring them down by establishing an overall regional ceiling on fluoride emissions and limiting existing phosphate plants to six-tenths of a pound of fluoride emissions per day for each ton of phosphate production capacity. New plants with newer control equipment were limited to only four-tenths of a pound. Existing phosphate plants were given a grace period for compliance until June of 1966. This moderate policy was not lax enough for the phosphate industry, however. At the January meeting where the new permit system was announced, an industry lawyer declared any such measure—and the very existence of the APCC as anything other than a wholly ineffectual research agency—to be illegal and unconstitutional under state law. As a sympathetic observer was forced to admit regarding the Florida phosphate operations throughout the 1960s, "Their strategy was to stall proceedings by legal actions and to deny any wrongdoing." As with Consolidated Edison in New York City, such legalistic obstructionism helped buy time for the phosphate corporations and helped keep the state control agency hesitant and hamstrung.[43]

Later, in September, 1965, the commission held a meeting in Lakeland to consider changes in the way of measuring fluoride concentrations in vegetation and to discuss the ramifications of the recent policy changes. The commissioners also briefly studied the extent of fluoride emissions both inside and outside the property boundaries of the phosphate companies—part of a new approach accepting that the pollution was chiefly a problem only if it fell on the property of others in significant amounts, while if the phosphate industry wished to allow very high concentrations on their own property, that was their business. Phosphate industry lawyers sought to exempt older, more polluting facilities from policies that would limit new plant construction and again questioned the legality of an overall emissions ceiling. In what would become a common complaint from industries nationwide, industry advocates also complained of the possibility of expensive changes in pollution control requirements when they were still adapting to earlier requirements.[44]

During this period, private citizens continued to demand action on dirty air. In February, 1966, the persistent Harriett Lightfoot, writing as head of both the Polk County Federation of Women's Clubs and the environmental committee of the

Florida Federation of Women's Clubs, presented to the Florida State Board of Health and the APCC a petition demanding action from responsible state authorities to control air contaminants in Polk and Hillsborough counties. Insisting that the local situation was still "critical," she noted recent tests of citrus groves in Polk County in January, 1966, that showed very high fluorine concentrations. By early February, three thousand local residents had signed the petition; within a few weeks, Lightfoot could claim two thousand more signatures, and several hundred more citizens would add their support during the following months.[45]

The APCC held another meeting in Tampa in April, 1966, to consider potential future changes in control policy and legislation, including possible new controls on sulfur oxide emissions to supplement older fluoride emissions standards. The commission heard other reports on new evidence of fluorine poisoning of citrus and on the need for taller smokestacks to disperse the phosphate plants' sulfur emissions more effectively. Commissioners then set a date for a public hearing on measures to control sulfur oxides, which passed over the objections of the two commission members representing the phosphate industry and an industry lawyer, who argued that there was insufficient scientific evidence of any need for control of sulfur emissions.[46]

At the April meeting, Harriett Lightfoot, annoyed by the foot-dragging of former APCC chairman Dr. Elwood R. Hendrickson on seeking more help from the federal government, promised to seek such assistance herself. She also read aloud a dramatic letter telling of her efforts on behalf of local citizens who were suffering from phosphate plant emissions and felt they were getting no help from control authorities. The residents in question were not affluent, educated, middle-class types with time on their hands to fuss over insignificant environmental problems, as environmentalists sometimes have been caricatured. As Lightfoot wrote, "These people are poor, they feel helpless, the fumes are making them physically and psychologically ill. . . . Mr. Lamb complains bitterly and keeps asking me what help he can expect from you?" In her letter, Lightfoot implied that Kay K. Huffstutler, director of the Polk-Hillsborough County Air Pollution Control District, had been unresponsive in answering these citizens' complaints. When Huffstutler later claimed to be unable to find any problem and tried to tell Lightfoot to mind her own business and let affected parties speak for themselves, at the next APCC meeting in June, 1966, Lightfoot justified herself by reading a signed letter from William Lamb complaining of insufferable conditions; repeated, severe fumigations; and official inaction. If Lamb's signed letter was not just a fabrication by the hard-driving Mrs. Lightfoot, it supports her charges that local control officials remained unresponsive to public complaints, as other residents had earlier alleged.[47]

However, the June APCC meeting was upstaged by hearings on June 2–3 about sulfur oxides and the effects of fluoride on citrus. While these were intended to be ordinary public hearings giving the public the chance to speak on the issues and to present any evidence for or against new regulations, the charged atmosphere in central Florida due to the long political and legal struggle between local citizens and the phosphate industry made such normality impossible. The frustration and resentment this long battle had engendered, and the presence and active participation of lawyers from both sides at the hearing, helped make what should have been ordinary proceedings effectively into a court case in all but the name, replete with legalistic quibbling about definitions and procedure and other such lawyerly pettifogging.

The legalistic tussling began immediately after APCC Chairman Dr. George F. Westbrook, from the Florida Department of Agriculture, took roll call. Westbrook announced that in addition to commissioners having the right to ask questions of individuals testifying for or against proposed new regulations, the lawyers present who represented either the pro- or anti-pollution control sides of the debate would also be able to ask questions, provided that these were not in the nature of "cross-examination, badgering or intimidating" the witnesses. Westbrook's warning was largely honored in the breach as lawyers on both sides of the issue turned the hearing into a replay of the actual court proceedings of the preceding few years. Chesterfield Smith, the lead attorney for the phosphate industry, promptly began quibbling about the order in which the two separate issues of fluorides and sulfur oxides would be heard as well as the rules regarding questioning of witnesses, leading the chairman to remind Smith, "This is not a court of law." The congeniality of the proceedings only deteriorated from there.[48]

The first two witnesses called by the "prosecution" were local citrus growers who complained of extensive damage from the phosphate industry's fluoride emissions over the preceding decade and a half. Despite Westbrook's earlier prohibition on such conduct, both industry lawyers and one of the industry's representatives on the commission cross-examined these witnesses so aggressively as to compel "prosecution" lawyers to demand that the chairman "limit the argumentative character of these questions." As with Floyd Bowen's earlier testimony before the Muskie subcommittee, the industry advocates sought to bring out every other potential cause of damage to citrus trees, to raise "reasonable doubt," and to confuse as much as possible the obvious and increasingly accepted pattern of fluoride damage.[49]

Most of the hearing, however, hinged on the testimony and interrogation of scientific experts regarding fluorine toxicity in citrus trees. Attorneys for the pro-regulation side called plant physiologist Dr. S. S. Woltz and citrus horticulturist Dr. Chester D. Leonard from Florida's state agricultural experiment stations; both testified that there was extensive evidence of fluorine poisoning of local citrus trees and corresponding economic injury to growers. In particular, Leonard described controlled greenhouse experiments he had conducted between 1962 and 1965, in which citrus trees were deliberately fumigated with varying concentrations of airborne fluorides, which produced stunting and reduced fruit yields. Thereafter, attorney Robert Murray, a sharp cross-examiner working for the phosphate industry, interrogated Leonard for more than two hours on details of his experiments and the state of knowledge in his profession, seeking to discredit Leonard's credentials, his familiarity with the literature in his field, and the adequacy of the controls, statistical analysis, and scientific method in his experiments. In defending himself as Murray threw an unending string of names of researchers and titles of publications at him, Leonard did bring out that Dr. R. S. Brewer, the leading national expert on citrus fluorosis from the California Citrus Experiment Station at Riverside, was surprised by the extent of visible fluorosis in Florida groves when he first visited the Polk-Hillsborough County area. Dr. Westbrook further interjected that one article Murray had quoted was merely a survey article rather than a new study itself. Meanwhile, the assembled audience grew more and more irritable.[50]

Following Dr. Leonard's long ordeal at the hands of the industry counsel, J. Hal Connor, an attorney for the citrus interests, opened a line of questioning seeking to expose Smith and Murray's clients. Since Murray had made an issue of the incompleteness of Leonard's greenhouse tests, Connor brought out that Leonard and his coworkers had hoped to continue their experiment until November, 1966, but the grove they were using had suddenly been bought out from under them by the Consumers' Cooperative Association, which Connor alleged was a front for the phosphate industry that employed Smith and Murray. The new owners ordered the ongoing experiment stopped by May 1 and allowed Leonard and his men "a few more days to dismantle the equipment." Leonard noted that the change of ownership happened just in time to deny the researchers "at least some additional information on yield." In this way, the pro-regulation attorney suggested that the research was undone by an underhanded conspiracy to suffocate the truth.[51]

The evening session on the first day of the hearing on fluorides and the citrus industry continued in the same vein. Edward A. Borsarge, an attorney for local plaintiffs against phosphate industry emissions, brought forward three further expert witnesses to discuss the horticultural and economic aspects of the issue: Dr. I. W. Wander, general manager of the Growers Fertilizer Cooperative in Lake Alfred and past director of graduate programs in soil science and horticulture at the University of Florida at Gainesville; Dr. James T. Griffiths, director of research for Cypress Gardens Citrus Products and a past professor of zoology and entomology at colleges in Alabama and Iowa as well as a member of the Citrus Advisory Committee to the Florida APCC; and Dr. Herman J. Reitz, professor at the University of Florida and director of the Citrus Experiment Station at Lake Alfred. All three of these experts with long experience of the local problem agreed that local groves were suffering significant economic and physical injury from fluoride emissions, while Murray did his best to undercut and cast doubt on their testimony. Griffiths, speaking as a member of the Citrus Advisory Committee, pointed out that this committee had unanimously concluded that there were grounds for emissions regulations based on fluoride content of citrus leaves.[52]

When the hearing reconvened the following morning, Murray presented a battery of imported expert witnesses to argue against the proposed fluoride standards. First were Dr. Leonard H. Weinstein and Dr. Delbert C. McCune, both from the Boyce Thompson Institute for plant research in Yonkers, New York. Weinstein and

McCune discussed how individual plants may have various different paths for the uptake of fluorine, some of them potentially more harmful than others based on what parts of the vegetation were exposed, such that the crucial factor in determining fluorine toxicity in plants was not the total level of fluorine but where this substance was concentrated within the different parts and cells of the plant. Murray's careful questioning revealed that the two New Yorkers had undertaken research specifically seeking to refute Leonard's study and to undercut its scientific assumptions and statistical basis, through experiments commissioned and funded by the phosphate industry. Like the resources to afford skilled and specifically focused legal talent, this ability to fund scientific counterstudies is an advantage large corporations have often had over ordinary citizens and communities in battles over environmental pollution.[53]

Following the New York visitors' testimony and various commissioners' polite but probing questions about their findings and the scientific details involved, hostile questioning from the citrus interests brought out that the visiting experts' conclusions were based solely on experiments conducted on corn plants, that their experiments on citrus plants were still incomplete, and that they were not particularly expert on the physiology of citrus trees. Moreover, the individuals directly superintending the conduct of the experiment were paid employees of the phosphate industry, some of whom had no special qualifications for participating in the study. Connor concluded his questions by quoting from Weinstein's earlier testimony on behalf of Armour and Company at the first major phosphate industry pollution trial in June, 1964, at which the scientist had admitted to being no authority on Florida citrus or "the effects of fluoride under field conditions in Florida on Florida citrus," and asking Weinstein whether his answers would be the same today. Weinstein said they would. Connor then asked the two visiting scientists whether they had ever testified for or against air pollution regulations before; they both said no. The pro-regulation attorney then asked one of the basic philosophical questions pertaining to many environmental regulations: given the fact that not all aspects of fluoride toxicity in plants were yet fully illuminated scientifically, did the scientists feel "that it is inappropriate to attempt to regulate, reasonably, air pollution in this manner?" Weinstein only answered noncommittally that it was "a dark area . . . crying for research." Further testimony and questioning followed similar lines.[54]

The question Connor raised is, unfortunately, a political, ethical, and economic question as much as a scientific one. As with the earlier debate over the safety of tetraethyl lead in gasoline, the decision to hold off on pollution regulations until the relevant scientific data is complete naturally benefits those in society who are creating potentially harmful pollution and would rather not clean it up, and it further harms those who may be suffering ill effects from such pollution but cannot scientifically demonstrate exactly how it is injuring them. The decision not to regulate pollution preemptively plays into the hands of polluters particularly because, given the nature of scientific exploration in the modern world, when a complex topic is involved, all the data never is in, and seldom are all possible points of scientific debate ever wholly laid to rest. Major polluters have often been happy to keep this scientific debate alive, funding experiments questioning the need for regulation while enjoying considerable economic benefits from not having to control their pollution and hiding their economic interests behind the pursuit of scientific truth.[55]

On the afternoon of June 3, the APCC moved on to the subject of sulfur oxides. This hearing was much tamer than the fluorides portion, since there was not as long a record of public anger and litigation over sulfur oxides, their local impacts were less well known, and comments from the general public were not accepted. A staff engineer from the Florida health department's Division of Industrial Wastes presented the evidence and arguments for why the state sought limits on sulfur oxide emissions, raising of industry smokestacks to a minimum of two hundred feet, installation of sulfuric acid mist collectors, and other new requirements. Industry lawyer Chesterfield Smith presented various witnesses for hire who argued that there was no need for such costly changes, including Louis C. McCabe. Like many other polluting industries throughout the nation, the phosphate industry found no pressing need to clean up their sulfur oxide emissions and expected to do this at their own pace.[56]

Following this hearing, there was a temporary glimmer of hope that state authorities would adopt a more aggressive posture on air pollution in the phosphate belt. On September 16, 1966, the APCC met again in central Florida and reached the unanimous conclusion that the "preponderance of evidence" from the June hearing had shown the need for regulations to protect citrus groves from air pollution. They had then set their policy committee to drafting such regulations. In addition to the expert testimony in favor of regulations at the hearing, the commission had received additional supportive testimony from Dr. Herman Wotz of the United States Department of Agriculture, while the expert testimony against further controls was discounted somewhat due to the witnesses' demonstrated lack of expertise regarding Florida citrus culture.[57]

However, when the time came for the APCC to vote on the proposed new regulations early in 1967, the agency proved to be as ineffectual as ever. Five commissioners—the chairman, Dr. George F. Westbrook of the Florida Department of Agriculture; Paul B. Huff of Florida Citrus Mutual; Ledley H. Wear, a cattle rancher of Bartow representing livestock interests; and two medical doctors, Dr. J. O. Bond, representing the state health officer, and Dr. R. E. Parks of Miami, representing the general public—readily voted to support the proposal. Three commissioners voted against the measure, which was not surprising, considering their background: two were commissioners representing the industry—Curtis A. Cox, manager of the Virginia-Carolina Division of Mobil Oil Company's phosphates branch, and Maywood W. Chesson, general manager of the Occidental Petroleum Company's phosphates division—and the third was a representative for the general public who happened to be from another mining company. Most important, two commissioners were absent from the proceeding. One was B. C. Thomas, executive secretary of the Florida Industrial Commission, an agency designed to make the state as attractive to industry as possible. Thomas's de facto negative vote was hardly a surprise.[58]

The other commissioner who was absent from this crucial vote was Dr. Elwood R. Hendrickson, a former air pollution engineering expert at the University of Florida. Since the hearings of June, 1966, Hendrickson reportedly had gone to work for the Florida branch of Louis McCabe's company, Resources Research Inc., which a local newspaper noted was "a scientific organization manned by the phosphate industry." Hendrickson's tacit siding with the phosphate interests on this crucial vote guaranteed official inaction, because, according to the rules under which the APCC was established, a quorum was sufficient to conduct ordinary business, but regulations could only be passed by a clear majority of the whole commission. As such, the regulation was to be sent back to the policy committee for reworking, but phosphate industry representatives, with fully half the commission on their side, vowed unending opposition to any regulation. A local reporter surmised that "one can only assume the matter is dead." Yet another hard-fought effort by ordinary central Floridians to gain protection from phosphate industry emissions had been frustrated.[59]

As evidence of the anger and frustration among local citizens in the wake of this important vote and perceived double cross, one local resident wrote an angry letter to President Lyndon B. Johnson and sent additional copies to Maine's Senator Muskie and Florida's Congressman Ray Mattox. In his letter, Robert H. Arnold of Winter Haven, Florida, fumed:

> I have always considered myself a conservative. Never before have I thought of advocating restrictive legislation. However, when the rights of others are so flagrantly transgressed upon so as to be almost criminal in nature, I feel that I must speak up.
>
> Picture, if you will, air pollution so severe that oak trees die, growing animals lose their teeth, citrus production is affected, and people cough, the telltale sign of emphesema [sic] and lung problems. This picture exists today in Central Polk County, Florida. It appears that our local county and state officials are helpless, although they recognize the need. . . .
> Won't you please help?

Arnold, who unsubtly identified himself as a "Registered Voter" and a Democrat, was probably not the only local resident who was forced to question traditional American (and southern) libertarian assumptions at the prospect of the phosphate industry so blatantly trampling the rights of ordinary citizens for so many years; he certainly was not the only one who was infuriated by the outcome of the vote.[60]

Florida saw further developments on air pollution control policy during the last few years of the 1960s. In 1967, the state legislature passed an Air and Water Pollution Control Act limiting aerial fluoride emissions in central Florida to 5,500 pounds per day and creating a new state agency, the Florida Air and Water Pollution Control Commission, to take over from the State Board of Health and enforce the new law. In 1969, this new agency was in turn superseded by a whole new department of the state government, the Department of Pollution Control, which elevated the issue to a new level of legitimacy in Florida. This department also stiffened emissions standards and penalties for polluters.[61]

In the meantime, Florida officials claimed to have made great strides in controlling air pollution in the phosphate belt. During the ten-year existence of the Air Pollution Control Commission between 1957 and 1967, total fluoride emissions over central Florida reportedly dropped from 40,000 pounds per day to slightly over 10,000 pounds per day and continued to drop, despite a tremendous increase in phosphate production. By the time of the new 1967 state pollution law limiting emissions to 5,500 pounds per day, the phosphate industry claimed to be emitting only 3,000 pounds a day, partly as a result of a slump in production and prices. By 1969, Vincent D. Patton, a longtime staffer of the Florida State Board of Health who was made first director of the new state Air and Water Pollution Control Commission, testified before United States senators that the region's fluoride air pollution problem had been solved to the limits of current technological capability. After many long years, Florida officials and phosphate companies happily declared victory in the war against fluorides.[62]

However, even accepting the control authorities' claims of progress—and the examples of Los Angeles and New York City demonstrate that such claims and statistics could sometimes be suspect due to agencies' desires to paint a rosier picture than the situation warranted—not everything had been satisfactorily resolved. For instance, the question of health effects, and the danger to which local citizens had been exposed over the preceding two decades from industrial fluoride and sulfur oxide emissions, remained unanswered. Throughout the late 1960s, industrial representatives and control officials continued to insist that there was no evidence of any impact upon human health from pollution levels in central Florida. Though this claim was based primarily on the lack of adequate research into either the individual or synergistic effects of local pollutants upon human health, the industry and its allies comfortably assumed that any fluorides ingested through low-level, long-term chronic exposure would pass harmlessly out in the urine. By 1973, on the other hand, the Florida Department of Pollution Control could state matter-of-factly: "Persons living near a source of fluoride pollution can suffer from eye irritation, respiratory inflammation and breathing difficulty" from the fluorides—significant harm satisfying the legal definition of nuisance. By this time, though, the federal government had stepped into intrastate air pollution affairs, and the phosphate industry had less veto power over state officials' pronouncements than it had once enjoyed.[63]

Similarly, the resolution of the question of economic damage to local cattle and citrus operations left something to be desired. In the absence of protection or action by state officials, residents of the phosphate belt increasingly turned to the courts to gain justice, and by the mid-1960s, they started to win some of their lawsuits against phosphate companies. In particular, some cattle ranchers were able to demonstrate damage persuasively enough to force phosphate companies to pay damages or to purchase the ranchers' land at prices well above what fluoride-poisoned pasturage would bring on the open market. By the mid- to late 1960s, the mere threat of legal action reportedly was often enough to make phosphate companies buy out nearby cattle or citrus operations, though the companies remained more likely to go to court against citrus owners, whose fluoride damage was perceived to be harder to prove than that of the ranchers. Through the use of private lawsuits, at least some property owners who felt they had been wronged were able to get justice and get around the slow-moving state bureaucracy with its excessive sensitivity to the desires of the phosphate industry. Ultimately, from the mid-1960s onward, even state authorities pushed phosphate companies to purchase the land of their neighbors for use as a dump for atmospheric fallout. By that time, many neighboring cattle ranchers, citrus growers, truck farmers, and commercial flower growers had already been driven away.[64]

It bears noting that whatever action happened at the state level inside or outside Florida during the later 1960s took place in a changed political context—namely, impending "federalization," or takeover of pollution control by the federal government, due to the demonstrated inadequacies of the existing policy of leaving it to state and local governments. As citizens in many parts of the United States grew progressively angrier and more frustrated by the unsolved pollution problems in their local areas, and as the federal government moved slowly but surely toward a greater involvement in environmental issues, state authorities and industry leaders grew aware that they would have to make a more respectable show of action to prevent a federal takeover, a prospect they viewed with horror. It is hard to say how much effective, spontaneous cleanup there would have been had there not been this looming threat; with the exception of unusually progressive cities such as Pittsburgh and Los Angeles, the record, in Florida and elsewhere, suggests there would have been rather little.

Despite a striking degree of spontaneous public mobilization to confront a serious environmental problem, residents of central Florida waited for almost two decades for any semblance of environmental justice. What perhaps could have been done by a strong control program in just a few years with existing technology and evidence in the mid-1950s took more than a dozen years due to the hesitance and foot-dragging of a weak, ineffectual state agency and a strong, recalcitrant industry. Moreover, help from the federal government was a long time coming.

# chapter 11

# The Perils of Federalism

## *The Intrastate Dilemma*

The story of atmospheric contamination in the phosphate belt of central Florida offers a good example of the weaknesses of the federalist approach to air pollution control in the decades before 1970. While angry, frustrated Floridians vainly pestered state and federal officials for meaningful action on the industrial pollution that was plaguing their communities, the federal government sat by mostly ineffectually, worrying over what minimal actions might represent an overstepping of the bounds of federalism. Federal air pollution control officials generally maintained the status quo regarding federal-state relations, adhering to a traditional view of the problem as merely a matter of science and engineering and keeping to a traditional federal role whereby federal scientists and experts were expected merely to give technical advice and keep out of the legal or political aspects of air pollution. Above all, federal authorities sought to avoid even the appearance of intrastate intervention without the invitation of state and local officials. In the face of persistent inaction by such officials on the problem, both in Florida and throughout the nation, these anachronistic concepts straitjacketed the early postwar federal control effort and largely precluded progress on reducing air pollution.[1]

Federal dealings with Florida also reveal some of the difficulties of trying to set policy by science alone. The complexities and uncertainties of air pollution chemistry and physics, the number of variables involved, and the difficulty of monitoring them offered policy makers a great temptation to continue conducting research into the problem without taking further action. Yet the air pollution damage in central Florida was serious and obvious enough to make at least some federal officials wish to take more immediate action. Such impulses were usually soon slapped down, however, as the overall federal control establishment hid out behind its customary justifications for inaction: scientific professionalism and federalism. As might be expected, this pursuit of pure science and avoidance of legal and political complications nevertheless had a political impact, helping to keep air pollution control policy effectively in the hands of polluting industries that benefited politically and economically from continued inactivity beyond the incessant research.

The federal government's participation in air pollution control efforts in Florida began tentatively in 1957, just a few years after the state government itself first began paying attention to the problem that had developed since the Second World War. That April, Dr. Harry Heimann, director of the Operational Research Section of the Air Pollution Medical Program of the United States Public Health Service (PHS), went to Jacksonville, Florida, to discuss health problems and other issues related to air pollution with Dr. John MacDonald of the Florida Board of Health's Division of Industrial Hygiene, then the lead state agency in air pollution control. Among other issues, such as pending legislation to promote air pollution research, the main topic of discussion was the state's most obvious air pollution problem, the phosphate industry of Polk County. In a subsequent report to his superiors in the federal bureaucracy regarding the seven triple superphosphate plants then operating around Bartow and Mulberry, Heimann reported "undocumented and documented . . . effects of fluoride on grazing animals, damage to conifers, and irritation of eyes and throat among the people" and suggested the Bartow-Mulberry area as a "'single-industry' type of community of the kind we were interested in studying at some future time." He also recommended that state or local public health officials look further into "other suggestive data on cardiac deaths" in the phosphate belt.[2]

In the wake of Heimann's visit, from May 28 to June 7, 1957, staff of the federal Occupational Health Program, a separate division of the PHS, joined Florida authorities in making a preliminary in-plant survey of work-related health problems in the phosphate industry. This study discovered information about the processes involved in manufacturing agricultural phosphates and compiled statistics on the workforce and the age and operating characteristics of individual plants. The state and federal researchers identified polluted air as the likeliest source of any health danger. In testing the levels of exposure of workers in different parts of the phosphate-processing operation, researchers found that some work areas had "extremely high concentrations of dusts and fluorides, both gaseous and particulate," and though such exposures were generally "brief and intermittent," the scientists recommended that "management study these operations" to "assure adequate protection of the worker" and to "define the extent and magnitude of . . . potential health hazards." While the visiting scientists found no clear evidence either of acute symptoms of fluoride poisoning or of health injury from subacute, chronic exposure, they did admit that the general lack of information available, such as workers' health records or previous studies of any significance, made it impossible to make a conclusive statement one way or the other about health risks.[3]

Later, in July, 1957, C. Stafford Brandt, a plant pathologist and an official in the PHS's Air Pollution Engineering Program, made a trip to Lakeland in Polk County, to observe reported air pollution damage to local citrus groves from the phosphate industry. Brandt saw the unique pattern of fluoride-caused chlorosis in local citrus trees, which Florida researchers had discovered and reproduced experimentally years earlier, though he noted that the diagnosis of fluoride chlorosis was complicated by the fact that various soil nutrient deficiencies could cause chlorosis. The federal scientist found the whole picture further complicated by the fact that the area's highly leached, sandy soils provided few nutrients and citrus growers had to add nutrients artificially through fertilizers and sprays, such that it was difficult to sustain the right balance of minerals needed; correspondingly, he reported, "deficiency and toxicity symptoms are common in many groves." He failed to observe that notwithstanding such considerations, Polk County was the center of the state's citrus industry, and citrus culture had been practiced extensively and profitably in the area for decades. While citrus growers alleged that fluoride chlorosis reduced their groves' yield and productivity, Brandt noted that no precise evidence or estimate of the extent of damage had yet been advanced, so state citrus experts were merely monitoring the problem. He concluded: "In view of the strictly local and minor importance of this symptom in relation to the entire citrus industry of the State, this attitude by the State Station is understandable and justifiable." In this statement and many others, Brandt showed the tendency of many contemporary professional scientists to dismiss the observations of laymen too hastily.[4]

The federal scientist (no veterinary expert) treated claims of damage to the local livestock industry with similar suspicion. Brandt explained the experimental and diagnostic difficulties in determining whether the phosphate industry was harming local cattlemen, emphasizing the various other local factors and variables that might be contributing to any apparent rash of fluorosis in cattle, such as high natural fluoride levels in water or forage from phosphatic local soils, or malnutrition and substandard care in an area on "the fringe of the cattle-producing area" where Brandt found operations generally "somewhat marginal." He concluded that "undoubtedly, fluorosis does occur" but declared that it would be difficult to distinguish air pollution impacts from the effects of natural fluoride levels. Brandt's lack of expertise in animal health did not prevent him from casting doubt on the validity of local ranchers' claims of unaccustomed damage.[5]

Despite his overall disdain for local citizens' allegations, apparently industrial pollution was serious and visible enough for Brandt to go on to criticize the local phosphate refiners for their nonchalant approach to the issue: "The industry cannot hide from the casual observer that there is a dust problem and that in certain areas, vegetation has been injured and even destroyed as a result of the dust or fumes. Cooperation among the individual companies in the approach to these problems would appear to be nonexistent." Noting how the overall phosphate industry was supporting some limited technical studies, Brandt remarked that only two companies were doing any work whatsoever on possible impacts on vegetation, and even these were not cooperating on research. Some of the phosphate refiners were also "operating producing [citrus] groves" near their plants in an effort to disprove citrus growers' claims of damage, but as Brandt complained, "The public relations possibilities of these company-operated groves has not been exploited except to a limited extent." As with many federal scientists of the day, Brandt's general tone suggests that he sympathized more with corporations and the corporate scientists with whom he worked than with ordinary citizens.[6]

In February, 1958, Brandt returned to central Florida along with a higher federal air pollution control official, Dr. Arthur C. Stern of the PHS's Robert H. Taft Sanitary Engineering Laboratory in Cincinnati, Ohio, to survey local conditions. The two men joined members of the Florida Air Pollution Control Commission (APCC), industrial representatives, and other interested citizens on a day-long field trip around Polk County to inspect alleged fluoride damage and control measures taken by the phosphate industry. On the morning of February 28, the air pollution experts reviewed emissions control efforts at three major local phosphate-processing facilities. Later, the visiting dignitaries were shown possible fluoride damage to citrus groves and livestock. Although Stern was an air pollution engineer whose primary expertise was in neither plant nor animal physiology, he showed little of the doubt of his companion as to the impact of fluoride emissions on regional agriculture. He noted that they saw numerous groves near various roads crossing the county, and that despite severe frost damage and signs of nutritional deficiency, "the distinctively chlorotic leaf pattern which is allegedly due to fluorides was evident." Noting the earlier experimental production of the same patterns through fluoride sprays and the confinement of this pattern to the phosphate-refining area, Stern contradicted Brandt's earlier hesitance about the diagnosis, concluding: "This evidence is quite sound and is probably sufficient." The federal visitors also saw evidence of fluorosis in local cattle, although Stern noted that complicating factors such as poor animal nutrition made it difficult to attribute all symptoms to fluorosis alone. After attending an APCC meeting, receiving accumulated data on local fluoride emissions, and discussing the potential role of the federal government in assisting the state's control efforts, the federal visitors left for Washington, D.C.[7]

The scientifically and politically complicated Florida air pollution problem that Stern and Brandt reviewed was developing at a time when the federal government had only begun to take rather tentative steps into the field of air pollution control, so its power remained limited and its role uncertain. On the 1958 field trip, Stern and Brandt revealed some of the different possible approaches federal officials could then take toward state or local atmospheric problems. First, within the limits set by existing assumptions regarding federalism, federal officials could be more or less responsive to the public in pollution-afflicted areas. Stern was at least faintly encouraging toward local citizens in their demands for action on their fluoride problem, informing the APCC how in his opinion, widespread and recurrent public complaints clearly indicated the need for an air pollution control district, and further explaining how the federal government could offer research and technical assistance within limits. Brandt, on the other hand, was relatively discouraging and seemingly wanted to deny the existence of a problem. For instance, he alleged that the cattle he had seen that day were poorly managed and hence were on the borderline of being malnourished relative to better-fed, better-tended animals in other states—an observation subsequently challenged by a veterinary expert. He also noted the possibility that natural fluoride in the water might account for some of the observed problems and concluded that the subject required more intensive study. Again, he appears to have dismissed wholly citizens' claims that the major problems began after the superphosphate plants started up, a technocratic attitude typical of a period when there was greater faith in the objectivity of science and scientists than in later decades.[8]

Besides revealing the emerging dichotomous tendencies between a more activist federal approach and a more quietist one, Stern and Brandt's performance in Polk County also displayed differences in the definition of science that would long hamper air pollution control efforts—the argument between those experts requiring almost impossibly strict standards of scientific proof before action was taken and advocates of looser standards and more action. Based on the visible situation in Polk County, even with an awareness of potential complicating factors, Arthur Stern, like most subsequent observers, was readily able to declare the existence of a significant problem requiring action. Brandt, by contrast, sought disproof, ignoring the obvious wider picture and dwelling on any variables that cast doubt upon the existence of an air pollution problem—a traditional, "strict-science" approach. This debate over sufficient scientific evidence to take action, within both the federal air pollution control establishment and wider scientific and engineering circles, tended to split observers into two camps: those who shared the public health concerns of the public and were more inclined to take action, and those who tended to favor the claims of industry that there was no demonstrated problem and who never felt there was sufficient evidence to act. During this period, most federal scientists, who shared the same training as industrial scientists and often came from the private sector themselves, tended toward the quietistic side, like Brandt. Such attitudes would long remain an impediment to a more meaningful federal role in air pollution control. Despite the scientists' desire to be purely apolitical, their tacit defense of the status quo of inactivity, as already noted, inevitably had economic and political ramifications for polluters and their victims.[9]

After the creation of Polk County's Air Pollution Control District in March, 1958, state control officials sought help from the federal government to stretch their inadequate budget, including help with research and the loan of costly, advanced air pollution sampling and monitoring equipment under the new federal air pollution law of 1955. In particular, Florida officials hoped that the PHS could assist in securing expert researchers to make initial veterinary and epidemiological studies to define the nature of their problem, including dental surveys to ascertain whether fluoride pollution was affecting the teeth of local children. However, Dr. Wilton M. Fisher, then director of the federal Air Pollution Medical Program (APMP), took a narrow view of the federal role and rejected most of these ideas, noting that the Florida situation was less involved with public health than with plant and animal health and regretting that while the PHS "could gain considerably by working with [Florida officials] in this new, almost unexplored, area of total community study," recent federal budget reductions made such cooperation impossible. While the federal government periodically agreed to loan equipment to the Polk County Air Pollution Control District, a process invariably complicated by the novelty and unreliability of sophisticated and sometimes untried new monitoring equipment, the federal officials declared that they were unable to loan scientists and technicians for any long period and offered only "technical assistance or consultation on a short-term basis." Regarding the state's desire for help conducting a full veterinary survey of the effects of fluoride emissions on livestock, Fisher politely suggested that the Florida authorities pester the United States Department of Agriculture (USDA) for a federal veterinarian to assist with their studies of impacts on animal health.[10]

The USDA was not very interested, either, so instead of a full veterinary survey, Polk County received a visit from Dr. Norman L. Garlick, a livestock inspector with the Animal Disease Eradication Division of the USDA's Agricultural Research Service; he made a brief inspection tour with state and local officials. Garlick—an expert on the matter, unlike earlier federal visitors—found that particular local animals "demonstrated the extreme maximum of dental fluorosis" as well as obvious exostoses on the leg and rib bones, and he noted that calves raised under such conditions would have shortened productive life spans. Besides affirming that local livestock were indeed being poisoned by airborne fluorides from nearby phosphate plants, Garlick, better able to judge than Brandt, observed that cattle ranching in Polk County was a significant and prosperous business: 110,000 head of beef cattle grazed on 900,000 acres of rangeland, plus a further 7,000 dairy cows and "an important beef cattle purebreeding industry." Thus, the USDA expert felt no doubt about the existence of a serious local air pollution problem, and he emphasized the potential for significant economic damage and urged further studies to ascertain the degree of damage and to determine safe exposure levels. But federal officials expressed little interest in the Garlick report.[11]

The problem would not go away, however. Federal officials were constantly reminded of the Florida situation by the letters and complaints that flowed in from angry local citizens, who remained convinced that their local air pollution was a health threat as well as a cause of economic damage and who were gradually despairing of getting real action out of their reluctant state officials. Federal control authorities also received worried letters and pleas for help or reassurance from local officials facing a firestorm of public protest over the phosphate industry's emissions. In late January, 1959, John H. Dewell, acting attorney for Polk County, sent a request for information to Harry G. Hanson, director of the PHS's Taft Laboratory, emphasizing local citizens' worries about possible effects of fluorides on human health. Dewell alleged that at a recent meeting of the APCC, a representative from the State

Board of Health had stated "as an established fact that there was no hazard to human health associated with the air pollution condition now existing in Polk County, Florida," and that "as a result of this fact the U.S. Public Health Service could not participate in any program directed to the elimination of the said air pollution condition." Dewell requested thorough scientific confirmation and evidence of such findings from the federal authorities to reassure "the numerous complaining citizens from whom we receive almost daily inquiries," since, as he continued, "a responsible group of vitally interested citizens of this County have in the past few years accumulated data which indicated that under U.S. Public Health Service standards human health was endangered," leading both citizens and county commissioners to have become "greatly alarmed."[12]

Dewell's politely alarmed letter touched off considerable discussion within the medical and engineering branches of the federal air pollution control program. In his initial draft response, Hanson protested that the state official at the recent meeting had probably misinterpreted what he had heard, since the PHS had "worked closely with the Florida State Board of Health in air pollution matters for the last several years." Professing his agency's interest in further efforts to control the region's fluoride emissions problem, Hanson noted the earlier visits by PHS staff and other federal officials for investigation and consultation as well as ongoing efforts to help the Florida county secure federal grant funds and to recruit a chemist skilled in the analysis of fluorides in vegetation to train chemists for the Polk County program. In his draft, Hanson did admit that "no definitive studies have been made in the area to determine whether a hazard to human health exists." However, he argued that Florida control authorities had adopted a sound approach by using their limited resources first to attempt to meet and control problems for which definite information existed, such as damage to cattle and vegetation, since livestock and plants were more sensitive to fluoride emissions than humans. If conditions were improved to the point that plants and animals were no longer afflicted, any potential risk to human health would likely have been removed. Hanson did not add that, given the basic complexities of air pollution science and epidemiological research, compounded by the limited budgets of both the Florida and federal control programs, any thorough research program would take a long while and still might not produce conclusive answers. The scientists of the PHS viewed problems in terms of such extended research programs and sought to be oblivious to the emotional and political aspects of issues such as air pollution, including the understandable desire of the local public for definite answers regarding health risks.[13]

Perhaps because Hanson's draft response was not sufficiently reassuring, Dr. Richard A. Prindle, then acting chief of the APMP, got the job of answering Dewell. Prindle offered a prepared statement of his staff's estimation of the possibility that airborne fluoride threatened human health, which the PHS medical experts found very unlikely. Nevertheless, Prindle added, this opinion would not preclude federal participation "in any program directed towards further study and appraisal of fluoride air pollution problems and their control," and he promised that the PHS would continue to work on their problems "to the extent that our funds and personnel will allow."[14]

Other agencies also wanted the PHS to tell them not to worry. The Florida State Board of Health had received a similar request for reassurance on the health issue from Polk County officials, and they turned to the federal government for a response. More pressingly, various individuals in central Florida had become frustrated enough by their local air pollution that they were threatening to sue the phosphate companies, and to help defuse this tense and politically problematic situation, state officials wanted "a statement from the Public Health Service saying that there is no evidence of human health effects as a consequence of fluoride community air pollution in the Polk County area." In an internal memorandum to his superiors on January 29, 1959, Richard Prindle observed that the earlier survey of in-plant fluoride exposure among phosphate workers by the PHS's Occupational Health Program had become "the subject of a bargaining problem between industry and the union" and that the local phosphate industry had retained Dr. Louis C. McCabe, formerly with the federal Bureau of Mines and one of the best-known American experts on air pollution, to represent them in their ongoing clash with workers over working conditions. These public worries, together with mounting evidence of significant damage to citrus and livestock growers, made an explosive situation. Hoping that the federal government could soothe fears as state and local officials evidently had been unable to do, the Florida Health Department suggested that the PHS should point out that the occupational health survey had found no evidence of fluorosis and that unless a state investigation of the teeth of local school children in February produced evidence of mottling and stains characteristic of fluorosis, citizens had no grounds for worry. The various branches of the PHS concerned with air pollution and occupational and dental health immediately got to work on this reassuring statement at the state's behest.[15]

The PHS statement on the human health hazard from airborne fluorides in Polk County, Florida, began by noting demonstrated damage to vegetation and livestock in the area from such pollution but went on to observe that many plants accumulated fluoride in much greater quantities than would be found in the ambient air, so that animals grazing on these plants would consequently face elevated exposure levels. Studies of livestock fluorosis in other areas of the United States had found symptoms when concentrations of fluoride in the air ranged from 0.01 to 0.4 micrograms per cubic meter of air—about one ten-thousandth of the threshold human occupational exposure limit of 2.5 milligrams of fluoride per cubic meter of air that the American Conference of Governmental Industrial Hygienists had declared a safe average concentration for workers to face "eight hours a day, five days a week, without adverse health effects." PHS officials thus concluded that "effects may become manifest in foraging cattle long before there are evident changes in humans from inhalation." Since the federal officials' mission was effectively to discount the possibility of serious health effects, they did not go on to observe that no one had yet properly studied or set standards for exposure to lower concentrations of aerial fluoride twenty-four hours a day, or for children, the elderly, and other sensitive individuals, patterns of exposure obviously different from workplace exposure of healthy adult males.[16]

Discussing the various ways by which humans could be exposed to fluorides, the report noted the earlier PHS occupational health survey that had found extremely high concentrations of fluorides in some work areas. However, despite some workers' complaints about health effects, the report's authors found that "no factual evidence has been provided to us that would indicate that health effects from air-borne fluorides have occurred in this area." Outdoor concentrations were of course generally much lower than in the workplace. Consequently, the authors concluded, "it is our opinion that at the present time the available data do not indicate the existence of a hazard to human health as a result of air-borne fluorides in the community." Urging further research, the report concluded that "the reduction of the fluoride levels in the community to the point where no further damage occurs to cattle would be a highly commendable aim" and "would assure levels with a great margin of safety for humans, as well as contributing to the solution of obvious economic problems." With this generally soothing disclaimer, the PHS passed the issue back to Florida officials, who found the report just what they had wanted. The federal authorities thus helped state officials and a polluting industry to sweep the matter under the rug for a time, reflecting an understanding of the federal program's mission as being to support state authorities uncritically while remaining aloof from unhappy citizens.[17]

About this same time, federal officials were arranging a visit to central Florida by a federal air pollution control expert as part of their program of limited consultation and technical assistance to state control agencies. This visit and its aftermath—the "Rossano affair"—brought unexpected tension between federal and state officials and further revealed some of the fundamental weaknesses of the federalist approach to air pollution control in the early postwar period.[18]

From March 30 to April 3, 1959, August T. Rossano, Jr., chief of the Office of State and Community Services in the federal Air Pollution Engineering Program, made a tour of inspection around Polk County with Harry Seifert, director of the Polk County Air Pollution Control District. On the way from the Tampa airport, they observed and took samples from orange groves that later showed very high fluorine concentrations in both fruit and foliage. Over the next two days, besides reviewing the Polk County District's offices and laboratory, Rossano had tours of two local phosphate plants. He also inspected a collection of deformed bones Seifert and Dr. Garlick of the USDA had accumulated, revealing fluorosis in local cattle, and saw areas of Polk County exposed to extensive plumes of emissions from the phosphate mills, "dense enough to seriously affect visibility on the highway." Rossano met with a local livestock veterinarian who had mapped the geographical distribution of the many diagnosed cases of fluorosis in cattle, which Rossano found "well interspersed among the phosphate plants." On the morning of April 3, Rossano flew to Washington, D.C., before returning to his usual post in Cincinnati.[19]

Unfortunately for Rossano, though, he had made statements before local journalists at press conferences prearranged by Seifert. Rossano told reporters from the *Lakeland Ledger* and *Tampa Tribune* that because fully three-quarters of the nation's phosphates were mined and processed in the area around the Florida county, he "assumed" based on similar situations that there could be potential trouble from pollution. Rossano warned of possible health impacts, in contrast to his own superiors in the PHS, who had recently helped Florida officials in glossing over the matter. Noting that "we don't know what effects the fluorides will have on children in their formative years," he declared, "We can't wait until fluorosis begins to damage humans before we act." He also offered some criticism of the existing state program and recommended a more activist approach. In particular, he found "a good deal of confusion between the State Board of Health and the State Air Pollution Control Commission," and he questioned the adequacy of the modest funding and leisurely pace of the current control program, given the significance of air pollution to economics and health in Florida and throughout the nation. The original allocation of sixty-five thousand dollars for two years he called "only lip service," and he encouraged residents of central Florida to make more of an effort at the county level as well as seeking federal grant money to help solve their problem. Considering the extensive evidence of underfunding, understaffing, and loose organization in Polk County's control program, which included only Harry Seifert and two assistants with inadequate equipment, Rossano's criticisms seem fairly gentle.[20]

However, Rossano's modest, brief foray into activism brought a storm of protest from Florida officials, who complained that Rossano had embarrassed them and harmed their control efforts through his "flagrant violation of Federal-States Relationship [*sic*]." The resentful state officials took steps to ensure that no more consultants would be brought into Florida without their visits first being cleared with top state officials, since Seifert had acted on his own in arranging the visit. The phosphate companies and their lawyers also complained about Rossano exceeding the traditional, limited federal advisory role. Rossano responded that he had been quoted selectively for inflammatory effect but that the facts demonstrated the truth of his statements about underfunding and mismanagement; other knowledgeable federal officials agreed that his statements were true, if unfortunate. Regardless of truth, though, and despite the PHS's best efforts to soothe the angry Florida officials, the matter snowballed until it reached the attention of U.S. Surgeon General Leroy E. Burney himself, who in late April formally apologized to Florida State Health Officer Dr. Wilson Sowder and gave assurances that "all of our personnel concerned will make special efforts to see that this does not happen again." Rossano was ultimately removed from the State and Community Services division of the federal program and was transferred (exiled?) to a position in California. Rather than dealing directly with the inadequacies of the Polk County control program or the hobbling limitations of the existing system of polite but ineffectual federalism in air pollution control, the PHS and the Florida agencies in effect killed the messenger who brought bad tidings.[21]

The Rossano affair shows the difficulty of conducting a meaningful federal air pollution control program at a time when federal powers in such areas remained limited and states jealously guarded their "states' rights" against federal incursions—except when they could get federal money with no strings attached. Federal officials could dangle modest incentives to try to stimulate cooperation and significant control efforts from the states, but they certainly could not tell state officials what to do, or even gently criticize disorganization and inactivity, without having state officials and industrialists throw a tantrum. This state of affairs suited the PHS well enough. According to their professional and institutional ethos, PHS officials were scientists, not politicians; they generally wished to keep their work strictly scientific and "above politics"; they assumed, incorrectly, that thorough scientific understanding would lead automatically and directly to effective control efforts; and it seemed not to bother most that their elaborate scientific investigations often produced little concrete action or progress. They were ill-equipped to respond when politics and emotionalism nevertheless intruded into their white-coated world of scientific, medical, and technical expertise, save to fall back on their standard policy of assiduously avoiding stepping on the toes of state governments.[22]

In the wake of the Rossano affair, Florida officials reorganized the state's air pollution control program. In late June, 1959, David B. Lee, director of the Florida State Board of Health's Bureau of Sanitary Engineering and secretary of the Florida APCC, informed a friend in the federal bureaucracy that at the commission's next meeting in July, Lee expected Sowder to propose that air pollution abatement activity be shifted from the health department's Division of Industrial Hygiene to Lee's own Bureau of Sanitary Engineering. It was probably no coincidence that Harry Seifert, whom state officials partly blamed for the Rossano affair, originally came from the Division of Industrial Hygiene. After this, Lee planned to request a PHS review of the technical and administrative aspects of the Polk County program to recommend changes in policy and operation, including an increase in staff, which still consisted only of Seifert, one "sanitarian," two chemists, and one secretary. Lee also announced his plans to establish a statewide air sampling network.[23]

Everything happened as Lee planned, and the PHS agreed to send Ralph Graber and Jean J. Schueneman, Rossano's replacement, to review and report on the Polk County program in early September, 1959. The Graber and Schueneman report stated, politely and privately, much of what Rossano had said publicly about a control program in disarray. They called for legislative changes to include "damage to property, and the reasonable enjoyment of property and life [in other words, the legal concept of nuisance], in the definition of air pollution in the Florida Air Pollution Control Law" and noted that the existing standard based only on demonstrated major damage to human, plant, or animal health was inadequate. Echoing Rossano, Graber and Schueneman also remarked: "It was apparent that considerable uncertainty has existed regarding the roles and responsibilities, and the lines of communication, among the several entities concerned with air pollution activities in the Florida State Board of Health." In obvious indirect reference to past disorganization, they recommended that such relationships should be workable, clear, and set forth in writing and that "once they are agreed to, it would be most desirable that they be adhered to by the entities concerned." The report also concluded that public education and information activities should be significantly improved, regularized, and expanded. In further agreement with Rossano's earlier conclusions, the two PHS officials noted that staffing of the Polk County program was so inadequate that professional employees were engaged in "washing of glassware" and other such "sub-professional activity"; as such, "staff should be so augmented [as] to release these personnel for the technical and public relations aspects of the activity."[24]

In a victory for Lee over Seifert, the Graber-Schueneman report stated: "The Director of the Winter Haven [Polk County] office should be responsible solely to the Director of the Bureau of Sanitary Engineering, who should set the policies and program within which the field office must operate." In an apparent indirect reference to Seifert, who had tried hard to be responsive to angry local citizens, the report continued: "In this way, the Winter Haven group would not be directly subject to wishes of official and citizen groups, and individuals, for modification of program and policy." Although there probably was a need for more centralized leadership and organization, the state control effort was to be insulated still further from the concerns of local citizens who were facing property damage and possibly worse. Graber and Schueneman also offered other criticism on experimental technique, data handling, and meteorological monitoring. Lee took advantage of this report and reorganization to squeeze out Seifert; he was soon replaced by Kay K. Huffstutler, who was careful to maintain good relations with the phosphate companies, if not always with concerned citizens.[25]

In the wake of Rossano's abortive effort at truth telling and stimulating action, the role of the federal air pollution control program in Florida reverted to near impotence and remained mostly that way through the end of the 1960s. As in its interactions with the many other states that were demonstrably not getting on top of their problems during this period, the federal government merely provided modest research grants to the state of Florida to help study the particular characteristics of the pollution developing in cities such as Miami, Jacksonville, and Tampa as well as in rural areas where the paper or phosphate industries operated. When desperate citizens contacted federal officials in a vain effort to go over the heads of their largely inactive state pollution control authorities, all that federal bureaucrats could do was to remind such individuals that existing federal law allowed a "program of research, technical assistance, and training in the field of air pollution" but provided that "the control of air pollution shall be primarily the responsibility of the states and communities." This was little consolation to residents of states where authorities were adamantly refusing to confront their pollution problems in a timely fashion.[26]

Even the weak federal laws of the 1960s were too strong for some Florida officials, who remained protective of their right to determine—or prevent—air pollution control efforts within the Sunshine State. In a letter to a federal official in December, 1963, David B. Lee complained, "I can truly say that there are no Federal-State relations in this program." He claimed that he had tried to contact Vernon G. MacKenzie and Arthur C. Stern in the new federal Division of Air Pollution (DAP) to discuss the situation, but, he sniffed, "they emulated . . . the new Federal government which will by-pass state agencies." Lee and Sowder had been horrified by the original draft of the federal Clean Air Act of 1963, which threatened "very rapid hearings without state people being notified"; they remained unhappy with the law as passed, which "still gives the Secretary [of Health, Education, and Welfare] direct connection . . . to the local areas without the advice and counsel of the states." He longed for the "smooth Federal-state program" in the earlier federal law of 1955, which had been even more toothless than the 1963 measure. However, Lee and his fellow Floridians were wrong to worry that the federal government might come galloping down to intervene in Polk County without their express permission; the limited federal enforcement provisions of the Clean Air Act of 1963 were much too weak and dilatory for that.[27]

The federal program did participate in some other activities in Florida during the 1960s, most of which were typical of their strictly limited role prior to 1970. For instance, in 1960, federal staffers joined Florida control officials and university scientists in putting together a *Report on Florida's Air Resources,* seeking to catalogue the available facts on all the known sources, varieties, and characteristics of air pollution problems throughout the state and to offer recommendations as to what to do about them. Jean J. Schueneman worked on this project along with many similar ones in other states and regions. While federal officials initially hoped that such a standard study might "significantly enhance the opportunity to secure desirable modification of existing legislation" in Florida, there is little indication that this accumulation of facts and statistics, released in early 1961, had much direct impact on control efforts in the state. The state also invited federal officials to join largely unproductive preliminary discussions of emissions standards for the phosphate industry. As usual, and contrary to the expectations of scientists and engineers, scientific research alone offered no solution for a complex legal, political, and economic problem.[28]

Meanwhile, Florida authorities continued fruitlessly studying and nibbling around the edges of the issue. A good example is their proposal late in 1960 to investigate pesticides containing fluoride as a possible cause of damage to gladioli grown commercially outside Tampa. Discussing a letter from Kay K. Huffstutler of the Florida Board of Health, PHS scientist C. Stafford Brandt, not one to jump to unwarranted, inflammatory conclusions, saw the proposed study as merely a diversionary tactic, observing impatiently: "I know of no one who has examined the area who does not accept the premise that the gladiolas [*sic*] of the Tampa region are affected by fluorides and that the fluorides are air-borne and not the result of management practices. I fail to see the point in continuing to dodge the issue." That the once suspicious Brandt found the issue this unmistakably certain to everyone with any knowledge of the subject implies that in undertaking such a study, Florida officials were struggling to come up with an alibi for local phosphate interests, whose emissions were by then causing serious economic damage in Hillsborough County as well as Polk County.[29]

One rare exception to the rule of federal avoidance of activism in central Florida occurred in relation to the first and only lawsuit brought by Florida state authorities against a phosphate plant during the 1950s or 1960s. On June 5, 1963, Vincent Patton of the Florida Board of Health telephoned the DAP to request federal assistance in providing "either direct testimony or deposition" regarding "the effects of $SO_2$, sulfuric acid mist and fluoride on people, animals, and vegetation" for legal action the board was taking against the Armour and Company fertilizer plant at Fort Meade. Vernon MacKenzie, director of the DAP and one of the more activist federal control officials, requested permission to make Drs. Harry Heimann and Stafford Brandt available for this purpose, arguing: "We believe that this request is in the best interest of the Public Health Service. It is both in an area of our activity and in an area in which we wish to give support to the Florida State Board of Health."[30]

Predictably, not all members of the federal air pollution program shared MacKenzie's eagerness to get involved in the Florida proceedings. Early in July, when he first found out about this potential new responsibility, Brandt fumed to MacKenzie about his neither having been "consulted prior to any commitment made for my services" nor having been "immediately informed of the full nature of the commitment and the nature of the action involved." Revealing the traditional self-image of PHS officers and their agency as existing to conduct scientific research, to offer technical assistance and advice apolitically, and to do nothing else, Brandt expressed horror at the prospect of the federal program taking one side of a political or legal issue. MacKenzie responded that it was unusual for the PHS to join such an action but affirmed that PHS rules and regulations did allow testimony in situations where state or local agencies were party to cases and the PHS foresaw the opportunity to advance program objectives. Encouraging the Florida authorities to take legal action—generally a rough indicator of the seriousness of a control program—represented such a situation. However, this case was settled out of court and did not signal any new trend for Florida control officials.[31]

After the sustained public outcry in central Florida led Polk and Hillsborough counties to be included on the schedule for the 1964 field hearings of Senator Edmund S. Muskie and his Senate Subcommittee on Air and Water Pollution, in mid- to late December, 1963, the DAP produced for the senators' use briefing booklets on some of the major local or regional air pollution problems in the United States, including one on central Florida. They also collected lists of the names of Florida citizens and officials who had participated in the thirtieth and thirty-first meetings of the APCC in May and August, 1963, as individuals who might wish to testify before the visiting U.S. senators. Beyond that, however, the hearings were basically left to Senator Muskie and the legislative branch of the federal government, which was at least slightly more activist regarding an expanded federal role and greater and more direct federal intervention on air pollution than was the executive branch throughout the 1960s. The many Floridians at this hearing who begged for the federal government to help resolve their seemingly hopeless situation simply did not understand the niceties of postwar federalism, to which most members of the executive and legislative branches of the federal government still steadfastly subscribed.[32]

The federal air pollution control bureaucrats continued to monitor policy developments in Florida through the rest of the 1960s. They received copies of hearings transcripts from the Florida APCC regarding the old problem in the phosphate belt as well as newer problems in cities such as Jacksonville and Miami, and they continued to receive angry or plaintive letters from local citizens in the region. Federal control officials were bemused to read Elwood Hendrickson's statement that federal authorities to whom he had spoken felt that Florida's program was "on the right track" when apparently no one in the federal program had ever said any such thing; they were similarly amazed to learn of Florida officials testifying that the sulfur oxides emission problem in central Florida was minimal when in fact ambient concentrations were "extremely high" and all available evidence pointed to a severe local sulfur oxide problem. In March, 1966, Vernon MacKenzie suggested that whether or not a visiting delegation of Florida state legislators wanted to hear it, federal officials should point out certain deficiencies of their current air pollution control law, such as divided authority and jurisdictional problems, inadequate funding, "exemption of certain industry emissions such as those which apply to the pulp and paper industry," and "conflict of interest on the part of present commission members." Later, federal officials even considered offering to help conduct a Florida study on the effects of sulfur oxides on human health and offered to meet with the Florida APCC to discuss implementation after the publication of new federal air quality criteria on sulfur oxides. The federal officials observed hopefully: "This situation would serve as the basis of an interesting intrastate abatement action, if an appropriate request for such action" were made; of course, by then they should have known better than to expect any such request to come from Florida.[33]

Instead, the federal air pollution control program had almost no role in Florida through the rest of the 1960s. Despite the pressing and highly visible need for meaningful control of atmospheric contamination in the phosphate belt, Florida officials moved in slow motion and avoided cooperation with higher authorities that might expose the inadequacies of their feeble control effort. Federal control officials, with no power to intervene in an intrastate situation unless they had a special request from the government of the state in question, sat idly by while the Sunshine State stubbornly resisted fully confronting its major air pollution problem. This did not bother the majority of federal air pollution scientists and engineers, who apparently cared less about results than about staying on good terms with state governments and private industry. In the years before 1970, federalism and scientific apoliticism went together to render the federal air pollution control program largely impotent. This demonstrated impotence would help fuel a growing nationwide cry for greater federal control efforts.

# chapter 12

# The Three-Thousand-Mile-Long Sewer

*Air Pollution as a National Problem*

In late 1966, E. F. Porter, Jr., the executive secretary of the Missouri State Air Conservation Commission, wrote a letter to Donald E. Nicoll, an assistant to Senator Edmund S. Muskie, describing the situation on the air pollution control front in Missouri in terms reminiscent of conditions or episodes in New York, Florida, or any number of other places. Porter reflected generally on the technical complexity of battling air pollution and the difficulties of building an adequate staff. As he observed, "The best brains in the business are working either for one of the jurisdictions in California or for the Public Health Service at salaries which a state like Missouri has difficulty meeting. Some of the large states had good luck raiding California and the PHS and we may do just that to fill our top technical positions. Most of the lower echelon positions, however, will have to be filled with bright, educable kids who can pick it up as they go along." Porter also chattily, and somewhat bemusedly, described the problems of trying to get help out of federal air pollution control officials, particularly the research-oriented, enforcement-averse Technical Assistance Branch, headquartered in Cincinnati:

> Dealing with the Division of Air Pollution Control [*sic*] of the PHS has been a Kafka-esque experience. It does not, as you are plainly aware, speak with one voice. Jean Schueneman thinks that Missouri, outside of St. Louis and Kansas City, is pretty small potatoes and doesn't merit a full-time consultant. I know because he told me so way back in March. On the other hand, Joe Fitzpatrick, the Air Pollution man in the PHS Regional Office in Kansas City, has been doing his damndest to get us a full-time consultant and for this the Technical

Assistance Branch has accused him of empire building. The regional office is plainly more closely aligned with the Abatement Branch—Smith Griswold's outfit, which the TAB considers, for its part, sloppy and unscientific.[1]

Porter then described his ongoing struggle to get a full-time consultant from the federal control program assigned to the State of Missouri to help guide abatement efforts. He characterized the situation of many states in depicting how far behind in its work and desperately in need of help in organization and planning the Missouri Air Conservation Commission was, confessing: "Like many agencies which should have been established long ago, but were not, we inherited a backlog of pressing problems. We have, therefore, been operating at a head run on a crisis basis since opening day. We have neglected the forest in favor of the trees. I hope our consultant on the other hand, will be able to devote some of his time to helping us take a look at the big picture."

Porter had someone in mind. For the preceding three years, a federal air pollution technician named Norman G. Edmisten had been assigned to St. Louis as part of a three-year, $750,000 research program to study the interstate air pollution problems of the Greater St. Louis area, where the city sprawls across the border between Missouri and Illinois. Porter observed, "He is aggressive, outspoken, technically well qualified, and good mannered. He likes Missouri and he wanted to stay here and help us." However, Edmisten, like August Rossano before him, was apparently a little too aggressive and outspoken. He gave clear answers to some of Missouri's questions about the interstate air pollution moving between St. Louis and southern Illinois, in particular the emissions from sulfuric acid plants that Missouri had asked Illinois to reduce, and so he became "persona non grata" with the state of Illinois and was "disciplined by the TAB for 'taking sides in an interstate dispute.'" His punishment was to be taken off the front lines and stuck in a safe desk job; Porter noted that he was kicked upstairs to Cincinnati. Porter complained that this was all "utterly unfair, because all he did was to answer our questions when we asked them," something he similarly would have done for Illinois, had they cared to find out anything about their interstate pollution problems.

Porter went on to complain about the uncooperativeness of the State of Illinois regarding interstate atmospheric contamination over the Greater St. Louis metropolitan area in terms many New Yorkers would readily have understood. Yet he candidly acknowledged that the blame for the situation was not all on one side. As he admitted, "The trouble with this thing is that we don't come into it with entirely clean hands. Jurisdictions on both the Illinois and Missouri sides of the Mississippi are sitting on their hands doing nothing awaiting the outcome and recommendations of this three-year study, which should be forthcoming in a few days. The Illinois Air Pollution Control Board has been in business for about two and one-half years and has accomplished very little." Porter did not bother to emphasize further how relatively little the state of Missouri had accomplished; he could not have known how little the two states would be able to achieve even after the release of the long-awaited report.

The window Porter's letter opens on the struggles of the State of Missouri sheds wider light on the general state of air pollution control in the United States during the mid- to late 1960s. By then, far from being merely an affliction characteristic of Los Angeles, New York City, central Florida, or a handful of other peculiar places, air pollution was recognized as a chronic problem causing economic damage and potentially threatening health in cities, towns, and even some rural areas through-out the nation. Yet most states and localities remained woefully unprepared to con-front this situation. Nationwide, funding and staffing for control agencies were minimal, where these agencies existed at all; legislation was generally inadequate and primitive, still rooted in prewar concepts; and scientific understanding, research, and technical capabilities devoted to particular local problems remained generally deficient at best. Many states and localities were in a position similar to Missouri's—just starting to grapple with air pollution and not sure how to go about it.

A number of these lower jurisdictions also shared with Missouri an uncertain relationship with the federal government on the matter of air pollution control. The fact that the federal government was participating in an interstate air pollution study at all, or that Porter could even dream of borrowing a full-time technical consultant from the federal government free of charge, indicated how far the federal effort had evolved from the nearly nonexistent program of the 1940s. The inconclusiveness of the activities in which federal air pollution control officials participated during the 1960s, and the overall lack of results following from most of them, in turn revealed how far the federal government and the nation in general still had to go to master the ever-expanding national air pollution problem. Inside the federal program itself, as Porter noted, debate still continued as to whether the federal effort should keep true to tradition and remain limited and chiefly research oriented in scope or adopt a more vigorous stance that might bring more actual progress. This situation—the unresolved air pollution crisis and the unresolved role of the federal government in facing it—would give way to growing federal involvement toward 1970, as Ameri-can citizens, increasingly environmentally conscious and alarmed over pollution, stridently demanded greater and faster action than feeble existing federal, state, and local programs could bring.

The nationwide extent of air pollution had grown increasingly evident by the late 1950s and early 1960s. Many coal-burning industrial cities had never gotten their traditional smoke and particulate emissions under control before they also began to experience newer sorts of atmospheric contamination, such as Los Angeles-style photochemical smog or the complex new emissions from modern chemical plants. Into the early 1960s, the automobile industry still denied that cars represented a pollution problem anywhere beyond Southern California and insisted that nowhere else was ever likely to experience choking, eye-stinging photochemical smog. Yet already in the mid-1950s, researchers found that such "widely separated cities as San

Francisco and New York have experienced smog-produced eye-irritation, low visibility, and crop damage," though such attacks remained infrequent. During that same decade, smog was already regularly visiting many other California cities far from Los Angeles, such as San Diego, Bakersfield, and even Fresno, while the LA smog itself was found to spread fifty miles to both the north and south of the Los Angeles city center, much farther than it previously had. Philadelphia apparently saw photochemical smog for the first time in 1957; by the early 1960s, Chicago, Washington, D.C., and even breezy Miami would see it too, as would many other places. During the field hearings of the Muskie Subcommittee on Air and Water Pollution in early 1964, the visiting senators heard about severe air pollution problems well beyond Los Angeles and New York City—in Chicago, Boston, Providence, and even Denver, a place once renowned for its pristine air that had come to suffer from too many uncontrolled industrial smokestacks and automobiles overloading its thin atmosphere. As the 1960s wore on, the list of chronically afflicted cities would only grow longer, including the sprawling new sunbelt city of Phoenix and the booming Pacific Northwest megalopolis of Seattle. By 1965, the United States Department of Agriculture reported: "Los Angeles no longer has, if it ever had, a monopoly on photochemical smog. . . . Characteristic symptoms have been found in almost every metropolitan area in the country." Even communities in Alaska and Hawaii would face the threat of dirty air. While photochemical air pollution was spreading in the United States, it would also soon appear in many other parts of the world; by the mid-1950s, characteristic vegetation damage was discovered in foreign cities such as London, Paris, Sao Paulo, and Bogota.[2]

As the problem spread, the resulting damage mounted ever higher. In 1949, at the First National Air Pollution Symposium, representatives of the federal Commerce Department estimated the annual costs of smoke damage in American cities to be from eight to twenty dollars per inhabitant. The U.S. Geological Survey found smoke damage just to merchandise and buildings to cost the nation at least half a billion dollars every year; the chairman of a New York City sanitation reform group guessed at an annual loss of $100 million in damage to fabrics, household furnishings, and structures in New York City alone. A few years later, in testimony at hearings on proposed federal legislation regarding polluted air, federal officials estimated $1.5 billion dollars in losses yearly from corrosion and deterioration of structures and materials and other damage. Such estimates would rise much higher during the 1960s.[3]

Dirty air also was causing significant damage to the nation's farmers. In 1957, an early study at the University of California's agricultural experiment station at Riverside found $5 million dollars in crop damage from dirty air in the Los Angeles area alone and a further million lost around the San Francisco Bay. A decade later, federal researchers at the Department of Agriculture's Air Pollution Laboratory estimated aerial contamination nationwide to be costing $325 million in crop damage and $175 million in stunting or slowed growth of livestock every year, while the spreading pollution increasingly drove agriculture away from urban hinterlands.[4]

The greatest concern of all—health risks—remained scientifically problematic, but scientific evidence, medical opinion, and public awareness generally moved toward greater concern or alarm from the 1950s through the 1960s as other citizens around the nation began to hear the sort of warnings Angelenos and New Yorkers had heard earlier. At the first National Conference on Air Pollution of late 1958, Dr. Lester Breslow of the California State Department of Public Health warned of unknown potential health damage from long-term, low-level, chronic exposure to air pollution, noting that some "harmful, even fatal, effects" might only surface after "years of exposure." He also reminded listeners that smog contained chemicals capable of causing cancer in animals. Surgeon General Dr. Leroy E. Burney commented on the statistical finding that lung cancer death rates were twice as high in large cities as in rural areas, possibly due to air pollution. In a report that same year, the U.S. Public Health Service (PHS) announced that it was "abundantly clear that acute air pollution can cause death among the aged and infirm and serious illness in the general population." Noting suspicions that pollutants caused damaging changes to tissues of the respiratory tract, they found "accumulating evidence" that polluted air contributed to various modern "'urban' diseases," such as heart conditions and "cancer of the lung, trachea, stomach, and esophagus." Such disturbing messages gradually got out to increasing numbers of American citizens through the print and broadcast media, as well as through a growing number of best-selling books published on the topic during the 1960s.[5]

By the mid-1960s, top federal officials showed an increasing sense of alarm regarding the health effects of polluted air. In June, 1966, Secretary of Health, Education, and Welfare John W. Gardner testified before the Muskie subcommittee: "We believe that air pollution at concentrations which are routinely sustained in urban areas of the United States is a health hazard to many, if not all, people." Later that month, another major figure in the federal government, Surgeon General Dr. William H. Stewart, echoed these ideas in a keynote speech for the fifty-ninth annual meeting of the Air Pollution Control Association in San Francisco, declaring: "There is little doubt that air pollution is at the very least a contributing factor to the rising incidence of chronic respiratory diseases—lung cancer, emphysema, chronic bronchitis and asthma." These statements came before the Thanksgiving Weekend air pollution episode of November, 1966, which did more than any public statement to hammer home these warnings about health effects and to increase public concern about the problem.[6]

By the end of the 1960s, health worries had grown and intensified among the American public. In particular, health scientists began following Dr. Breslow in raising disturbing questions about the possible long-term health effects from exposure to relatively low levels of pollutants—a matter that had drawn little research because it was so difficult to set up experiments to study it. Epidemiological evidence suggested that death rates rose above statistical norms from concentrations of sulfur dioxide that were far below alert levels and were fairly common in the New York

City area. Dr. Samuel S. Epstein, chief toxicologist for the Children's Cancer Research Foundation in Boston, warned of potential long-term mutagenic and carcinogenic effects from chronic low-level exposure to hydrocarbon-based pollutants and in 1968 attributed the "dramatic increase in mortality from lung cancer . . . now approaching epidemic proportions" at least partly to polluted air. By this time, the evidence of links between tobacco smoke and lung cancer had developed to where smoking was seen as the single main culprit for the increase in lung cancer in the western world, but, as one observer wryly noted, that did not explain why the birds and mammals at the Philadelphia Zoo, surrounded by heavy automobile traffic, had experienced a sixfold increase in deaths from lung cancer since 1902—they certainly did not smoke. The nation's best-known living ecologist, Barry Commoner, echoed his famous predecessor Rachel Carson in his best-selling book, *Science and Survival,* warning that uncontrolled human intrusion in the environment was subjecting humans to "a huge experiment on ourselves" that might reveal dire public health consequences in future generations.[7]

While knowledge of air pollution and its dangers and damage was growing, most state and local governments were still sitting on their hands. In 1961, less than half of the urban areas with major or moderate air pollution problems were spending more than five thousand dollars a year to cope with them, while of the total of eight million dollars spent by eighty-five local agencies throughout the nation, 55 percent was spent in California and more than 40 percent in Los Angeles alone. Most municipal air pollution authorities were so underfunded and understaffed as to be almost wholly ineffectual.[8]

The story was not much different at the state level. In 1961, only seventeen states devoted more than five thousand dollars a year to the fight against atmospheric pollution, and California spent 57 percent of the total of two million dollars. As late as 1967, when the federal government passed yet another law to try to stimulate state action without federal compulsion, fully fourteen states had no money and no personnel allocated toward air pollution control in their budgets, notwithstanding the lure of federal grants under the Clean Air Act of 1963. Another twenty-two states had fewer than ten employees working on air pollution issues; only six had more than twenty. Seventeen states still had no state air pollution control law by 1967, while only eighteen had such laws before 1963—not including some states containing major emissions problems, such as Illinois, Missouri, and Texas. In many states, major industrial polluters had representatives on the state pollution control boards. As one observer commented at the end of the 1960s, "Until quite recently the interest of most state governments in controlling air pollution was minimal or nonexistent." Despite the major postwar expansion of knowledge about air pollution and its control, most states and localities were barely at the stage of initial research into the problem, let alone taking significant abatement action, while many of the minimal state programs that were started during the 1960s probably represented in part token efforts to forestall further federal intervention.[9]

Yet while state and local governments were almost sitting still, public interest and concern raced forward, and not only in a few polluted major cities. During the 1960s, air pollution grew to be a public concern throughout the nation as more and more individuals and organizations began seeking information on the topic. By 1965, the Kiwanis Club noted its increased emphasis on the "prevention of air and water pollution" and made these major topics of discussion at its 1966 convention. That same year, when the Johnson Administration briefly balked at proposed national motor vehicle emissions standards and regulations then under consideration in Congress, an anonymous federal staff member commented: "I don't believe the people at the White House realize how great the public support for this kind of legislation is, or the extent to which many people in industry are ready to go along with some kind of air pollution control." In 1966, even before the Thanksgiving Weekend episode, the Junior Chamber of Commerce of Linden, New Jersey, formed a clean air committee and sought whatever educational materials it could find on the subject. Immediately after that frightening incident, Region 1 of the Boy Scouts of America (the Northeast) proposed a new merit badge for "air conservation." Around the same time, a group of Missouri housewives organized to fight air pollution, while thousands of high school students from around the nation sought information on the problem for school projects. Countless other citizens wrote to federal authorities with questions about air pollution. During the mid-1960s, the tone of such letters was usually inquisitive; the outright alarm mostly came later.[10]

As in California earlier, many Americans offered quick solutions to the problem. E. H. Sullivan of Tennessee proposed spraying "chemicals such as Lysol . . . into the outdoor air to neutralize the adverse effects of air pollutants," a suggestion that federal authorities politely explained was "novel, but impractical." Referring to a newspaper article discussing the federal government's growing interest in controlling vehicular emissions, Kenneth R. Brown of Atlanta announced: "Realizing the need for a device as out linned in the inclosed Artical, I have several different devices ON PAPER that can be produced cheaply and installed easyly on any make car or truck. As for industrial uses on sulphourous pollutions there is no problem." Brown went on to warn naysayers against rejecting his proposal too soon:

You might give thought to the fact that many times, advancement is held back, but for no other reason then expearmentation and investigation is placed in the hands of trained monkeys with a hand full of clipboards and a head full of nothing Lets be honest about this thing—give a problem to a man who has proven himself—I mean, a man with dirt under his finger nails who has supported his family by doing, rather then telling some one to do. Tell him what you need and he will come through with a workable solution. . . . If you've any faith in the common-sence, Jack of all trades type american, take a gamble, furnish me with a few hundred dollars for materials, give me thirty days, and I'll serve up your answer to pollution.

A federal public information official politely reminded this latter-day populist that the federal control program could do no selection or marketing of a proprietary device, and he referred Brown to the Atlanta office of the Small Business Administration.[11]

By the later 1960s, with the explosion of general public environmental awareness and particular concern about air pollution, the volume of public complaints and letters to elected officials naturally grew. Many of these were directed toward federal officials and legislators, complained of nonfeasance by state and local authorities, and begged for federal help. Senator Muskie, as the best-known congressional figure involved with air pollution control, was a favorite shoulder to cry on. Henry A. Kreutzer, chairman of the Swannanea Valley (North Carolina) Association for Protection of Health and Property, described his community's suffering at the hands of a local chemical plant. After many letters sent to state and local authorities without results, Kreutzer's group had concluded that control officials cared more about preserving "payrolls and tax revenue" than stopping air pollution and protecting "the health of human beings, their lives and property." Kreutzer further complained that while the laws passed and the millions of dollars spent by the federal government implied that it "should be able to give poor people a hand in stopping pollution and protecting their civil and constitutional rights," he and his neighbors were "beginning to wonder if we have any rights."[12]

Thomas A. True of New Iberia, Louisiana, similarly wrote to report, "The black soot is almost upon us in Southwest Louisiana. I have written my state and federal representatives, but have received no help." True and his neighbors were suffering from heavy particulate emissions from local sugar refineries, and sugar was "a big money crop" supporting "much of the local power structure," though he claimed that it benefited the "vast majority of the citizenry . . . not at all." True noted that "a local woman filed an injunction to stop the air pollution; but the courts threw out her case claiming the parish depended on the cane and nothing could be done. We, the little people, have submitted dozens of petitions with very little results." Asking Muskie for help and advice, True concluded, in a statement expressing the anti-corporate sentiments spreading among the public in the late 1960s, "Well over three-fourths of the town wants clean air, but public apathy and lack of funds make it almost as hard to fight as challenging General Motors."[13]

Anti-air pollution activism also cut across class lines. Rose Owen of Philadelphia had an even worse tale of local and state authorities' nonfeasance to tell, as well as a story of local mobilization in a poor neighborhood. As she complained to Muskie:

I am writing this letter in hope that you will try to help us in our desperate fight against Air Pollution. We have been in and out of court since last Sept. 27th 1968. Our problem is Coke Dust, Fumes, & Noise from the Phila. Coke Company. . . . My main concern is the health of my six children. They eat, sleep and drink coke dust. My children as well as many others have had serious eye infections and lingering colds. We also have been sick and have burning eyes from the coal fumes. The Noise from the Company has wakened my children at 3 and 4 o'clock in the morning and scared them. We had called the State into this mess, because the city wasn't doing its job. We picketted the plant in late September and the Coke Company got an injunction against us. At Court, Judge Sloane on Sept 27th with the State, gave Phila. Coke certain requirements to do by March 1st. They said if certain Air Pollution [*sic*] wasn't put in by March 1st, he himself would close the plant. Well, we never went back to Court and the state never lived up to their promises. . . . The people are desperate and feel that it will take a . . . death to make the City Officials wake up to what is going on here, we are living in a Hell. Please help us![14]

Years earlier, even before the major surge in popular environmental awareness, another working-class mother, Ann Belcher, had mobilized her community, the relatively poor Talleyrand neighborhood in Jacksonville, Florida, to fight back against noxious and corrosive sulfuric acid, sulfur dioxide, and dust pollution from the local industrial plants and generating stations. On October 8, 1963, Belcher took to the city council a petition to take immediate action against air pollution in the city, along with the names and signatures of a thousand neighbors in Talleyrand and residents of nearby communities. Belcher also led a delegation of a hundred women from the Talleyrand area to complain to the city council in person about the way the "air in their neighborhood destroyed their homes, their cars, their clothing, and their health." Belcher brought a stained sheet and a sickly potted plant as exhibits to demonstrate the effects of polluted air on vegetation and previously clean laundry. Other residents offered further complaints of damage both to property and personal health. A neighbor of Belcher's, Stanley Charles Carter, threatened to take direct citizens' action against the polluters: "We can be very mandatory. . . . We'll lie down in front of trucks [at the local paper pulp mill] and keep them from moving." Another local citizen to call for cleaner air was Ulysses Cook, a spokesman for black residents in the neighborhood, who angrily told how the pollution had caused one of his children to develop chronic respiratory trouble and had corroded his sister's gas furnace so badly as to make it unusable.[15]

When their efforts brought no results from city officials, Belcher began writing letters, first to the state governor then to Senator Muskie, seeking help and action to control the local air pollution—despite the fact that writing clearly was a struggle for her. That people from a working-class neighborhood such as Carter, Cook, and Belcher and her legion of angry housewives should have gone out of their way to appear before the Jacksonville City Council to demand cleaner air, in addition to writing letters, goes against the common assumption that environmentalism is and always was strictly a pastime of middle-class professionals. The presence of Ulysses Cook at the Jacksonville City Council meeting along with other unhappy citizens from Talleyrand—particularly at a time when most of the South remained in the grip of racial segregation—also demonstrates that at least at times, contrary to later assumptions, early environmental activism could cross racial divisions and was not exclusively a white phenomenon. However, since the predominant social ideal of the time was a color-blind society in which race and ethnicity truly would be ignored, African Americans who wrote letters to their elected representatives or federal officials may not have advertised their ethnicity.[16]

In some other cases, though, there is less doubt. For instance, in September of 1966, Emma Kai of New York City wrote to Senator Muskie after seeing the television broadcast "Poisoned Air." She recounted:

> I've been concerned about "Air-Pollution" for a very long time. I've written [New York] Senator [Jacob K.] Javits about Air P. in the past. Now I write you also. Please do something constructive and immediate about ridding N.Y.C. of Air P. We need Quick Action, not something two three years later. Human lives are at stake. If it means some personal sacrifice on our part to be made I am more than willing, God Willing. I cannot support this problem financlly, as my husband only makes $68.00 dollars a week. . . . We have a family to support in Kowloon Hong Kong, besides regular monthly bills. I suffer from Bronchitis which I got this year. I cannot afford to leave N.Y.C. for Arizona or some warmer climate because my husband can only find work in N.Y.C. chinatown. He is a cook in a Restaurant in chinatown in N.Y.C. Please find time to do what ever you can to rid N.Y.C. of Air pollution.

Some years later, José Martinez of Gary, Indiana, wrote to Muskie with a cry from Mother Earth:

> "Dear Senator: I'm in desperate need of HELP! Your industrial society is killing me. The steel industries and chemical laboratories are polluting all of my air and waterways. The rise in mercury in my water streams has literally ended all swimming and fishing because of that lethal element. Please lend a helping hand I'm in desperate need of your help.—Earth"

> Dear Sir, The above letter was my impression of what mother earth would be pleading for if she could speak. We need help around here, because our local government does nothing about these problems; they're nothing but a political machine that only cares about itself. This is why I am writing to you, because you can lend a helping hand to pass stiffer pollution legislation.

Although Kai and Martinez may have been unusual cases, their efforts to express their concerns about environmental issues nevertheless show that even back then, environmental worries were not exclusively a white, Anglo-Saxon, middle-class fixation.[17]

More surprising, given later presumptions of hostility between the labor and environmental movements, the battle against air pollution got militant support from some labor unions and union members during the late 1960s. The United Auto Workers (UAW) sent representatives to testify in favor of various environmental measures, including nationwide emissions standards and federal enforcement powers on air pollution, from 1967 onward. In 1970, the UAW formally joined several environmental groups in calling for vehicular emissions standards so tough as to banish the internal combustion engine by 1975. The United Steelworkers also demanded action on air pollution, and in 1970, some steelworkers around Peoria, Illinois, threatened to strike over the atrocious air pollution outside, not inside, their local steel mill. Even the larger, more conservative AFL-CIO branded air pollution a threat to life on earth. In St. Louis in 1969, members of locals of the relatively conservative and usually not very pro-environmental Teamsters Union formed a delegation to complain to Senator Muskie's Subcommittee on Air and Water Pollution regarding the health effects and property damage from the local air pollution, and they ridiculed the ineffectual efforts at cooperative interstate control, calling for federal enforcement. Kenneth L. Worley sent a letter on behalf of the Greater St. Louis Council of the UAW wondering what was the use of worrying about wages, contracts, medical insurance, civil rights, nuclear disarmament, moon exploration, or ending the Vietnam War "if we continue to poison and destroy the life supports of the world?" Noting how his union had put air pollution "at the top of its list of priority problems of man," Worley stridently continued, "We demand that uncompromising and irreversible standards and controls be established to preserve our environment, no matter what the cost, no matter how great the violation of property rights, no matter what the effect on dividends and no matter what the effect on our own bold plans for collective bargaining." While such workers may have been atypical, they nevertheless show that environmental concerns had taken root among at least some portions of the working class.[18]

In the face of such strident and broad-based public demands for air pollution control and the overall evidence of state and local incapacity or unwillingness to address the problem, the federal government ultimately moved to take action and expand its authority to regulate aerial contamination throughout the nation. Yet this was a decidedly gradual process. Far from rapidly reaching beyond the boundaries of traditional federalism to seize state power arbitrarily, both the legislative and executive branches of the federal government assiduously sought to maintain air pollution control as a state and local responsibility through the end of the 1960s—to the dismay of activists hoping to see more rapid and decisive action.

The significant expansion of postwar federal attention to polluted air began abruptly with the lethal Donora air pollution disaster of October, 1948. The shocking news from western Pennsylvania was reported widely in the United States, briefly drawing unprecedented national attention to air pollution. After the incident, local and state authorities in Pennsylvania sought to determine what exactly had happened and what had caused the deaths. However, local citizens, who had long resented the heavy, uncontrolled emissions and resulting damage from a local branch of the United States Steel Corporation, feared that the industry-oriented state government would merely try to cover up the incident and absolve the major local polluter from responsibility. As such, they demanded an inquiry by the federal government, and the Public Health Service was sent in to investigate the matter. However, the PHS, also, was worried about keeping good relations with industry and with state governments. It ultimately joined the state government and local industry in glossing over the role of industrial pollution, instead mostly labeling the episode a freak disaster and focusing on meteorological conditions and the need for further research into the causes and effects of air pollution. Thus, from the very outset, the postwar federal air pollution control effort opted for a safe, quietist approach of scientific research and technical assistance, carefully avoiding stepping on the toes of state or local governments or industry.[19]

The next few years saw a slight expansion of the federal government's limited involvement in air pollution issues. In 1949, years after the earlier Trail Smelter case was left unresolved, the PHS became involved in another international air pollution dispute between the United States and Canada, regarding pollution from ships, boats, and industrial plants crossing between Detroit, Michigan, and Windsor, Ontario. That same year, four unsuccessful bills were introduced in Congress, two of them calling for the Public Health Service to study the health effects of air pollution. Then, in December, 1949, President Harry S. Truman requested that the Department of Interior organize an interdepartmental committee to plan the first United States Technical Conference on Air Pollution. Truman did this reluctantly, as a modest answer to the Donora affair, and reflecting the predominant localist presumptions of the time, he warned that he did "not contemplate . . . programs which will commit the Federal Government to [significant] material expenditures . . . since the responsibilities for corrective action and the benefits are primarily local in character." The resulting conference, in 1950, brought together more than 750 representatives of governmental agencies and industry to hear papers on a wide variety of mostly technical subjects, and the participants concluded that the federal government should "help identify air pollution problems and develop the technology to combat them." Beyond this limited role of research and technical assistance, however, the conference assumed that the federal government would steer well clear of any unsolicited intervention in local affairs.[20]

After some additional years of little action and unsuccessful proposals for federal legislation, the federal government took further significant steps toward a role in air pollution control during the mid-1950s. In 1954, Republican senators Thomas H. Kuchel of California, who was familiar with the situation in Los Angeles, and Homer Capehart of Indiana, who had learned of developing air pollution problems in Gary and Indianapolis, sought to include an amendment to that year's federal housing bill offering federal aid for air pollution research and the installation of abatement equipment. This effort failed, but President Dwight D. Eisenhower did respond to a suggestion from the two senators by establishing a standing interdepartmental committee on "Community Air Pollution," bringing together representatives from potentially concerned departments, such as Interior, Agriculture, Commerce, and Health, Education, and Welfare (HEW), and other federal agencies, such as the Atomic Energy Commission and the U.S. Weather Bureau, to recommend what further action the federal government should take on the issue. Echoing a familiar theme, this committee called for a broad, general program of federal research and technical assistance.[21]

In 1955, Senator Kuchel proposed another bill to spend three million dollars of federal money yearly for four years to study the problem of air pollution and to assist the states with research projects, professional training, and general technical expertise most states did not possess. In the bill's passage through Congress, the yearly sum was raised to five million dollars. The bill envisioned no outright federal role in control or enforcement; Kuchel himself declared that control authority would remain just "where it ought to remain"—at the state and local level—and officials within the executive branch of government claimed that "instances of troublesome interstate air pollution are few in number" and that the issue was "essentially a local problem." In 1959, the untitled law, sometimes referred to as the federal Air Pollution Control Act, was renewed for another four years, reemphasizing the fact that this federal commitment was not only a modest, limited one but a temporary one as well, until the states and cities could take care of their own problems. In late 1958, HEW held the first National Conference on Air Pollution, but while this conference explored many different aspects of the problem, it faithfully reaffirmed the standard line regarding problems being local, the federal role relatively minor. With the exception of limited federal activities in Congress and the executive branch to investigate the role of the automobile in polluting the air, this was the extent of federal air pollution control during the 1950s.[22]

The next leap forward in federal involvement came with the passage of the Clean Air Act in 1963. In 1961, the new president, John F. Kennedy, had called for a somewhat greater federal role on the issue, declaring: "We need an effective Federal air pollution control program now." The following year, he requested that Congress pass a law to promote interstate conferences regarding shared air pollution problems, a proposal that had passed the Senate previously in 1960 and 1961 but had died each time in the House Subcommittee on Health and Safety because of the opposition of both subcommittee chairman Kenneth Roberts and the timid, technically fixated Public Health Service—not to mention the industrial interests, which steadfastly resisted any extension of federal authority. To oppose these forces, in 1962, an urban lobby composed of the United States Conference of Mayors, the American Municipal Association, and the National Association of Counties mobilized to promote an expanded federal role on air pollution, and their draft proposal became the basis for the Clean Air Act, introduced in the Senate in early 1963. In late 1962, the

Second National Conference on Air Pollution again affirmed fundamental state and local responsibility, but more calls for greater federal involvement were heard. By 1963, even Congressman Roberts changed his views and supported expanded federal air pollution abatement authority for particularly serious problems. As a result, in both houses of Congress, the urban lobby and conservationist interests were able to overcome the arguments of industrialists and the PHS, and the Clean Air Act became law in December, 1963.[23]

The new federal law differed from the 1955 law in various ways. First, rather than being a temporary federal effort requiring reauthorization every few years, under the 1963 Clean Air Act the federal program became permanent, with a hugely expanded budget authorization of ninety-five million dollars over three years. Also, besides conducting research and offering technical assistance, the federal program gained the authority to make grants to state and local control agencies, using federal money directly to help initiate or improve state and local control programs. Reflecting the growing awareness of automotive air pollution by the early 1960s, the act directed federal officials to encourage the development of devices or techniques to reduce vehicular emissions and to make regular progress reports to Congress. The law further called for HEW to develop federal air quality criteria, though these would be nonbinding recommendations rather than actual mandates for state and local authorities. Finally, federal officials gained some mostly token enforcement powers: they could intervene in an intrastate air pollution situation only with the invitation of the governor of the state in question, and they could only take action on an interstate air pollution problem endangering citizens' health and welfare after an elaborate, lengthy procedure including consultation with state and local officials, formal federal recommendations, public hearings, and further recommendations. Only if the situation persisted unabated after all these preliminary steps had been taken could federal officials commence actual abatement proceedings. This extensive set of preconditions was intended to prevent the federal government from recklessly interfering in states' affairs; it more than accomplished this purpose, precluding any meaningful federal enforcement. Nevertheless, the federal abatement provisions represented a significant precedent for the future. Despite all the significant changes from the 1955 law, the 1963 law remained limited and firmly committed to the policy of primary state and local responsibility.[24]

In 1964, as we have seen, following passage of the Clean Air Act and 1963 hearings on the extent and severity of the nation's air pollution, Senator Edmund S. Muskie, chairman of the new Senate Subcommittee on Air and Water Pollution, held field hearings on air pollution in various cities of the United States to ascertain what particular problems or research needs the federal government could best help answer. Two recurrent suggestions from witnesses were automobile exhausts, a technically complex problem produced by a nationwide industry engaged in interstate commerce and thus arguably subject to federal purview, and solid waste disposal. The latter was suggested as potentially a partial responsibility of the federal government, since various places, especially large eastern and midwestern cities, were becoming inundated with large volumes of heavy construction and demolition debris as federal urban redevelopment programs tore down lower buildings to replace them with high-rise federal housing projects. The traditional method of dealing with such building debris was to burn it in smoky open dumps, a source of significant air pollution and public complaints. In 1965, responding to these suggestions, Muskie proposed a three-part bill calling for enforceable federal vehicular emissions standards, research into more benign means of solid waste disposal, and further study of sulfur oxide emissions. Following some initial hesitance by the executive branch regarding federal compulsion of the auto industry, the 1965 amendments to the Clean Air Act were signed in October, 1965. The following year, the 1963 law was further amended to extend and expand the federal grants-in-aid program to help start or improve state and local air pollution control programs, and the Muskie subcommittee held hearings on the health and environmental risks of automotive lead emissions and experimental fluorine- or beryllium-powered rocket motors for missiles. Also in 1966, President Lyndon B. Johnson signed Executive Order 11202, requiring federal agencies to develop plans for the installation of air pollution control equipment at all federal facilities. Senators Muskie and Jennings Randolph proposed a further bill to promote federal research into low-emission alternative vehicles, but this died in committee.[25]

In spite of the increasing federal attention to air pollution after 1963, public pressure for an even greater federal role in controlling atmospheric contamination grew steadily during the 1960s as most state and local governments still proved reluctant to confront the problem. The alarming Thanksgiving Weekend incident helped pave the way for a more radical mood at the Third National Conference on Air Pollution, coincidentally scheduled to take place the following month, in December, 1966. In the keynote address to the four thousand participants, HEW Secretary John Gardner showed how far federal policy and the public mood had moved away from the stubborn localism of 1955 when he sharply criticized the states for failing to take initiative and failing "to establish the regional approaches demanded by a problem that ignores traditional state boundaries." He called for uniform national air quality and emissions standards. Also addressing the conference was Senator Muskie, who had become the clear leader in Congress on pollution issues due to his subcommittee chairmanship and was well on the way to establishing his national reputation as "Mr. Clean," which would later help to make him both a vice-presidential and a presidential contender. While he strongly favored the interstate regional cooperation approach, Muskie argued against "fixed national emission standards for individual sources of pollution" except in the case of motor vehicles. As a senator from a mostly rural state still eager to attract industry and not too heavily afflicted with pollution, Muskie worried that national emissions standards would inhibit development in such states and argued that the federal government should focus its efforts on those places that were already seriously polluted.[26]

This was the backdrop for the struggle over the 1967 Clean Air Act amendments, known as the Air Quality Act of 1967. In his special message to Congress on "Protecting Our Natural Heritage" in January, 1967, President Johnson mostly ignored the preferences of Muskie and the nation's industry and called for uniform national emissions standards for major industries; for the establishment of a control policy based on "regional airsheds" crossing state boundaries; and for federal assistance in state vehicle inspection programs, federal registration of motor fuel additives, and further federal research into both automotive and stationary-source emissions. As the subsequent administration bill went through congressional process, Muskie reduced the proposal for national emissions standards to only a two-year federal study of the feasibility of such standards. While he kept the provision that would have HEW designate regional interstate or multistate air quality control zones, Muskie sought to restore to the states the primary initiative and responsibility for creating such regions and setting air quality standards for them. Like the earlier air and water pollution control legislation Muskie had shepherded through Congress, the Air Quality Act expected the states to fulfill their responsibilities on their own, with the federal government having the right to step in after a potentially lengthy process only if states failed to meet deadlines. Also, Senator Jennings Randolph from West Virginia—chairman of the Public Works Committee of which the Air and Water Pollution Subcommittee was a part and a long-time champion of the coal industry—took measures to make certain that the law would not reflect proposed federal air quality criteria for sulfur oxides, which threatened to exclude most coal from many major urban energy markets. He inserted in the bill requirements that HEW reconsider its proposed sulfur oxides criteria, that all future criteria be issued along with viable recommended control methods, and that advisory committees reflecting the views of industry be created to guide federal policy. Randolph added extra federal money for research into the control of sulfur oxides. In this form, the act became law in November, 1967. It was the federal government's last effort to avoid major intervention in state and local air pollution control.[27]

At the very end of the 1960s, a knowledgeable observer commented that it was "clear that the Air Quality Act will shape the nature of air pollution control [in the United States] at least for the next several years." This prediction would prove inaccurate. Although the new law was resoundingly passed by both houses of Congress and was received with hope and optimism by most of the concerned public, it soon came under heavy criticism as public agitation over dirty air raced ahead while the moderate Air Quality Act, with its built-in delays and reliance on state and local initiative, failed to show results fast enough to assuage public concern. Those few modest efforts at interstate cooperation on air pollution control that were undertaken during the 1960s—between Missouri and Illinois, between Ohio and West Virginia, and between New York and New Jersey—still mostly failed to bear any fruit and remained tangled up in bureaucratic and jurisdictional uncertainties through the end of the decade. The sole example of the federal government initiating abatement proceedings in an interstate air pollution situation under the terms of the Clean Air Act of 1963—against a poultry rendering plant in Bishop, Maryland, that dumped noxious emissions on Selbyville, Delaware—had only led to inconclusive court proceedings and further appeals by the end of 1969, after four years of mandated conferences and delays; as of early 1970, the pollution continued unabated. Most state control programs remained rudimentary and ineffectual. To many concerned Americans, the need for direct federal involvement in air pollution abatement and enforcement was only too obvious.[28]

Indeed, public impatience with federal and state officials' gradualist approach to confronting air pollution even led more radical control advocates to criticize Senator Muskie sharply by the late 1960s. Muskie, who had grown accustomed to being celebrated as Mr. Clean, was blamed for what many had come to perceive as an excessively moderate bill; some even branded him a sellout to corporate America for his efforts to obstruct federal enforcement and keep policy at the state and local level. The radical young law and medical students of Ralph Nader's Study Group on Air Pollution in particular criticized Muskie for being a conciliator rather than a fighter and blasted his vision of "creative [or cooperative] federalism" in air pollution control, given the demonstrated failure of most states to act. Reflecting the public's new sense of urgency about environmental matters, "Nader's Raiders" condemned the perpetuation of the old system of delays before the federal government could intervene, noting: "Do we have twenty, thirty, or forty years before major areas will be literally uninhabitable? Can we wait for the mechanisms of the Air Quality Act to be implemented? If we can wait, will the Act work when it is implemented?" With some minor nods of respect to Muskie for past accomplishments, the radical students nevertheless branded his Air Quality Act as lifted straight from the pages of a typical dilatory industrial statement on air pollution abatement published by the Manufacturing Chemists' Association in 1952—charges that stung a politician with presidential ambitions and a career built on pollution control.[29]

By 1969 and 1970, the overall radicalization of public environmental attitudes was having an influence on other federal legislators and officials besides Muskie. In the Senate, Gaylord Nelson of Wisconsin and Henry Jackson of Washington State threatened to take the mantle of environmental leadership away from Muskie after their successful promotion of the National Environmental Policy Act of 1969, which for the first time formally committed the federal government to the pursuit of environmental soundness and sustainability and required the federal government to consider environmental impacts before taking any actions. In August, 1970, Senator Nelson even followed California's example by proposing a bill to ban the internal combustion engine throughout America by 1975. Meanwhile, in the White House, the master political strategist Richard M. Nixon, who had shown little interest in the environment previously but realized that it had become an immensely popular issue which Republicans could not afford to abandon to the Democrats, sent a message to Congress in early 1970 calling for legislation to reduce automotive emissions by 90 percent by 1980 and to increase federal research on alternative vehicles. These leaders and others were increasingly responsive to growing public demands for direct federal intervention on air pollution control.

In the end, Senator Muskie abandoned his earlier commitment to "creative federalism" and proposed the Clean Air Act Amendments of 1970, which finally gave federal officials ultimate authority, responsibility, and direct enforcement power for air pollution problems throughout the nation and provided for the setting of federal emissions and air quality standards that the states would have to meet, although the federal government would still leave normal day-to-day operations in the hands of lower authorities. Further major air pollution incidents in Chicago in 1969 and on the East Coast late in 1970 helped increase support for the measure, which was signed into law on December 31, 1970. The greatly expanded, more powerful federal program would be run by the new Environmental Protection Agency, which brought together most federal programs related to pollution and many of those concerning other environmental issues that had previously been divided in bits and pieces among various federal departments and agencies. Recognizing the inadequacy of this division of authority regarding inextricably interconnected environmental issues, as some states and cities had already done, the Nixon Administration, under Reorganization Plan Number 3 in December, 1970, created the larger, more powerful federal umbrella agency covering most environmental matters.[30]

While various legislative changes formally expanded the federal role in air pollution control through the 1950s and 1960s, evolution occurred at a slower rate within the executive branch, where a conservative institutional culture that had long resisted any expanded role gradually had more responsibility and authority dropped in its lap. As Porter of the Missouri air conservation authority had observed, the federal officials did not always speak with one voice, and it was difficult even to get a straight answer out of them, let alone action. As already discussed, the federal control program had been divided from early on between two camps, one favoring the status quo of harmonious relations with states and industries, and a limited research and advisory mission, the other favoring more activist federal involvement. The latter view started out small in the late 1940s and gradually expanded through the end of the 1960s, but the former view remained solidly entrenched throughout the whole period, particularly among the scientists and engineers who wished to think only of science and technology and to ignore the inescapable social, political, legal, and economic aspects of the problem. The divide between activists and quietists persisted through the various reorganizations of the federal air pollution control program within the Department of Health, Education, and Welfare.[31]

Among the most important activists were Vernon G. MacKenzie, the federal sanitary engineer who began pushing for greater research into air pollution in the late 1940s and who later became head of the federal Division of Air Pollution from 1960 until 1966, and Thomas F. Williams, whom MacKenzie brought in as chief information officer for the federal effort in 1957. MacKenzie was instrumental in getting

Senators Kuchel and Capehart to propose a federal involvement in air pollution in 1954 and went on to work closely with control advocates on subsequent legislation during the 1960s, helping to draft both the 1963 Clean Air Act and the 1965 amendments, while MacKenzie and Williams together worked to increase public awareness and concern about air pollution throughout the nation.[32]

Activists long remained in the minority within the federal control bureaucracy, however. For every activist like Williams and MacKenzie, there were probably at least two quietist scientists or technicians fearful of antagonizing state or local governments and industry, such as plant pathologist C. Stafford Brandt. Much of the quietism remained centered at the federal Technical Assistance Branch laboratory in Cincinnati, where so many of the technicians were, and where they expected to go on working harmoniously with polluting industries on research projects. It was partly because of such quietist attitudes on the part of so many PHS scientists and technicians that in 1966, after too many years of PHS staffers deciding not to act but only to conduct further research, the federal water pollution control program was wrested from the PHS and HEW altogether and was placed in the Department of Interior, where it proved much more active. During the late 1960s, the new National Center for Air Pollution Control (later the National Air Pollution Control Administration) similarly would come under ever-increasing criticism for its sluggish quietism and persistent research orientation. As one of the all-time great activist control officials, S. Smith Griswold, observed, "They are studying the hell out of everything but they haven't removed one ounce of crud from the air." The replacement of the NCAPC/NAPCA and the Air Quality Act of 1967 with the new Environmental Protection Agency and the Clean Air Act Amendments of 1970 brought to the war on air pollution a larger, better-funded, more vigorous agency with greater power and authority, though the EPA must have inherited a good many hesitant technicians and bureaucrats from the earlier agencies.[33]

Although both the legislative and executive branches of the federal government long resisted direct federal intervention and enforcement on air pollution and only finally moved in that direction after the explosion of public anxiety about the environment during the late 1960s, another major presence in American society and politics resisted that transition even more stubbornly—business and industry. Industry's basic set of arguments and tactics remained mostly unchanged through the whole period before 1970: to minimize or deny air pollution problems; to argue that not enough was known about the matter to take any action, offering the prospect of virtually limitless delay through further research; to insist that air pollution was totally different in every locality throughout the land and so had to be studied and understood anew in every situation, and that it was a strictly local problem in which the federal government had almost no role; to claim that industry knew best how to handle its own pollution problems and should be left to do this itself on a voluntary basis; and to warn that requiring atmospheric cleanup at any rate faster than what the polluting industries selected for themselves would lead to plant closings, job losses, higher prices for consumers, and other frightful economic hardships. Industries also liked to point out, correctly, that they were not the only polluters, but that everyone contributed to the problem, and they often proudly announced the very considerable sums of money they had spent on pollution control, though usually neglecting to offer these statistics as percentages of gross capital expenditures, profits, or the sums actually necessary to cure the problem. They would generally accept further studies of air pollution, but were often ready to twist arms in local, state, or federal legislatures to prevent more aggressive action if necessary.

Already in 1952, the Manufacturing Chemists' Association (MCA) had offered some of these key ideas in its booklet *A Rational Approach to Air Pollution Legislation:* "Comparatively little is known about the effects of air pollution . . . but research . . . is gradually lifting the veil of ignorance. In the meantime it is illogical to impose arbitrary or uniform restrictions." Implicit in this statement is the assumption that those using the atmosphere as a sewer had the right to continue doing so, as much as they wished, until someone proved incontrovertibly that they were causing harm. Also, the chemical industry booklet pointed out that the atmosphere was a convenient sewer and should be used to maximum capacity short of causing undeniable harm to human health: "It is logical . . . to regard the atmosphere as one of the many natural resources which are being and should be used to technical and economic advantage." Later, in the 1960s, a favored industry catchphrase, "Dilution is the solution to pollution," sought to emphasize that any amount of pollution would be acceptable if there were adequate ventilation to disperse it and that it might be better and cheaper to seek improved means of diluting air pollution, such as ever-taller smokestacks, rather than limiting its production—a policy oblivious to the warnings of ecologists like Barry Commoner that, in fact, there ultimately is no "away" in which to throw harmful pollutants.[34]

The chemical industry in 1952 also minimized the seriousness of atmospheric contamination, noting how the Donora disaster had to be viewed in its "proper perspective," for "in spite of highly concentrated air-polluting operations in many localities there has never been a similar occurrence elsewhere in this country"—although not long thereafter, the city of London, England suffered its Great Smog of 1952, taking an estimated four thousand lives prematurely. Of course, as the title of the booklet suggests, any position differing from that of the Manufacturing Chemists' Association and other industries was by definition irrational, emotional, and groundless, even "hysterical"—a typical way for self-proclaimed experts to dismiss public criticism. As a later speech prepared by the MCA for delivery by plant managers throughout the nation read, "*Sensation* may sell books on this subject. But only *information* can produce the public awareness needed to gain support for effective and realistic air pollution control programs." The experts in industry naturally claimed to have a monopoly on "information" that gave them alone the power to define what was "effective" and "realistic."[35]

Industry representatives would hammer away at the same themes in countless subsequent appearances before federal and state legislatures, where they sought to prevent or delay the stiffening of air pollution control laws. During the hearings on Senators Kuchel and Capehart's proposed amendments to the 1954 federal housing bill relating to air pollution, industrial lobbies did not take the issue seriously enough to send actual representatives, but the National Association of Manufacturers, the National Coal Association, and the National Aviation Trade Association did submit statements for the record. In a typical argument, the National Coal Association commended the goals of the proposal but found much wanting in the approach, noting: "The proper solution to a problem within an industry can best be found by people within the industry itself who know every point of the operations" and who could solve the problem more cheaply and efficiently than "thru [sic] the avenue of government bureaucracy." The following year, industry representatives testified in favor of the proposed air pollution control act of 1955, once they were sure it was nothing but a tame, toothless act to promote federal research and not enforcement; nevertheless, they all reaffirmed their support for industry self-regulation and control at the lowest possible level of government, where local industries' power was greatest, and they sought to keep the federal commitment from growing any larger. Further hearings on proposed extensions of the federal control program invariably brought forth the same arguments through the late 1950s and 1960s.[36]

Considering the potential political power of American industry united, it is perhaps surprising that industry was not more successful in preventing the federalization of air (and water) pollution control entirely. Some observers have noted that the whole issue caught industry slightly off guard and threw it on the defensive up through the early 1970s. Industrial lobbies, of course, could not actively support air pollution but could only accept its undesirability while trying to minimize its significance, a weak position at a time when control advocates were growing steadily more radical. Also, though resisting costs, industry, like governments and ordinary citizens, initially thought the problem would prove far easier to solve than it ultimately did.[37]

By the later 1960s, though, as industrialists felt their control over the process of formulating air pollution control policy steadily slipping away, industry adopted a more combative stance in resisting further stiffening of control policies and federal encroachment. This happened earliest with one of the industries that felt most threatened—the coal industry. In its issue of September 16, 1966, *Time* magazine ran a full-page advertisement paid for by an organization calling itself "Coal for a Better America" that read:

> If you want an instant end to air pollution . . . stop driving your car—then turn off your oil burner, brick up your fireplace, bundle your leaves, box your trash, refuse delivery of anything by truck, boycott airplanes, busses [sic] and cabs. Don't use anything which requires oil, gas, coal, or atomic energy in its manufacture—such as electricity, steel, cement, clothes, food, newspapers,

babies' rattles, and on and on and on and on and on and on and on . . . OR—let's face the fact that any combustion generates pollutants . . . and that any "instant end" to air pollution brakes our civilization to a halt. Coal is a minor cause of this contamination, but the coal industry is working hard to clean the air. After all, we're breathing it, too.

A furious Norman Cousins, editor of *Saturday Review* magazine and chairman of the New York City Mayor's Task Force on Air Pollution, wrote to Senator Muskie to observe, "I have seldom seen a more intellectually dishonest public statement. One of its unfortunate implications, in effect, is that it puts your Senate Subcommittee in the position of dismantling the American way of life." Muskie responded that this was something they had seen a lot of, "reminiscent of the automobile industry campaign against the Clean Air Act Amendments of 1965"; he hoped that American industry would start doing more voluntary research and cleanup and less diversionary public relations activity.[38]

Whatever Muskie may have wished, American industry instead stepped up its public relations campaign against further regulations on air pollution. By May of 1971, even before the onset of the really severe economic trouble of the 1970s, John J. Coffey, a spokesman for the United States Chamber of Commerce, warned that pollution control laws were a serious economic threat that could wipe out entire industries faced with air or water pollution problems and insufficient technology or funds with which to confront them. Coffey specifically offered as an example the automobile industry, which had suddenly come to seem perhaps even more endangered than the coal industry, given the various state and federal proposals to ban the internal combustion engine and to require only low-emission vehicles. For his part, Henry Ford II declared in a speech that the federal government had set "an impossible task" in proposing under the Clean Air Act Amendments of 1970 that urban air must be cleaned up by 1975 even if it meant restrictions on the use of automobiles. Ford informed the American people that, like it or not, the nation was now built around the automobile, such that "to call for a massive switch to public transportation is to call for a massive rebuilding of metropolitan areas that would take generations to complete if it is possible at all." He further argued that the federal air quality standards were "based on very questionable evidence" that vehicular fumes were harmful, and that the federal government was very unkindly putting the auto industry to great expense "to achieve goals for which there is little or no demonstrated need." Predicting all manner of dire consequences if the federal government went ahead with its control plans, Ford also helpfully suggested that atmospheric decontamination came far below problems of poverty, education, housing, and crime—issues that did not impinge directly on the profits of the auto makers—on any reasonable list of domestic priorities. To buttress the claim of excessive and unjustified regulation, the auto industry also funded research finding that motor vehicles were "responsible for less than 10 per cent of the total U.S. air pollution problem," that the federal carbon monoxide standards were far too strict, and that they were unattainable without a "major economic convulsion." Although the term "environmental extremist" would not become commonplace until a decade later, the anti-environmental rhetoric of American industry was already moving in that direction.[39]

Another trick the automobile industry learned to use in its public relations campaign against further regulation or alternative vehicles was to declare victory in the war on smog. The most notable example of this was a 1969 speech given by Charles M. Heinen, Chrysler's chief engineer for emissions control, entitled, "We've Done the Job—What's Next?" The first line declared, "The main battle against automotive air pollution has been won"; the rest made bold claims for the industry's success in eliminating pollutants—claims that sounded hollow when it was revealed how few cars measured up to the standards of the specially tuned and monitored prototypes that the auto makers were allowed to submit to federal air pollution control officials for model certification. In 1971, Lee Iacocca, then president of the Ford Motor Company, similarly declared: "By the middle of this decade, though we aren't sure yet how to do it, we expect that every new car in the United States will be virtually pollution-free. Despite the litany from our critics on safety and emissions, most of the job is already done." Apparently he was off by at least a few decades, though American cars by the 1990s were certainly much less polluting than they had once been.[40]

Despite industry's best efforts at public relations, or in some cases, outright misinformation, during the late 1960s and early 1970s the American public kept moving toward differing, more radical conclusions regarding environmental problems and how to deal with them. As statements from individuals like José Martinez and Kenneth Worley clearly demonstrate, public concern about environmental problems and people's distress at the sluggish existing efforts to abate them had helped lead to the spread of ecological attitudes among the American public. Related to this partial transformation in attitudes—partly a cause and partly a reflection of it—was a striking change in the statements and rhetoric of officials and public figures as ecological concepts crept into public discourse during the 1960s, thanks in no small measure to many Americans' first exposure to ecological concepts through Rachel Carson's best-selling exposé of dangers and abuses in the handling of pesticides, *Silent Spring*, published in 1962. Air pollution, one of the most salient environmental problems in the public mind, naturally became one of the chief topics of ecological thought and discussion as many, though certainly not all, Americans began considering the need for controls and limitations on human behavior, rather than just the introduction of new technology to sweep problems aside.

Ecological awareness first surfaced among life scientists such as Rachel Carson, but by the mid-1960s, the new ecological rhetoric was appearing frequently in the mainstream press, as various journals began covering stories chiefly on *human* ecology, or humans' interrelationship with the rest of the biosphere, and the dangers people potentially faced from ignoring ecological considerations. More strikingly, such ideas also crept into the realm of government and policy. In his Special Message to Congress on Natural Beauty of February 8, 1965, President Lyndon Johnson called for specific proposals to end air and water pollution and spoke more generally of "a new conservation" concerned "not with nature alone, but with the total relationship between man and the world around him"—decidedly ecological rhetoric, and an early effort to give a name to the new set of ideas and attitudes that would later be known as environmentalism. Johnson quoted Rachel Carson when signing the Clean Air Act amendments of 1965. Later, in July, 1965, Richard N. Goodwin, a presidential aide and speechwriter for both the Kennedy and Johnson administrations, identified "the decontamination of the air Americans breathe and the water they use" as one of the four principal objectives of Johnson's Great Society; he also noted the problems of abundance and "progress" and warned: "The creations of our genius—science and inventive skills—are poisoning our rivers and our air, devastating the nature that once fed the spirit." While the liberal Goodwin may have been more sensitive to environmental issues than many other members of the Johnson Administration or the American public at that time, he was nevertheless a representative of the federal government who was using alarmist ecological rhetoric and suggesting that technology might be the problem rather than the solution. This was new to the American political milieu, but it was a sign of things to come.[41]

By December, 1966, following the Thanksgiving Weekend episode, the outlook on air pollution control and the American environment was even darker. In his keynote address before the Third National Conference on Air Pollution, HEW Secretary Gardner was somber: "The truth is . . . that we are actually losing ground in the fight against air pollution—the smog continues to grow more dense even as we talk about it. . . . There is not a major metropolitan area in the United States without an air pollution problem today." Offering a glimmer of hope if major action were taken immediately and warning that "certainly we cannot afford further delay," Gardner excoriated existing state, local, regional, and industrial efforts at controlling air pollution as grossly inadequate. Without immediate, major action, Gardner foresaw a science-fiction nightmare: "Our choices are narrow. . . . We can remain indoors and live like moles for an unspecified number of days each year. We can issue gas masks to a large segment of the population. We can live in domed cities. Or we can take action to stop fouling the air we breathe."[42]

In an editorial reflecting on the conference, the *New York Times* editors were equally somber and alarmed about the prospects facing the twentieth-century metropolitan citizen: "The air he breathes is slowly poisoning him, defacing his buildings and destroying his environment." Reflecting on Gardner's stark vision of the future, the *Times* editors concluded: "The passage of time only narrows the choice. Each day brings more people and more vehicles, but no more air."[43]

This new, worried, broadly ecologically inspired rhetoric—this sense of being besieged by more people, more cars, more smokestacks, and more pollution, reflecting a new suspicion of the technological, economic, and demographic growth that

Americans had traditionally worshipped unreflectively—would only increase in force and frequency among public figures, journalists, and ordinary citizens and students through 1970, as the environmental movement arose to challenge the status quo, to stress people's inescapable connection with the natural world, and to call for new limits on human activity for the sake of long-term survival. Many politicians would empathize with Nixon cabinet secretary John C. Whitaker's later recollection: "When President Nixon and his staff walked into the White House on January 20, 1969, we were totally unprepared for the tidal wave in favor of cleaning up the environment that was about to engulf us." Indeed, by 1970, even President Richard M. Nixon, who had ignored the environment until the very end of the presidential campaign of 1968, felt compelled to proclaim in his February, 1970, State of the Union Address: "The decade of the 1970s absolutely must be the years when America pays its debt to the past by reclaiming the purity of its air, its waters and our living environment. It is literally now or never." On April 22, 1970, the first Earth Day commemoration represented probably the largest nationwide demonstration of the whole decade of the 1960s—its principal organizer later claimed it was "the largest organized demonstration in human history"—with as many as fifteen hundred colleges, ten thousand schools, and an estimated twenty million people around the nation participating in various activities, marches, and demonstrations, including the closing of New York City's Fifth Avenue to automobile traffic so that a hundred thousand people could hear speeches on environmental topics. Beneath the festive mood was a surprising earnestness and sense of urgency about the state of the environment: as New York City's Mayor John V. Lindsay informed the crowd gathered in Manhattan, "Beyond words like ecology, environment and pollution there is a simple question: do we want to live or die?" Similar rhetoric echoed throughout the nation. Even if most Americans' understanding of ecology, environmental issues, and the interrelationships between them remained limited, and even if most citizens never entirely perceived or questioned the environmental impacts of their overall consumer lifestyle and society, nevertheless the spark of a new idea challenging established assumptions had emerged in the American mind.[44]

By the early 1970s, with a new, sterner, nationwide policy on air pollution control, a federal government committed to intervening in and solving this and other environmental problems, and a blossoming, aggressive national environmental movement picking up the rhetoric of ecology, it appeared that the old cycle of environmental policy for air pollution and other issues—the cycle of temporary surges of local public concern being stifled by government, industrial, and public inertia, and the issues then being forgotten all over again—had been broken in a number of ways. Finally, air pollution had been acknowledged to be no mere local matter but a nationwide affliction deserving the attention of the highest level of government, which would not cater to individual states' desires to ignore the problem. Thanks largely to the pioneering efforts of Los Angeles County and the State of California, as later built upon by the federal government, a significant body of scientific research and knowledge had been accumulated such that the matter could no longer so easily be eternally passed over as needing more research before action could be taken. The shift in the popular mood toward more anti-corporate sentiments during the late 1960s and early 1970s gave promise that the old notion of leaving the problem to the experts in industry to fix voluntarily would henceforth be banished; from that time on, the industries would have a powerful and persistent governmental and public presence policing their pollution and helping to keep them on the straight and narrow. Above all, a new, potentially powerful social and political movement enjoying wide popular support was coalescing to ensure that air pollution and other environmental issues would always remain on the policy agenda and that the public would no longer forget them. The lessons and attitudes of the newfound science of human ecology offered hope that humans might discover a new, more harmonious and sustainable way of living and of interacting with the natural systems that made their lives possible—and with one another. The 1970s, it was widely assumed, would be the "environmental decade."

Yet "ecotopia"—the dream some contemporary environmental activists had of an ecological paradise harmoniously balancing and blending the human and natural realms—was not so easily called into being. If the traditional air pollution cycle had been shaken out of its customary course by the environmental upheaval around 1970, its cyclical nature nevertheless remained, and some of the forces traditionally affecting that cycle remained powerful. One of the most obvious of these was the economy, an inescapable influence with cycles of its own. During the 1970s, a raft of economic woes befell the United States, including severe inflation, recession, the major revival of foreign competition, and most of all, a decade-long energy crisis that revealed dangerous latent weaknesses and inefficiencies in the American economy.[45]

As in earlier decades, such serious economic troubles tested and exposed the limits of Americans' commitment to environmental goals. From the late 1970s through the 1980s, the environmental movement lost much of its momentum and the almost unquestioned aura of righteousness it had enjoyed at the time of the first Earth Day, as critics sharply attacked environmentalists for perceived shortcomings and excesses. With the movement thrown on the defensive, some hostile business leaders increasingly branded environmentalism a dangerous, radical, even communistic threat to economic growth and jobs, and many Americans came to share this belief. Many frightened workers in troubled industries felt threatened along with their employers, and they sometimes joined their employers in blaming environmental regulations for plant closings, as with the Anaconda Copper Company workers in Montana who protested 1980 smelter shutdowns with signs reading, "Our Babies Can't Eat Clean Air." Certain writers characterized environmentalists as affluent white elitists seeking to preserve special privileges for themselves at the expense of poor and minority communities; others went further, branding environmentalists irrational, fanatically opposed to science and progress, and hostile to modernity itself. Some religious fundamentalists in America saw environmentalism as sacrilegious, paganistic nature worship, while some conservatives attacked environmentalism as one of the prime examples of liberal absurdity. As one prominent arch-conservative fundamentalist, James G. Watt, former secretary of interior, observed in 1990, "If the troubles from environmentalists cannot be solved in the jury box or at the ballot box, perhaps the cartridge box should be used." Such rhetoric was not sheer fancy, either, as various environmental activists were physically attacked or threatened during the 1980s and 1990s. During the early 1980s, conservative intellectual and political pundit David Gergen introduced the phrase *environmental extremists,* which became a popular label.[46]

By the 1990s, faithful listeners could regularly hear popular conservative radio talk-show host Rush Limbaugh deliver tirades against "environmental wackos." Jokes of the late 1970s and 1980s about the snail darter, a small, endangered fish that almost stopped a major dam-building project in Tennessee in the mid-1970s—"Save the economy, eat a snail darter"—gave way to similar 1990s jokes about the northern spotted owl, which impeded wholesale clearcutting of old-growth timber in the Pacific Northwest—"Shoot an owl, save a job." The villain in the popular 1980s comic movie *Ghostbusters* was a meddling EPA official. In 1998, members of a community group in Southern California seeking to stop a major luxury-home development that would ruin a biologically diverse and recreationally rich local mountain area took offense at being labeled "environmentalists" in a sympathetic piece in the *Los Angeles Times;* rather, claimed the group's president, they were concerned citizens seeking to maintain "quality of life"—precisely what environmental historian Samuel P. Hays has identified as the essence of environmentalism. To say the least, during the three decades after 1970, environmentalists became safe, easy targets for criticism, while for many, *environmentalist* became a dirty word, meaning a professional busybody preoccupied with abstract "nature" and unconcerned with human problems.[47]

Along with environmentalism, federal environmental policies of the early 1970s also drew sharp criticism, as economists, policy analysts, business leaders, and conservative theorists pointed out various alleged flaws and misfirings. Reflecting the resurgent conservatism of the 1980s and 1990s, many conservative, pro-development lawmakers longed to slow down or reverse environmental reform along with other reforms of the 1960s and 1970s. In particular, they wished to end or reduce federal environmental mandates and restore authority over environmental policy to the states, which would then have greater flexibility to promote economic development by bending, bypassing, or ignoring environmental regulations. In these attitudes, such officials could draw on the examples of Ronald W. Reagan, the popular conservative, anti-environmental president during most of the 1980s, and of Representative Newt Gingrich, primary leader of the conservative, Republican-dominated 104th Congress. This assault on federal environmental policy was blunted by overall public resistance, but during the 1980s and 1990s, the idea steadily advanced that environmental goals must be balanced against the sorts of economic goals that traditionally had always weighed more heavily in policy decisions; through the end of the 1990s, the demand that states be freed from federal environmental mandates continued to resound. Although the American public maintained overall support for many environmental policies, such support was often lukewarm, moderate, and conditional, depending on the state of the local economy. In many areas, environmentalists remain on the defensive, having to explain and justify themselves to their many aggressive critics.[48]

Such developments since the 1970s have perhaps caused many American lawmakers and citizens to forget why environmentalism and environmental policy took shape as they did. Those who lambaste the federal government's seizure of state authority over environmental issues have apparently forgotten the often pathetic inactivity and incapacity of state governments on issues such as air pollution prior to 1970. Those who view environmentalists narrowly as professional conservationists or as a screaming lunatic fringe clearly fail to understand that the environmental movement was created—and is continually renewed—by a mass mobilization of ordinary Americans from many walks of life confronting a wide range of threats to human quality of life and to the natural world in their communities, their states, and their nation. Critics in industry often complain of the crisis-driven policy process and coercive command-and-control environmental regulations of the late 1960s and early 1970s; they now recommend a newer, more flexible style of control based on market forces and economic incentives, which many concerned business leaders and moderate environmentalists hope will bring more rapid progress at lower cost. Yet such arguments ignore the long historical record of the battle against air pollution, which suggests that there would have been little progress on atmospheric decontamination if an alarmed public and their elected officials had not wrested policy on control from industrialists accustomed to no controls and confronted them in earnest with powerful, coercive regulations that made market-based controls seem an attractive alternative. In general, when critics carp about the current state, and failings, of environmentalism and environmental policy, they often cavalierly take for granted the significant achievements of the environmental movement and fail to recognize how much worse conditions might have been without it.[49]

Whatever the relative strengths or faults and the wins or losses of postwar environmentalism in the United States, people in America and throughout the world face a variety of new environmental threats and challenges as we begin a new century and new millennium. The globalization of economic activity has accelerated the globalization of environmental problems and ecological damage, as market demand in one nation can promote deforestation or species extinction in another nation on the other side of the world. Local atmospheric problems such as smelter fumes or vehicular exhaust have been joined by global atmospheric concerns such as ozone depletion and global warming, both subject to ongoing—and sometimes politically or economically motivated—scientific debate, just as the earlier air pollution issues were. Although federal laws and national standards in the United States helped to limit the ability of American corporations to relocate domestically from states with stricter pollution controls to states with laxer regulations, global free trade now makes it easy for polluting industries to find relative freedom from environmental regulations, along with low-cost labor, in developing nations desperate for jobs and investment, while still selling their products to the affluent citizens of the developed world. In this way, developed nations effectively export their pollution to underdeveloped nations. While Americans in general have gained more progressive environmental sensibilities than we as a nation had in 1945, we still freely consume and waste more natural resources per capita than any other people in the world, putting stress on the natural systems supplying our wants. Meanwhile, improvements in nutrition, control of epidemic diseases, and reductions of infant mortality have brought skyrocketing growth in the global human population, which places further stress on natural systems, especially as residents of developing nations seek the sort of middle-class consumer lifestyle that Americans enjoy.[50]

Thus, citizens and environmentalists in America and throughout the world have our work cut out for us as we confront the new millennium. Whether we will be able to learn enough from the past, and build upon the successes of earlier environmental activists such as America's postwar air pollution control crusaders while avoiding their mistakes, remains to be seen.

# Notes

## Chapter 1. Introduction

1. "Pollution," in Lehrer, *Too Many Songs by Tom Lehrer,* pp. 112–15.
2. Philip Shabecoff, *A Fierce Green Fire: The American Environmental Movement,* pp. 116–20; Kirkpatrick Sale, *The Green Revolution: The American Environmental Movement, 1962–1992,* pp. 14–23; John C. Esposito and Larry J. Silverman, *Vanishing Air: The Ralph Nader Study Group Report on Air Pollution,* p. 299; J. Clarence Davies III and Barbara S. Davies, *The Politics of Pollution* 2e, pp. 86–87. Such headings as "Environment," "U.S. Environment," or "Environmentalism" do not appear in the *New York Times Index* until 1965 or later, and the term *environmentalist* probably did not gain currency until 1967 at the earliest. Some far-sighted conservationist leaders, such as Howard Zahniser of the Wilderness Society, were active on pollution control from the 1950s onward.
3. Alan P. Carlin and George E. Kocher, *Environmental Problems: Their Causes, Cures, and Evolution, Using Southern California Smog as an Example,* p. 3. *Webster's Ninth New Collegiate Dictionary,* published in 1983, supports Carlin and Kocher's impression, defining *environmentalism* as "advocacy of the preservation or improvement of the natural environment; *esp:* the movement to control pollution" (416).
4. Davies and Davies, *Politics of Pollution* 2e, pp. 81–84. Notable alarmist works from the period include Howard R. Lewis, *With Every Breath You Take: The Poisons of Air Pollution, How They Are Injuring Our Health, and What We Must Do about Them;* Donald E. Carr, *The Breath of Life;* Edward Edelson and Fred Warshofsky, *Poisons in the Air;* and Esposito and Silverman, *Vanishing Air.* Among the more dramatic examples of water pollution harming aquatic life were the shocking Mississippi River fish kills of the early 1960s, the Santa Barbara Channel oil spill of 1969, and the gradual death of Lake Erie. Because it more often stretched outside of urban areas and visibly harmed wildlife, water pollution had more obvious conceptual links to wilderness conservation than did air pollution, which was generally construed more directly as a public health concern.
5. Shabecoff, *Fierce Green Fire,* pp. 111–21; Sale, *Green Revolution,* pp. 11–28; Terrance Kehoe, "'You Alone Have the Answer': Lake Erie and Federal Water Pollution Control Policy, 1960–1972," *Journal of Policy History* 8, no. 4 (1996), pp. 440–69. For a brief overview of the evolution of federal pollution control legislation up to 1974, see Davies and Davies, *Politics of Pollution* 2e, pp. 26–57. The National Environmental Policy Act of 1969 (NEPA) included just two short paragraphs affirming that the federal government henceforth would first consider potential environmental impacts before taking proposed actions. The law stipulated no procedure, but its general statement of principle became the basis for required federal environmental impact statements (EISs). For examples of state enactments, see the monthly

summaries of state air pollution control legislation in the *Journal of the Air Pollution Control Association.*

6. For sharp leftist critiques of traditional wilderness conservation efforts as affluent white male elitism, as well as examples of the typical but historically inaccurate assumption that the early environmental movement was identical with the preexisting conservation movement, see generally Mark Dowie, *Losing Ground: American Environmentalism at the Close of the Twentieth Century;* Marcy Darnovsky, "Stories Less Told: Histories of US Environmentalism," *Socialist Review* 22, no. 4 (Oct.–Dec. 1992): 11–54. For more typical right-wing or economistic critiques of environmentalists as overprivileged elitists opposed to progress or as irrational fanatics, see generally William Tucker, *Progress and Privilege: America in the Age of Environmentalism;* Aaron Wildavsky, "Aesthetic Power or the Triumph of the Sensitive Minority over the Vulgar Masses: A Political Analysis of the New Economics," in *America's Changing Environment,* ed. Roger Revelle and Hans H. Landsberg, pp. 147–60; Ronald Bailey, *Eco-Scam: The False Prophets of Ecological Apocalypse;* Wallace Kaufman, *No Turning Back: Dismantling the Fantasies of Environmental Thinking;* Dixy Lee Ray and Lou Guzzo, *Environmental Overkill: Whatever Happened to Common Sense?*

7. The cavalier dismissing of wilderness and wildlife conservation as frivolous self-indulgence or undue attention to nonhuman interests indicates that some liberal and conservative critics of environmentalism still have not gotten the point about overall ecological relationships: that protected wildlife and habitat are not necessarily just cute, dispensable recreational or aesthetic resources but are integral components of interconnected living systems, upon which we humans ultimately depend for our own survival and the parts of which we remove at our own peril. Although there were abortive efforts early in the twentieth century to extend the conservation movement to include a wide range of urban environmental and social problems beyond wilderness preservation or resource conservation in rural areas, urban environmentalism and traditional conservation basically remained on separate tracks until the 1960s. For background on this, see generally Robert Gottlieb, *Forcing the Spring: The Transformation of the American Environmental Movement.*

8. Shabecoff, *Fierce Green Fire,* pp. 203–30; Sale, *Green Revolution,* pp. 48–56. In his dissertation on the early postwar federal efforts to combat air pollution from 1945 to 1960, Shawn Bernstein in particular tends to portray the issue as something cooked up by control advocates in the federal executive branch. See Bernstein, "The Rise of Air Pollution Control as a National Political Issue: A Study of Issue Development" (Ph.D. diss., Columbia University, 1982).

9. Regarding the difficulty of getting air pollution on the policy agenda as an issue and the powerful forces opposing this, see Matthew A. Crenson, *The Un-Politics of Air Pollution: A Study of Non-Decisionmaking in the Cities.* Regarding the cyclicality of American public opinion and policy regarding the environment and other issues, see generally Arthur M. Schlesinger, Jr., *The Cycles of American History,* especially pp. 23–48; W. Douglas Costain and James P. Lester, "The Evolution of Environmentalism," in *Environmental Politics and Policy: Theories and Evidence* 2e, ed. Lester, pp. 16–18.

10. In their very helpful though not historically oriented study of pollution control policy, Clarence and Barbara Davies observe, in a typical statement, "The lack of state interest in air pollution [before the 1960s] was due in part to the problem being considered nothing more than an unaesthetic nuisance." See Davies and Davies, *Politics of Pollution* 2e, p. 158.

11. For general information on the politicization of science during the postwar period, see Joel Primack and Frank von Hippel, *Advice and Dissent: Scientists in the Political Arena*. A good, brief, and commendably even-handed discussion of the politics of science with regard to environmental issues from the 1960s through the 1980s may be found in Samuel P. Hays, *Beauty, Health, and Permanence: Environmental Politics in the United States, 1955–1985*, pp. 329–62. Further valuable case studies of the instrumental use of science to avoid regulation or to sidestep liability from environmental damage and injury to public health may be found in David P. Rosner and Gerald R. Markowitz, eds., *Dying for Work: Workers' Safety and Health in Twentieth-Century America*, pp. 121–59; Lynne Page Snyder, "'The Death-Dealing Smog over Donora, Pennsylvania': Industrial Air Pollution, Public Health, and Federal Policy, 1915–1963" (Ph.D. diss., University of Pennsylvania, 1994). Statements about scientists' connections with the corporate industrial power structure apply even more strongly and directly to their fellow technologists, the engineers. On this topic, see David F. Noble, *America by Design: Science, Technology, and the Rise of Corporate Capitalism*.

12. Although it is beyond the scope of this study to give full coverage to the economic aspects of air pollution, for more information on this crucial topic, see James E. Krier and Edmund Ursin, *Pollution and Policy: A Case Essay on California and Federal Experience with Motor Vehicle Air Pollution, 1940–1975*, pp. 24–37; Robert W. Crandall, *Controlling Industrial Pollution: The Economics and Politics of Clean Air;* Ronald G. Ridker, *Economic Costs of Air Pollution: Studies in Measurement;* V. Kerry Smith, *The Economic Consequences of Air Pollution*.

13. Hays, *Beauty, Health, and Permanence*, pp. 2–5, 22–24, 34–39, 541–43. Distinguished historian Samuel P. Hays seeks to answer the various anti-environmental critics of the 1970s and 1980s in their own terms by arguing that environmentalists have been predominantly pro-progress, pro-science, and pro-technology; and that environmentalism is a perfectly natural and economically rational outgrowth of the shift in the nation's economy and culture from a producer orientation to a consumer orientation from the 1920s through the 1960s, particularly after the Second World War, when America's affluence became so much more broad-based. In his compendious and insightful book, Hays offers much convincing evidence to support his argument. However, by focusing on the more moderate, mainstream, consumeristic aspects of environmentalism, and also by focusing mostly on events after 1970, Hays tends to overlook some more radical aspects of early environmentalism and the challenges it raised, at least in principle if not always in practice, to mainstream American economics, society, and consumer culture. Also, although Hays depicts environmentalism as a broad-ranging popular movement, his focus on environmentalism as the pursuit of mere "environmental amenities" to make life more enjoyable, as opposed to demands for environmental necessities such as air fit to breathe, still tends to leave environmentalists open to the typical and oft-repeated charges of self-indulgent elitism on the part of affluent middle-class whites denying opportunity to poorer whites and ethnic minorities. For a perceptive brief critique of the currently fashionable notion of consumerism as the solution (rather than the cause) for environmental ills, see generally Alan Thein Durning, *How Much Is Enough? The Consumer Society and the Future of the Earth*.

14. For good, brief, general background on expanding federal involvement in issues such as unemployment, poverty, education, civil rights, and senior citizens' issues through the mid-1960s, see James L. Sundquist, *Politics and Policy: The Eisenhower, Kennedy, and Johnson Years*. Regarding conceptual links between environmentalism and other social issues or movements

in which the federal government became involved from the 1960s onward, see Victor B. Scheffer, *The Shaping of Environmentalism in America*, pp. 16–28; Sale, *Green Revolution*, pp. 11–13; Shabecoff, *Fierce Green Fire*, p. 116.

15. For good examples of this recent tendency toward "eco-deconstruction," see the various essays in William Cronon, ed., *Uncommon Ground: Rethinking the Human Place in Nature*.

16. Lewis D. Eigen and Jonathan P. Siegel, eds., *The Macmillan Dictionary of Political Quotations*, p. 538.

## Chapter 2. Recurring Cycles

1. Peter Brimblecombe, *The Big Smoke: A History of Air Pollution in London since Medieval Times*, pp. 2–4.

2. Ian G. Simmons, *Environmental History: A Concise Introduction*, pp. 24–25, 28; Brimblecombe, *Big Smoke*, pp. 6, 11–14.

3. Leslie A. Chambers, "Classification and Extent of Air Pollution Problems," in *Air Pollution* 3e, ed. Arthur C. Stern, 1:7; Brimblecombe, *Big Smoke*, pp. 5–7, 10–13, 30.

4. Brimblecombe, *Big Smoke*, pp. 16–17, 26–27, 30, 34; William H. TeBrake, "Air Pollution and Fuel Crises in Preindustrial London, 1250–1650," *Technology and Culture* 16, no. 3 (July, 1975): 337–59.

5. Brimblecombe, *Big Smoke*, pp. 92–93, 8–9; Martin V. Melosi, *Garbage in the Cities: Refuse, Reform, and the Environment, 1880–1980*, pp. 12, 26–27, 60–61, 80–82.

6. Henry C. Perkins, *Air Pollution*, p. 5; Brimblecombe, *Big Smoke*, p. 9. Brimblecombe finds no primary reference to an early air pollution control law reputedly passed in England in 1273 and frequently mentioned in twentieth-century sources.

7. Brimblecombe, *Big Smoke*, pp. 16–17, 14–15, 22, 30, 34; Chambers, "Classification and Extent," p. 7.

8. Eric Ashby and Mary Anderson, *The Politics of Clean Air*, p. 1; Brimblecombe, *Big Smoke*, pp. 30–33, 39–52; John Evelyn, *Fumifugium*, reprinted in *The Smoak of London: Two Prophecies*, ed. James P. Lodge, Jr., pp. 14–20.

9. Brimblecombe, *Big Smoke*, pp. 63–69, 71–73, 113; Lewis Mumford, *The City in History*, p. 471.

10. Brimblecombe, *Big Smoke*, pp. 93, 99–100.

11. Ashby and Anderson, *Clean Air*, pp. 55–56; Brimblecombe, *Big Smoke*, pp. 59–60, 122–26, 149, 165, and generally pp. 108–31; Sir Hugh E. C. Beaver, "The Growth of Public Opinion," in *Problems and Control of Air Pollution*, ed. Frederick S. Mallette, p. 4.

12. Brimblecombe, *Big Smoke*, p. 90; Brian William Clapp, *An Environmental History of Britain since the Industrial Revolution*, pp. 14–15, 19–23; Carlos Flick, "The Movement for Smoke Abatement in 19th-Century Britain," *Technology and Culture* 21, no. 1 (Jan., 1980): 29.

13. Ashby and Anderson, *Clean Air*, pp. 1–15; Beaver, "Public Opinion," p. 3; Brimblecombe, *Big Smoke*, pp. 101–102; Clapp, *Environmental History*, pp. 32–33; Flick, "Smoke Abatement," pp. 30–34.

14. Ashby and Anderson, *Clean Air*, pp. 20–76.

15. Ibid., pp. 56–60, 81–91; Brimblecombe, *Big Smoke*, pp. 103–106, 115–16, 148–50; Beaver, "Public Opinion," pp. 3–4; Flick, "Smoke Abatement," pp. 33–39. For further perspective on local battles against air pollution and their relation to the national government in Victorian En-

gland, see Peter Brimblecombe and Catherine Bowler, "Air Pollution in York 1850–1900," in *The Silent Countdown: Essays in European Environmental History,* ed. Peter Brimblecombe and Christian Pfister, pp. 182–95.

16. Ashby and Anderson, *Clean Air,* pp. 12, 54–64, 92–95, 104–53; Brimblecombe, *Big Smoke,* pp. 103–105, 162–77; Beaver, "Public Opinion," pp. 4–6; Clapp, *Environmental History,* pp. 43–64.

17. Brimblecombe, *Big Smoke,* pp. 46, 64, 90; Flick, "Smoke Abatement," p. 30; Robert Dale Grinder, "The Anti-Smoke Crusades: Early Attempts to Reform the Urban Environment, 1893–1918" (Ph.D. diss., University of Missouri–Columbia, 1973), p. 26.

18. See E. C. Halliday, "A Historical Review of Atmospheric Pollution," and Albert Parker, "Air Pollution Legislation: Standards and Enforcement," pp. 39–48 and 376–77 respectively, both in World Health Organization, *Air Pollution.* On the Meuse Valley incident, see Harry Heimann, "Effects on Human Health," in World Health Organization, *Air Pollution,* pp. 163–64; John R. Goldsmith and Lars T. Friberg, "Effects of Air Pollution on Human Health," in *Air Pollution* 3e, ed. Arthur C. Stern, 2:470. For further information on air pollution and control efforts in Britain, Germany, and France from the 1800s through the post–World War II period, see Albert Weale, Timothy O'Riordan, and Louise Kramme, *Controlling Pollution in the Round: Change and Choice in Environmental Regulation in Britain and Germany;* Jeremy Bugler, *Polluting Britain: A Report;* Friends of the Earth, *How Green is Britain?;* Chris Rose, *The Dirty Man of Europe: The Great British Pollution Scandal;* Helmut Weidner, *Clean Air Policy in Great Britain: Problem-Shifting as Best Practicable Means;* Franz-Josef Brüggemeier, "The Ruhr Basin 1850–1980: A Case of Large-Scale Environmental Pollution," in *Silent Countdown,* eds. Brimblecombe and Pfister, pp. 210–27; E. Schramm, "Experts in the Smelter Smoke Debate," in *Silent Countdown,* eds. Brimblecombe and Pfister, pp. 196–209; Gerd Spelsberg, *Rauchplage: Hundert Jahre Saurer Regen;* Ilja Mieck, "Luftverunreinigung und Imissionsschutz in Frankreich und Preussen zur Zeit der frühen Industrialisierung," *Technikgeschichte* 48, no. 3 (1981): 239–51.

19. William Cronon, *Changes in the Land: Indians, Colonists, and the Ecology of New England,* pp. 48–49; Richard Lillard, *The Great Forest,* p. 85.

20. Martin V. Melosi, *Coping with Abundance: Energy and Environment in Industrial America,* pp. 18–24.

21. Melosi, *Coping with Abundance,* pp. 19–20.

22. Joseph M. Petulla, *American Environmental History* 2e, pp. 149–51; Melosi, *Coping with Abundance,* pp. 17–34; Grinder, "Anti-Smoke Crusades," pp. 12, 125–26.

23. Grinder, "Anti-Smoke Crusades," pp. 27–29; Petulla, *American Environmental History,* pp. 54, 151–52; Melosi, *Coping with Abundance,* pp. 22, 28–29, 30–34.

24. Grinder, "Anti-Smoke Crusades," pp. 29–30; Marvin Brienes, "The Fight against Smog in Los Angeles, 1943–1957" (Ph.D. diss., University of California–Davis, 1975), pp. 17–18; Clarence A. Mills, *Air Pollution and Community Health,* pp. 155–56.

25. William Cronon, *Nature's Metropolis,* pp. 9–12.

26. J. Clarence Davies III, *The Politics of Pollution,* p. 33; Charles W. Gruber, *Air Pollution Control in Cincinnati,* p. 24; City of Cincinnati et al., *The Bridge to Clean Air,* p. 1; Brienes, "Fight against Smog," pp. 15, 18; Grinder, "Anti-Smoke Crusades," pp. 27, 32, 45, 47, 93, 118–19, 125–127, 130–31, 139–41; Christine Meisner Rosen, "Businessmen against Pollution in Late Nineteenth Century Chicago," *Business History Review* 69, no. 3 (autumn, 1995): 351–

97; Harold L. Platt, "Invisible Gases: Smoke, Gender, and the Redefinition of Environmental Policy in Chicago, 1900–1920," *Planning Perspectives* 10, no. 1 (Jan., 1995): 67–97.

27. Grinder, "Anti-Smoke Crusades," pp. 7–11, 18–20.

28. Ibid., pp. 20–21, 23–24.

29. Ibid., pp. 50–59; Brienes, "Fight against Smog," p. 16; Petulla, *American Environmental History*, pp. 417–18; Gruber, *Air Pollution Control*, p. 24.

30. Grinder, "Anti-Smoke Crusades," pp. 50–52. For a condensed version of Grinder's dissertation, see Grinder, "The Battle for Clean Air: The Smoke Problem in Post–Civil War America," in *Pollution and Reform in American Cities, 1870–1930*, ed. Martin V. Melosi, pp. 83–103.

31. Grinder, "Anti-Smoke Crusades," pp. 90–95, 109.

32. Ibid., pp. 22, 33, 95–102; Platt, "Invisible Gases," generally. Regarding chiefly middle-class women's role in early environmental reform efforts, see also Maureen A. Flanagan, "The City Profitable, the City Livable: Environmental Policy, Gender, and Power in Chicago in the 1910s," *Journal of Urban History* 22, no. 2 (Jan., 1996): 163–90; Carolyn Merchant, "Women of the Progressive Conservation Movement, 1900–1916," *Environmental Review* 8, no. 1 (spring, 1984): 57–85; Raymond W. Smilor, "Toward an Environmental Perspective: The Anti-Noise Campaign, 1893–1932," in *Pollution and Reform*, ed. Melosi, pp. 135–51; Suellen M. Hoy, "'Municipal Housekeeping': The Role of Women in Improving Urban Sanitation Practices, 1880–1917," in *Pollution and Reform*, ed. Melosi, pp. 173–98; Melosi, *Garbage*, pp. 117–24. For more general information on the ideology of civic motherhood and women's participation in reform movements of the late nineteenth and early twentieth centuries, see Carl N. Degler, *At Odds: Women and the Family in America from the Revolution to the Present*, pp. 279–361; Kathleen D. McCarthy, "Parallel Power Structures: Women and the Voluntary Sphere," in *Lady Bountiful Revisited: Women, Philanthropy, and Power*, ed. Kathleen D. McCarthy, pp. 1–23; and Anne Firor Scott, "Women's Voluntary Associations: From Charity to Reform," in *Lady Bountiful Revisited*, ed. McCarthy, pp. 44–46 and generally.

33. Grinder, "Anti-Smoke Crusades," pp. 20, 38–39, 51–52, 66–67; Rosen, "Businessmen against Pollution," pp. 355–65.

34. Grinder, "Anti-Smoke Crusades," pp. 36–39, 81–82. Regarding early twentieth-century control technology, see Joel A. Tarr, "Railroad Smoke Control: The Regulation of a Mobile Pollution Source," in Tarr, *The Search for the Ultimate Sink: Urban Pollution in Historical Perspective*, p. 274. Regarding the engineering profession and its corporate employers, see Edwin T. Layton, Jr., *The Revolt of the Engineers: Social Responsibility and the American Engineering Profession*, pp. 1–19 and generally; Noble, *America by Design*.

35. Grinder, "Anti-Smoke Crusades," pp. 34–36, 68.

36. Ibid., pp. 58–59, 68–69, 80, 93, 105–108, 118–19, 122–34; Rosen, "Businessmen against Pollution," pp. 375–80.

37. Grinder, "Anti-Smoke Crusades," pp. 14, 27–28, 87–89, 103–105, 131–44; Randall B. Ripley, "Congress and Clean Air: The Issue of Enforcement, 1963," in *Congress and Urban Problems*, ed. Frederic N. Cleaveland, p. 228; Bernstein, "Air Pollution Control," pp. 64–68, 82–83.

38. Grinder, "Anti-Smoke Crusades," pp. 16–17, 34–36, 68; Halliday, "Historical Review," p. 16.

39. Grinder, "Anti-Smoke Crusades," p. 12; Melosi, *Coping with Abundance*, pp. 155–56.

40. Grinder, "Anti-Smoke Crusades," pp. 101, 130–31; Utah Legislative Council Air Pollution Advisory Committee, *Air Resources of Utah*, pp. 8, 15; New Jersey Air Pollution Commission, *Report to the New Jersey Legislature on Air Pollution in New Jersey and Recommendations for Its*

*Abatement,* pp. 35, 44; Melosi, *Coping with Abundance,* p. 156; Petulla, *American Environmental History,* p. 338.

41. Duane A. Smith, *Mining America: The Industry and the Environment, 1800–1980,* pp. 11–12, 45–46, 75–80, 94–100.

42. Smith, *Mining America,* pp. 96–98; James C. Cobb, *Industrialization and Southern Society, 1877–1984,* pp. 122–25; Grady Clay, "Copper Basin Cover-up," *Landscape Architect,* July, 1983, pp. 49–55 ff.; Rick Schuurmans, ed., *A History of Air Pollution Control in Chattanooga and Hamilton County,* pp. 3–4, 6; Gene B. Welsh and Thomas E. Kreichelt, *Clean Air for Chattanooga,* p. 41; James D. Williams and Norman G. Edmiston, *An Air Resource Management Plan for the Nashville Metropolitan Area,* pp. 10–11; Mills, *Air Pollution,* p. 124–27.

43. Bernstein, "Air Pollution Control," pp. 64–66; Joseph A. Holmes, Edward C. Franklin, and Ralph A. Gould, *Report of the Selby Smelter Commission, with Accompanying Papers.*

44. Keith A. Murray, "The Trail Smelter Case: International Air Pollution in the Columbia Valley," *BC Studies* 15 (autumn, 1972): 68–85; Robert Gilkey Dyck, "Evolution of Federal Air Pollution Control Policy" (Ph.D. diss., University of Pittsburgh, 1971), pp. 18–19.

45. David Rosner and Gerald Markowitz, "'A Gift of God'?: The Public Health Controversy over Leaded Gasoline during the 1920s," in *Dying for Work: Workers' Safety and Health in Twentieth-Century America,* ed. Rosner and Markowitz, pp. 121–24; Seth Cagin and Philip Dray, *Between Earth and Sky: How CFCs Changed Our World and Endangered the Ozone Layer,* pp. 34–45.

46. Rosner and Markowitz, "'A Gift of God'?," pp. 121, 124–36; Cagin and Dray, *Between Earth and Sky,* pp. 42–55; California State Department of Public Health, *Lead in the Environment and Its Effects on Humans,* pp. 2, 11–13.

47. For a very good brief overview of the politics of science with regard to environmental issues, see Hays, *Beauty, Health, and Permanence,* pp. 329–62.

48. Oscar Hugh Allison, "Raymond R. Tucker: The Smoke Elimination Years, 1934–1950" (Ph.D. diss., St. Louis University, 1978), pp. 1–15; Brienes, "Fight against Smog," pp. 17–26.

49. Allison, "Raymond R. Tucker," pp. 16–21.

50. Ibid., pp. 23–46, 55–61, 79–99, 117–28.

51. Tarr, "Changing Fuel Use Behavior and Energy Transitions: The Pittsburgh Smoke Control Movement, 1940–1950," in Tarr, *Search for the Ultimate Sink,* pp. 227–61. Regarding Pittsburgh's later struggles with air pollution, see Charles O. Jones, *Clean Air: The Policies and Politics of Pollution Control.*

52. Marshall I. Goldman, *The Spoils of Progress: Environmental Pollution in the Soviet Union,* pp. 122–50; Boris Komarov, *The Destruction of Nature in the Soviet Union,* pp. 20–31; Murray Feshbach and Alfred Friendly, Jr., *Ecocide in the USSR: Health and Nature under Siege,* p. 184; F. W. Carter and D. Turnock, eds., *Environmental Problems in Eastern Europe;* Edelson and Warshofsky, *Poisons in the Air,* p. 4; Howard R. Lewis, *With Every Breath You Take,* pp. 162–63; Ian Burton, Robert W. Kates, and Gilbert F. White, *The Environment as Hazard,* pp. 71–73.

## Chapter 3. Trouble in Paradise

1. Brienes, "Fight against Smog," pp. 1–3, 28; Krier and Ursin, *Pollution and Policy,* p. 44.

2. Brienes, "Fight against Smog," p. 4; Krier and Ursin, *Pollution and Policy,* p. 45.

3. Brienes, "Fight against Smog," p. 4; Krier and Ursin, *Pollution and Policy*, p. 45.

4. Brienes, "Fight against Smog," pp. 4–5; Krier and Ursin, *Pollution and Policy*, p. 46.

5. Among the many, slightly varying descriptions of the Southern Californian atmospheric inversion layer are Brienes, "Fight against Smog," pp. 7–8; Krier and Ursin, *Pollution and Policy*, pp. 42–43; Harold W. Kennedy, *The History, Legal and Administrative Aspects of Air Pollution Control in the County of Los Angeles: Report Submitted to the Board of Supervisors of the County of Los Angeles, May 9, 1954*, pp. 3–5; Richard Wagner, *Environment and Man*, pp. 179–81; and Louis J. Battan, *The Unclean Sky: A Meteorologist Looks at Air Pollution*, pp. 96–97.

6. Brienes, "Fight against Smog," pp. 33–35; Krier and Ursin, *Pollution and Policy*, pp. 52–53; Dyck, "Federal Air Pollution Control," p. 19.

7. Brienes, "Fight against Smog," pp. 32–33, 35–53; Robert Adam Doty, "Life Cycle Theories of Regulatory Agency Behavior: The Los Angeles Air Pollution Control District" (Ph.D. diss., University of California–Riverside, June, 1978), p. 45.

8. Brienes, "Fight against Smog," pp. 58–83. The misnomer *smog*, as a label for photochemical air pollution rather than the traditional mixtures of coal smoke and fog found in London or Pittsburgh, came into widespread use by 1944 and later became a general label for all modern urban air pollution.

9. Brienes, "Fight against Smog," pp. 88–100; Los Angeles County Office of Air Pollution Control, *First Annual Report of Office of Air Pollution Control, 1945–1946*, pp. 10–12.

10. Brienes, "Fight against Smog," pp. 100–108; Kennedy, *Aspects of Air Pollution Control*, pp. 6–7, 21; *New York Times*, Nov. 3, 1946, sec. 4, p. 12, col. 5. (The *New York Times* is hereafter cited as *NYT*, and page and column numbers for microfilmed editions of the newspaper are given in the format used by the *New York Times Index:* "p. 12:3" for "p. 12, col. 3.")

11. Brienes, "Fight against Smog," pp. 114–18; Kennedy, *Aspects of Air Pollution Control*, pp. 7–8.

12. Brienes, "Fight against Smog," pp. 17–21, 114–25; Doty, "Life Cycle Theories," pp. 48–52; Kennedy, *Aspects of Air Pollution Control*, pp. 5, 12–13; Krier and Ursin, *Pollution and Policy*, pp. 57–60; Raymond R. Tucker, *The Los Angeles Smog Report*.

13. Brienes, "Fight against Smog," pp. 125–27; Kennedy, *Aspects of Air Pollution Control*, pp. 7–8; Krier and Ursin, *Pollution and Policy*, pp. 60–63.

14. *NYT*, Oct. 26, 1947, sec. IV, p. 6:4; Brienes, "Fight against Smog," pp. 127–32; Doty, "Life Cycle Theories," pp. 55–59; Kennedy, *Aspects of Air Pollution Control*, pp. 10–14, 16, 20; Krier and Ursin, *Pollution and Policy*, pp. 60–61.

15. Brienes, "Fight against Smog," pp. 132–33, 169–72; Krier and Ursin, *Pollution and Policy*, pp. 68–71.

16. Brienes, "Fight against Smog," pp. 132–33, 169–72; Krier and Ursin, *Pollution and Policy*, pp. 68–71; Kennedy, *Aspects of Air Pollution Control*, pp. 14–15.

17. Brienes, "Fight against Smog," pp. 132–33, 169–80; Kennedy, *Aspects of Air Pollution Control*, pp. 72–74; Krier and Ursin, *Pollution and Policy*, pp. 68–71.

18. Brienes, "Fight against Smog," pp. 176, 215, 223–24; Krier and Ursin, *Pollution and Policy*, pp. 66–67; *NYT*, Nov. 28, 1948, p. 6:2.

19. Brienes, "Fight against Smog," pp. 191–92; Kennedy, *Aspects of Air Pollution Control*, pp. 46–48, 64; George H. Hagevik, *Decision-Making in Air Pollution Control: A Review of Theory and Practice, with Emphasis on Selected Los Angeles and New York City Management Experi-*

*ences,* pp. 82–83. Regarding various specific APCD regulations, see Robert G. Lunche et al., *Air Pollution Engineering in Los Angeles County,* p. 5.

20. Brienes, "Fight against Smog," pp. 169–76, 228–32, 256; Hagevik, *Decision-Making,* pp. 84–87; Kennedy, *Aspects of Air Pollution Control,* pp. 14–15, 51–52; *NYT,* Sept. 26, 1955, p. 12:1; *NYT,* Oct. 2, 1955, p. 66:1; *NYT,* Apr. 7, 1959, p. 77:1.

21. Brienes, "Fight against Smog," pp. 175–76, 237–42; Krier and Ursin, *Pollution and Policy,* p. 83; *NYT,* Feb. 21, 1951, p. 29:8.

22. Los Angeles County Office of Air Pollution Control, *First Annual Report,* pp. 10–11; Brienes, "Fight against Smog," pp. 89, 96, 105, 155–63; Kennedy, *Aspects of Air Pollution Control,* p. 48.

23. Brienes, "Fight against Smog," pp. 124, 161–64, 192; Krier and Ursin, *Pollution and Policy,* pp. 59, 73–75.

24. Krier and Ursin, *Pollution and Policy,* pp. 75–76.

25. Brienes, "Fight against Smog," pp. 180–83; *NYT,* Nov. 13, 1952, p. 16:5.

26. Brienes, "Fight against Smog," pp. 183–87; Krier and Ursin, *Pollution and Policy,* pp. 78–80; Arie J. Haagen-Smit, C. E. Bradley, and M. M. Fox, "Ozone Formation in Photochemical Oxidation of Organic Substances," *Industrial and Engineering Chemistry* 45, no. 9 (Sept., 1953): 2086–89.

27. Brienes, "Fight against Smog," pp. 187–204, 256; Krier and Ursin, *Pollution and Policy,* pp. 80–86; *NYT,* Aug. 1, 1954, p. 59:1; Lunche et al., *Air Pollution Engineering,* pp. 5–6.

28. Krier and Ursin, *Pollution and Policy,* p. 80; Andrew Rolle, *Los Angeles: From Pueblo to City of the Future* 2e, pp. 158–62.

29. Brienes, "Fight against Smog," pp. 192–95, 200–202; Kennedy, *Aspects of Air Pollution Control,* pp. 48–51; Krier and Ursin, *Pollution and Policy,* pp. 86–89; Kenneth F. Hahn, *Smog: A Factual Record of Correspondence between Kenneth Hahn and Automobile Companies* 6e.

30. Brienes, "Fight against Smog," pp. 224–42.

31. Krier and Ursin, *Pollution and Policy,* pp. 11–13.

32. Brienes, "Fight against Smog," pp. 250–55.

33. Ibid., pp. 252–53; Hagevik, *Decision-Making,* pp. 114–24.

34. Kennedy, *Aspects of Air Pollution Control,* p. 66; Los Angeles County Air Pollution Control District (hereafter APCD), *Annual Report, 1954–1955,* sections on "The New Look," "The Control Program," and "Enforcement."

35. Los Angeles County APCD, *Annual Report,* sections on "The New Look" and "The Control Program."

36. Ibid., section on "The Alert System"; Doty, "Life Cycle Theories," pp. 69–71; Los Angeles County Office of Air Pollution Control, *First Annual Report,* p. 10; Bernstein, "Air Pollution Control," pp. 32–33.

37. *NYT,* May 15, 1956, p. 33:6; *NYT,* Sept. 29, 1957, p. 48:1; *NYT,* Oct. 13, 1957, p. 75:3; Lunche et al., *Air Pollution Engineering,* pp. 5–6; Doty, "Life Cycle Theories," p. 73, 84–85; Scott H. Dewey, "Part Cause, Part Cure: The Changing Relationship between Air Pollution and Aviation in the United States, 1927–1973," in *The Meaning of Flight in the 20th Century: Proceedings of the National Aerospace Conference at Wright State University, Dayton, Ohio, October 1998,* p. 347.

38. Hagevik, *Decision-Making,* pp. 85–114; *NYT,* Sept. 27, 1959, p. 35:3; *NYT,* July 27, 1966, p. 49:4; Doty, "Life Cycle Theories," pp. 73–74.

39. *NYT,* July 29, 1966, p. 16:1.
40. *NYT,* Apr. 15, 1968, p. 88:1.
41. Ibid.; *NYT,* Jan. 8, 1968, p. 92:1.
42. Doty, "Life Cycle Theories," pp. 77–79; Esposito and Silverman, *Vanishing Air,* pp. 23–25; Benjamin Linsky, *Fifty Years Backward and Forward in Air Pollution Control; NYT,* May 30, 1962, p. 20:1.
43. *NYT,* July 29, 1966, p. 16:1.

## Chapter 4. Smog Town vs. the Motor City

1. Part of the information in chapter 4 was used in Scott H. Dewey, "'The Antitrust Case of the Century': Kenneth F. Hahn and the Fight against Smog," which appeared in the *Southern California Quarterly* 81, no. 3 (fall, 1999): 341–76. I gratefully acknowledge permission from the *Southern California Quarterly,* Historical Society of Southern California. See also "APCD Completes Job—All except Auto Exhaust," *Montebello News,* Nov. 7, 1957, reprinted in Los Angeles County APCD, *Crossing the Smog Barrier: A Factual Account of Southern California's Fight against Air Pollution;* and testimony of Robert Chass, quoted in California State Assembly Interim Committee on Public Health, Subcommittee on Air Pollution and Radiation Protection, *Air Pollution, Its Health Effects, and Its Control,* p. 28.
2. "APCD Completes Job."
3. Neil Goedhard, *Report No. 6: Basic Statistics of the Los Angeles Area,* pp. 9, 11, 17.
4. "APCD Completes Job"; Goedhard, *Basic Statistics,* pp. 23–24, 30.
5. *NYT,* Jan. 5, 1954, p. 78:4; Hahn, *Smog: A Factual Record;* Krier and Ursin, *Pollution and Policy,* pp. 86–87; Briones, "Fight against Smog," pp. 200–202.
6. Kennedy, *Aspects of Air Pollution Control,* pp. 49–50; Krier and Ursin, *Pollution and Policy,* pp. 98–99.
7. Krier and Ursin, *Pollution and Policy,* pp. 87–89; Ronald Schiller, "The Los Angeles Smog," *National Municipal Review* 44, no. 11 (Dec., 1955): 563.
8. Hahn, *Smog: A Factual Record;* California State Chamber of Commerce, *Proceedings of the Third Southern California Conference on Elimination of Air Pollution: Moving Closer to Clean Air, Featuring If's, And's and But's of Automobile Exhaust Control,* pp. 61–62.
9. Wallace Linville to Gordon P. Larson, Nov. 18, 1954, file 3.6.1 "Air Pollution Control District 1952–1954," box 266, in Papers of Kenneth F. Hahn, Henry E. Huntington Library, San Marino, Calif. (hereafter Hahn MSS, with file names also abbreviated); Schiller, "The Los Angeles Smog," p. 563; *NYT,* Apr. 15, 1958, p. 35:5; *NYT,* May 25, 1960, p. 25:4.
10. *NYT,* Jan. 4, 1959, p. 28:1; *NYT,* Oct. 15, 1961, sec. 10; Thomas Raymond Roberts, "Motor Vehicle Air Pollution Control in California: A Case Study in Political Unresponsiveness" (baccalaureate honors thesis, Harvard College, 1969), pp. 28–30; California State Assembly Interim Committee on Public Health, Subcommittee on Air Pollution, *Motor Vehicle Created Air Pollution: A Control Program for California,* pp. 28–30.
11. Krier and Ursin, *Pollution and Policy,* pp. 102–12.
12. Ibid., pp. 112–15.
13. Ibid., pp. 116–18; Roberts, "Motor Vehicle Air Pollution," pp. 27–29; *NYT,* Nov. 3, 1958, p. 73:3; *NYT,* Jan. 1, 1959, p. 20:5; *NYT,* June 14, 1959, p. 120:6; California State Assembly, *Motor Vehicle Created Air Pollution,* pp. 18–19.

14. *NYT,* Apr. 7, 1960, p. 22:1; *NYT,* Apr. 10, 1960, sec. VI, p. 6:1; Krier and Ursin, *Pollution and Policy,* pp. 137–40; Roberts, "Motor Vehicle Air Pollution," pp. 30–31; California State Assembly, *Motor Vehicle Created Air Pollution,* pp. 31–48.

15. *NYT,* Jan. 4, 1959, p. 28:1; *NYT,* Jan. 15, 1959, p. 29:7; *NYT,* Feb. 22, 1959, p. 40:5; *NYT,* Aug. 3, 1959, p. 1:4; *NYT,* Oct. 23, 1959, p. 23:2; *NYT,* Nov. 5, 1959, p. 30:6.

16. *NYT,* June 14, 1959, p. 120:6; *NYT,* Aug. 14, 1960, sec. III, p. 1:3.

17. *NYT,* May 1, 1960, sec. III, p. 15:7; *NYT,* Oct. 25, 1960, p. 56:8; *NYT,* Feb. 10, 1961, p. 42:5; *NYT,* May 11, 1961, p. 57:4; *NYT,* May 6, 1960, p. 44:3; *NYT,* June 30, 1962, p. 23:3; *NYT,* Dec. 19, 1962, p. 10:6; *NYT,* Oct. 22, 1957, p. 35:6; Roberts, "Motor Vehicle Air Pollution," pp. 31–32.

18. *NYT,* Nov. 5, 1959, p. 30:6; *NYT,* Sept. 20, 1961, p. 31:2; *NYT,* Oct. 15, 1961, sec. 10; *NYT,* Dec. 7, 1961, p. 1:3; *NYT,* Dec. 19, 1962, p. 8:6; *NYT,* Sept. 26, 1963, p. 52:3; *NYT,* Feb. 7, 1965, sec. IV, p. 7:2; Krier and Ursin, *Pollution and Policy,* pp. 146–51; Roberts, "Motor Vehicle Air Pollution," p. 31; Esposito and Silverman, *Vanishing Air,* p. 50.

19. *NYT,* Nov. 5, 1959, p. 30:6; *NYT,* Sept. 20, 1961, p. 31:2; *NYT,* Oct. 15, 1961, sec. 10; *NYT,* Dec. 7, 1961, p. 1:3; *NYT,* Dec. 19, 1962, p. 8:6; *NYT,* Sept. 26, 1963, p. 52:3; *NYT,* Feb. 7, 1965, sec. IV, p. 7:2; Krier and Ursin, *Pollution and Policy,* pp. 146–51; Roberts, "Motor Vehicle Air Pollution," p. 31; Esposito and Silverman, *Vanishing Air,* p. 50.

20. Dyck, "Federal Air Pollution Control," pp. 33–35; Krier and Ursin, *Pollution and Policy,* pp. 110–11, 169–70; *NYT,* July 15, 1959, p. 4:6.

21. Dyck, "Federal Air Pollution Control," pp. 33–35; Krier and Ursin, *Pollution and Policy,* pp. 110–11, 169–70; *NYT,* July 15, 1959, p. 4:6; *NYT,* Dec. 2, 1958, p. 39:6; *NYT,* Dec. 31, 1958, p. 6:3; *NYT,* June 25, 1959, p. 59:3; *NYT,* Feb. 24, 1960, p. 74:5; *NYT,* Feb. 25, 1960, p. 31:4; *NYT,* Aug. 8, 1961, p. 31:1; *NYT,* Dec. 7, 1961, p. 1:3; *NYT,* Dec. 18, 1961, p. 34:2.

22. *NYT,* Mar. 14, 1962, p. 21:2; *NYT,* Mar. 18, 1962, sec. IV, p. 6:4.

23. Krier and Ursin, *Pollution and Policy,* pp. 153–57; Roberts, "Motor Vehicle Air Pollution," p. 32.

24. *NYT,* June 18, 1964, p. 35:1; Krier and Ursin, *Pollution and Policy,* pp. 157–58; Roberts, "Motor Vehicle Air Pollution," pp. 32, 62; Drew Pearson, "US Freezes Grand Jury Quiz on Smog Controls," in unidentified newspaper, Part II, May 23, 1966, file 3.6.12 "APCD 1965–66," box 266-3, in Hahn MSS.

25. *NYT,* July 15, 1965, p. 35:8; *NYT,* Apr. 10, 1966, sec. 12 (Auto Expo), p. 27:1; Krier and Ursin, *Pollution and Policy,* pp. 158–59; Roberts, "Motor Vehicle Air Pollution," p. 32; Esposito and Silverman, *Vanishing Air,* p. 51.

26. Krier and Ursin, *Pollution and Policy,* pp. 160–63.

27. Ibid.; *NYT,* Jan. 1, 1966, p. 1:4; Roberts, "Motor Vehicle Air Pollution," pp. 90–92.

28. Krier and Ursin, *Pollution and Policy,* pp. 147, 154–55, 165–66; Roberts, "Motor Vehicle Air Pollution," pp. 33–34.

29. Krier and Ursin, *Pollution and Policy,* pp. 164–67; Roberts, "Motor Vehicle Air Pollution," pp. 36–37, 84–88; Doty, "Life Cycle Theories," pp. 81–82; *NYT,* Jan. 19, 1967, p. 12:3; *NYT,* Feb. 14, 1967, p. 29:1; *NYT,* Mar. 9, 1967, p. 79:3.

30. Krier and Ursin, *Pollution and Policy,* pp. 141–44, 167–68; Roberts, "Motor Vehicle Air Pollution," p. 70.

31. Krier and Ursin, *Pollution and Policy,* pp. 141–44, 167–68; Roberts, "Motor Vehicle Air Pollution," p. 70; Doty, "Life Cycle Theories," pp. 78–80; California State Motor Vehicle Pol-

lution Control Board, *Our Exhausted Air: California's Fight against Automobile Smog—The Story of the California Motor Vehicle Pollution Control Board.*

32. *NYT,* Aug. 11, 1967, p. 80:2; Krier and Ursin, *Pollution and Policy,* pp. 177–79; Roberts, "Motor Vehicle Air Pollution," pp. 37–38, 87–88.

33. Dyck, "Federal Air Pollution Control," pp. 33–34, 38–40, 51–52.

34. Ibid., pp. 53–58; Davies and Davies, *Politics of Pollution* 2e, pp. 45–47.

35. Davies and Davies, *Politics of Pollution* 2e, pp. 47–48; Dyck, "Federal Air Pollution Control," pp. 59–63; *NYT,* Jan. 5, 1965, p. 19:4.

36. Dyck, "Federal Air Pollution Control," pp. 65–68; *NYT,* Apr. 5, 1965, p. 1:2; *NYT,* Apr. 6, 1965, p. 1:3; *NYT,* Apr. 8, 1965, p. 34:6.

37. Davies, *Politics of Pollution,* pp. 48–49; Dyck, "Federal Air Pollution Control," pp. 65–70; *NYT,* Apr. 6, 1965, p. 1:3; *NYT,* June 22, 1965, p. 32:5; *NYT,* July 1, 1965, p. 1:7.

38. *NYT,* June 1, 1966, p. 59:5.

39. Krier and Ursin, *Pollution and Policy,* pp. 181–83; Roberts, "Motor Vehicle Air Pollution," p. 36–38, 56, 71–72; Doty, "Life Cycle Theories," pp. 82–83; *NYT,* Apr. 9, 1965, p. 38:6.

40. Krier and Ursin, *Pollution and Policy,* pp. 181–83; Roberts, "Motor Vehicle Air Pollution," p. 36–38, 56, 71–72; Doty, "Life Cycle Theories," pp. 82–83; *NYT,* Apr. 9, 1965, p. 38:6; Davies, *Politics of Pollution,* pp. 49–52; Hays, *Beauty, Health, and Permanence,* p. 298.

41. Esposito and Silverman, *Vanishing Air,* pp. 26–33; General Motors Corporation, *Progress of Power: A General Motors Report on Vehicular Power Systems,* presented at the General Motors Technical Center, Warren, Michigan, May 7–8, 1969.

42. Krier and Ursin, *Pollution and Policy,* pp. 184–89; Roberts, "Motor Vehicle Air Pollution," pp. 39–40, 68, 123 ff.

43. *NYT,* Jan. 11, 1969, p. 1:3; Krier and Ursin, *Pollution and Policy,* pp. 87–88. For fuller coverage of this story, see Dewey, "'Antitrust Case of the Century'"; Esposito and Silverman, *Vanishing Air,* pp. 41–47.

44. Esposito and Silverman, *Vanishing Air,* p. 42; Krier and Ursin, *Pollution and Policy,* pp. 88 and 356 n.64; "Auto Industry's Headaches Multiply: Mr. Hahn Keeps on Writing Letters," *New York Herald Tribune,* Apr. 10, 1966, newspaper clipping in file 3.6.12 "APCD 1965–66," box 266-3, Hahn MSS; Kenneth F. Hahn to Attorney General John N. Mitchell, Sept. 4, 1969, file 3.6.16 "APCD 1969-I," box 266-4, Hahn MSS.

45. Esposito and Silverman, *Vanishing Air,* pp. 42–44; Pearson, "US Freezes Grand Jury Quiz on Smog Controls"; Hahn to Mitchell, Sept. 4, 1969; *NYT,* Jan. 11, 1969, p. 1:3.

46. Hahn to Mitchell, Mar. 7, 1969, file 3.6.16 "APCD 1969-I," box 266-4, Hahn MSS; memorandum, John D. Maharg to Los Angeles County Board of Supervisors, Aug. 8, 1969, file 3.6.18 "APCD 1969," box 266–267, Hahn MSS; *NYT,* Sept. 6, 1969, p. 24:3.

47. Esposito and Silverman, *Vanishing Air,* pp. 44–47; *NYT,* Sept. 12, 1969, p. 1:2; *NYT,* Oct. 29, 1969, p. 28:1; "Smog at the Bar," *Newsweek,* Nov. 10, 1969, p. 67; Dewey, "'Antitrust Case of the Century,'" pp. 356–63, 366.

48. *NYT,* Nov. 18, 1969, p. 49:1; *NYT,* Mar. 17, 1970, p. 28:2; *NYT,* Nov. 27, 1973, p. 22:4; Dewey, "'Antitrust Case of the Century,'" pp. 364–66.

49. *NYT,* Nov. 23, 1966, p. 42:8; *NYT,* Apr. 21, 1968, p. 63:3; California Air Resources Board, Los Angeles County APCD, and Western Oil and Gas Association, *Gasoline Modification: Its Potential as an Air Pollution Control Measure in Los Angeles County;* "The Advocates: Should

the Sale of the Internal Combustion Engine Be Banned in California by 1975?" (transcript of *The Advocates* episode on Los Angeles public television station KCET, 7:00–8:00 P.M., Pacific Standard Time, Oct. 5, 1969); *NYT,* Sept. 8, 1965, p. 49:8; *NYT,* June 9, 1966, p. 49:6; *NYT,* June 10, 1966, p. 90:5.

50. Roberts, "Motor Vehicle Air Pollution," pp. 53–55.

51. Ibid., pp. 53–55, 72–77, 112, 123; Judith Louise Lamare, "Urban Mass Transportation Politics in the Los Angeles Area: A Case Study in Metropolitan Policy-Making" (Ph.D. diss., University of California–Los Angeles, 1973), pp. 70–75, 114–16, 138–46, 151–52, 202–208, 234.

52. Regarding allegations of auto makers underperforming or cheating on emissions compliance, see Esposito and Silverman, *Vanishing Air,* pp. 59–61; *NYT,* Oct. 6, 1971, p. 20:1; *NYT,* Dec. 2, 1971, p. 23:1; *NYT,* May 23, 1972, p. 1:7; *NYT,* Sept. 27, 1972, p. 83:1.

53. Jeffry Fawcett, *The Political Economy of Smog in Southern California,* p. 185. The *New York Times,* Apr. 10, 1966, sec. 12 (Auto Expo), p. 27:1, noted how even with the "proliferation of people, automobiles and industry, smog hasn't gotten worse." Regarding California policy on motor vehicle emissions during the 1980s and 1990s, see Wyn Grant, *Autos, Smog and Pollution Control: The Politics of Air Quality Management in California.*

# Chapter 5. Folklore

1. Brienes, "Fight against Smog," pp. 32–37, 39–40; Krier and Ursin, *Pollution and Policy,* pp. 52–54.

2. Brienes, "Fight against Smog," pp. 69–70.

3. Ibid., pp. 70–73.

4. Ibid., pp. 12, 23–24, 48–53, 79–81, 134–36.

5. Ibid., pp. 73–74, 76, 90–92, 94–100, 107, 173.

6. Ibid., pp. 102, 105–108, 132.

7. Ibid., pp. 116–17; Kennedy, *Aspects of Air Pollution Control,* p. 7. The threat to tourism was seen as real and serious. In 1949, area hotel owners were warned that smog would probably reduce tourist revenue by as much as $73 million a year. See Richard D. Cadle and Henry C. Wohlers, "Smog Lore," *Air Repair: Journal of the Air Pollution and Smoke Prevention Association of America* 1, no. 4 (May, 1952): 32.

8. Brienes, "Fight against Smog," pp. 118–25; Krier and Ursin, *Pollution and Policy,* pp. 57–61.

9. Brienes, "Fight against Smog," pp. 134–36, 175–76, 217, 222–30, 238–42; Roberts, "Motor Vehicle Air Pollution," p. 26; *Los Angeles Examiner,* Dec. 10, 1954, p. 1; Larson to Los Angeles County Supervisors, Mar. 31, 1954, file 3.6.1 "APCD, 1952–54," box 266, Hahn MSS; *NYT,* June 28, 1955, p. 2:4.

10. H. F. Johnstone, *Technical Aspects of the Los Angeles Smog Problem: A Report Prepared for the Los Angeles County Air Pollution Control District,* p. 28.

11. New Jersey Air Pollution Commission, *Report to the New Jersey Legislature,* pp. 38–39, 65–67.

12. Los Angeles County Office of Air Pollution Control, *First Annual Report,* p. 12; California State Assembly Interim Fact-Finding Committee on Water Pollution, *Interim Report on Air Pollution Research,* pp. 11–12.

13. Doty, "Life Cycle Theories," p. 64; Dr. Robert B. Hope, M.D., to County Supervisor Kenneth F. Hahn, Oct. 9, 1953; Dr. Boris Arnov, M.D., to Hahn, Oct. 8, 1953; Dr. Samuel Ayres,

Jr., M.D., to Hahn, Oct. 9, 1953; Dr. Paul H. Reed, M.D., to Hahn, Oct. 14, 1953; S. C. Glassman, M.D., to Hahn, Oct. 10, 1953; Joseph B. Stevens, M.D., to Hahn, Oct. 9, 1953; all in file 3.6.1 "APCD, 1952–54," box 266, Hahn MSS.

14. Larson to Kenneth F. Hahn, Sept. 24, 1953, file 3.6.1 "APCD 1952–54," box 266, Hahn MSS.

15. *NYT*, Nov. 24, 1953, p. 28:1; *NYT*, Apr. 20, 1955, p. 35:4; Lester Breslow, *Air Pollution: Effects Reported by California Residents*, p. 7.

16. California State Assembly, Subcommittee on Air Pollution and Radiation Protection, *Air Pollution, Its Health Effects, and Its Control*, pp. 25–27; Los Angeles County Medical Association, *Physicians' Environmental Health Survey: A Poll of Medical Opinion*, p. i.

17. *NYT*, Apr. 17, 1962, p. 28:1; *NYT*, July 26, 1962, p. 11:1; *NYT*, Jan. 19, 1963, p. 7:2; *NYT*, Mar. 23, 1963, p. 3:4. *NYT*, Oct. 16, 1964, p. 76:7; *NYT*, Mar. 3, 1966, p. 26:1; *NYT*, Mar. 5, 1966, p. 22:3; *NYT*, Mar. 18, 1966, p. 41:4.

18. Los Angeles County APCD, *Testimony of S. Smith Griswold, Dr. Leslie A. Chambers, and Hoyt R. Crabaugh before the Select Committee on Small Business of the House of Representatives at Los Angeles, California, May 18 and 19, 1956*, p. 8; *NYT*, Feb. 16, 1959, p. 31:5.

19. Brienes, "Fight against Smog," pp. 223–24.

20. *Arcadia Tribune and News*, Sept. 17, 1953, p. 24, file 3.6.3 "APCD 1952–54," box 266, Hahn MSS.

21. Brienes, "Fight against Smog," p. 241; *Los Angeles Examiner*, Oct. 21, 1954, p. 1; *Los Angeles Times*, Oct. 21, 1954, pp. 1–2; *Pasadena Independent*, Oct. 21, 1954, p. 1.

22. *Los Angeles Times*, Oct. 21, 1954, pp. 1–2; *Los Angeles Daily News*, Oct. 16, 1954, p. 1; *NYT*, July 8, 1954, p. 16:3; *NYT*, Oct. 16, 1954, p. 22:6.

23. Brienes, "Fight against Smog," pp. 239–40; Krier and Ursin, *Pollution and Policy*, pp. 113–15; *Los Angeles Daily News*, Oct. 16, 1954, p. 1; *Los Angeles Examiner*, Oct. 20, 1954, p. 1; *Pasadena Independent*, Oct. 21, 1954, p. 1; *NYT*, Mar. 1, 1953, p. 64:1; *NYT*, Oct. 16, 1954, p. 22:6; *NYT*, Oct. 17, 1954, p. 34:1; *NYT*, Oct. 21, 1954, p. 1:3; *NYT*, Oct. 23, 1954, p. 1:3; *NYT*, Oct. 24, 1954, p. 61:1; *NYT*, Oct. 31, 1954, p. 73:8; *NYT*, Feb. 22, 1955, p. 9:1.

24. Brienes, "Fight against Smog," pp. 240–41; *Los Angeles Daily News*, Oct. 15, 1954, p. 1; *Los Angeles Mirror*, Oct. 21, 1954, p. 1; *NYT*, Oct. 16, 1951, p. 22:6; *NYT*, Oct. 31, 1954, p. 73:8; *NYT*, Dec. 5, 1954, p. 78:3.

25. *NYT*, Sept. 1, 1955, p. 21:2; *NYT*, Sept. 7, 1955, p. 35:2; *NYT*, Sept. 14, 1955, p. 37:3; *NYT*, Sept. 15, 1955, p. 40:2; *NYT*, Sept. 16, 1955, p. 12:1; *NYT*, Sept. 21, 1955, p. 21:5; *NYT*, Sept. 23, 1955, p. 16:6; *NYT*, Sept. 25, 1955, p. 40:1; *NYT*, Sept. 27, 1955, p. 39:6; *NYT*, Oct. 2, 1955, p. 66:1; *NYT*, Nov. 20, 1955, p. 124:3; Schiller, "The Los Angeles Smog," p. 563.

26. Brienes, "Fight against Smog," pp. 253–55; Los Angeles County APCD, *Annual Report, 1954–1955*, section on "Public Information and Education."

27. Doty, "Life Cycle Theories," pp. 71–72.

28. Brienes, "Fight against Smog," pp. 253, 255–57; Harold Lahn Sims, "The Emergence of Air Pollution as a Political Issue in Southern California: 1940–1970" (Ph.D. diss., University of California–Riverside, 1973), pp. 105–34.

29. Roberts, "Motor Vehicle Air Pollution," pp. 48–50; Lucy Kavaler, "Today's Dirty Air Need Not Bring Gas Masks Tomorrow: How Los Angeles Women Are Fighting Smog—and Winning," *Family Circle*, September, 1968, pp. 55, 99–100.

30. Roberts, "Motor Vehicle Air Pollution," pp. 48–50.

31. Such suggestions had been offered from the very beginning of the smog sieges in the mid-

1940s but increased in number through the 1950s. See Cadle and Wohlers, "Smog Lore," pp. 30–35.

32. Brienes, "Fight against Smog," p. 12; Krier and Ursin, *Pollution and Policy,* pp. 93–94; Cadle and Wohlers, "Smog Lore," pp. 33–34.

33. Brienes, "Fight against Smog," p. 12; Krier and Ursin, *Pollution and Policy,* pp. 93–94; Cadle and Wohlers, "Smog Lore," pp. 33–34.

34. Krier and Ursin, *Pollution and Policy,* p. 94; *NYT,* Apr. 21, 1955, p. 30:1.

35. Von Braun to Kenneth F. Hahn, Aug. 23, 1964, file 3.6.10 "APCD 1964," box 266-2, Hahn MSS.

36. J. James to Kenneth F. Hahn, Oct. 13, 1964, ibid.

37. See, for example, Barkelew to Kenneth F. Hahn, Sept. 15, 1953, file 3.6.1 "APCD 1952–54," box 266, Hahn MSS; Mrs. Harry Barkelew to Hahn, undated, file 3.6.10 "APCD 1964," box 266-2, Hahn MSS; Raymond L. Woolley to Hahn, Feb. 14, 1967, file 3.6.13 "APCD 1967," box 266-3, Hahn MSS; Krier and Ursin, *Pollution and Policy,* pp. 94–95.

38. Krier and Ursin, *Pollution and Policy,* pp. 95–97.

39. Brienes, "Fight against Smog," pp. 228–34, 256.

40. Krier and Ursin, *Pollution and Policy,* pp. 148–50; Roberts, "Motor Vehicle Air Pollution," pp. 76–78.

41. Krier and Ursin, *Pollution and Policy,* pp. 146–48.

42. Ibid., pp. 147–50; W. G. Abbott to Kenneth F. Hahn, Jan. 3, 1964; Lee H. Hanson to Hahn, Jan. 7, 1964; Walter L. Smith to Hahn, Oct. 8, 1964; Irving Thomas Byrne to Hahn, Oct. 11, 1964; letters all from file 3.6.10 "APCD 1964," box 266-2, Hahn MSS.

43. Krier and Ursin, *Pollution and Policy,* pp. 150–51; Van Meter to Kenneth F. Hahn, Jan. 24, 1965; W. M. Raines to Hahn, Dec. 30, 1964; Ray Goodson to Hahn, Jan. 2, 1965; O. Hodges to Los Angeles County Air Pollution Control District, Aug. 12, 1964; all letters (and many others) in file 3.6.11 "APCD 1965–66," box 266-3, Hahn MSS.

44. Krier and Ursin, *Pollution and Policy,* pp. 151–52; *NYT,* Feb. 7, 1965, sec. IV, p. 7:2.

45. Krier and Ursin, *Pollution and Policy,* pp. 151–52; *NYT,* Feb. 7, 1965, sec. IV, p. 7:2; Lester A. Snyder to Kenneth F. Hahn, Jan. 21, 1965; Petersen to Hahn, Jan. 21, 1965; both letters in file 3.6.11 "APCD 1965–66," box 266-3, Hahn MSS.

46. Pagel to Kenneth F. Hahn, Dec. 31, 1963; Elster to Hahn, Oct. 27, 1964; both in file 3.6.10 "APCD 1964," box 266-2, Hahn MSS; Valleley to Hahn, Feb. 14, 1967, file 3.6.13 "APCD 1967," box 266-3, Hahn MSS; Sims, "Emergence of Air Pollution," pp. 105–34.

47. *NYT,* Aug. 7, 1965, p. 49:4; *NYT,* Aug. 12, 1968, p. 53:1; *NYT,* Dec. 2, 1968, p. 49:1.

48. University of California Air Pollution Research Center, *The Polluted Air: Smog Research at the University of California;* Sims, "Emergence of Air Pollution," pp. 105–33; *NYT,* Aug. 30, 1967, p. 20:6; *NYT,* Oct. 13, 1967, p. 10:6; Doty, "Life Cycle Theories," pp. 82–83; Krier and Ursin, *Pollution and Policy,* pp. 181–82; Roberts, "Motor Vehicle Air Pollution," pp. 56, 71–72.

49. Roberts, "Motor Vehicle Air Pollution," pp. 50–51; Hemmer to Hahn, undated (1969), file 3.6.18 "APCD 1967," box 266–267, Hahn MSS; Sims, "Emergence of Air Pollution," p. 246.

50. Doty, "Life Cycle Theories," pp. 91–92.

51. Ibid., pp. 92–94. In early 1970, Los Angeles County Supervisor Kenneth F. Hahn reaffirmed Slade's charge of lax enforcement. See Press Releases from Kenneth F. Hahn, Feb. 5, 1970, and Mar. 2, 1970, file 3.6.23 "APCD 1970," box 267, Hahn MSS.

52. Doty, "Life Cycle Theories," pp. 94–95.

53. Ibid., pp. 96–127; Fawcett, *Political Economy of Smog*, pp. 84–88, 105 note 40; City of Riverside, California, Office of Disaster Preparedness, *Air Pollution–Smog: An Existing Health Hazard, an Imminent Disaster; NYT*, Aug. 3, 1972, p. 30:1; Robert A. Doty and Leonard Levine, *Profile of an Air Pollution Controversy: Volume II*, pp. 1–8.

54. Regarding a rising tide of criticism of regulatory agencies and "experts" by the early 1970s, see Doty, "Life Cycle Theories," pp. 19–28.

## Chapter 6. Reinventing the Wheel

1. This discussion of events in New York City relies heavily on newspaper accounts, because relevant agency records are not available. The present air pollution control agency in New York City, the Bureau of Air Policy and Programs within the city Department of Environmental Protection, has kept no records from before 1970, since it has been relocated five times just since 1970, along with the various reorganizations and relocations that occurred before then. Earlier records were given to the municipal archives, but the New York City Department of Records and Information Services has been unable to locate them (letter from Antonia Bryson, Bureau of Air Policy and Programs, New York City Department of Environmental Protection, Apr. 25, 1995, in author's personal possession).

2. *NYT*, Jan. 30, 1946, p. 27:6.

3. *NYT*, Jan. 30, 1946, p. 27:6.

4. *NYT*, Jan. 9, 1947, p. 25:3; *NYT*, Jan. 10, 1947, p. 23:3; *NYT*, Jan. 11, 1947, p. 15:4 and p. 18:3; *NYT*, Feb. 4, 1947, p. 27:8; *NYT*, Feb. 8, 1947, p. 16:3.

5. *NYT*, Apr. 4, 1947, p. 5:5.

6. *NYT*, Oct. 23, 1947, p. 27:2.

7. *NYT*, Oct. 24, 1947, p. 25:1; *NYT*, Oct. 29, 1947, p. 6:6.

8. *NYT*, Mar. 24, 1948, p. 22:4; *NYT*, Apr. 9, 1948, p. 18:3.

9. *NYT*, June 3, 1948, p. 27:2; *NYT*, June 15, 1948, p. 26:3; *NYT*, June 4, 1948, p. 22:3; New York City Council, *Proceedings of the Council of the City of New York, January–June 1948*, vol. 1, June 15, 1948, pp. 1331–34. Unfortunately, records of actual debates and discussions of the New York City Council are rarely kept.

10. *NYT*, Oct. 19, 1948, p. 55:4; *NYT*, Oct. 21, 1948, p. 29:7; *NYT*, Oct. 22, 1948, p. 27:1; *New York Herald Tribune*, Oct. 22, 1948, p. 18:7; *NYT*, Oct. 23, 1948, p. 14:2; *NYT*, Oct. 28, 1948, p. 28:6.

11. *NYT*, Oct. 31, 1948, p. 1:2; *New York Herald Tribune*, Oct. 31, 1948, p. 1:4; *NYT*, Nov. 1, 1948, p. 1:4. A good general account of the Donora incident is found in Lewis, *With Every Breath You Take*, pp. 188–202. See also Snyder, "Death-Dealing Smog," 117–39.

12. *NYT*, Nov. 5, 1948, p. 8:8; *NYT*, Nov. 6, 1948, p. 7:7; *NYT*, Nov. 8, 1948, p. 20:3; *NYT*, Nov. 15, 1948, p. 27:3; *NYT*, Nov. 19, 1948, p. 48:7; *NYT*, Dec. 3, 1948, p. 27:5.

13. *NYT*, Jan. 17, 1949, p. 12:3; *NYT*, Jan. 19, 1949, p. 29:8; *NYT*, Feb. 2, 1949, p. 35:1; *NYT*, Mar. 1, 1949, p. 27:1; New York City Council, *Proceedings, January–June 1949*, vol. 1, Feb. 1, 1949, pp. 88–90.

14. *NYT*, May 25, 1949, p. 21:5; *NYT*, July 7, 1949, p. 27:6; *NYT*, May 26, 1949, p. 28:3; *NYT*, July 8, 1949, p. 18:2.

15. *NYT,* July 9, 1949, p. 28:5; *NYT,* Aug. 6, 1949, p. 18:3; *NYT,* Aug. 16, 1949, p. 25:1; *NYT,* Aug. 17, 1949, p. 22:3; *NYT,* Sept. 20, 1949, p. 31:1; *NYT,* Oct. 5, 1949, p. 31:4.

16. *NYT,* Jan. 26, 1950, p. 30:1; *NYT,* Feb. 4, 1950, p. 14:3.

17. *NYT,* Apr. 23, 1950, p. 54:3.

18. *NYT,* May 3, 1950, p. 34:7; *NYT,* June 5, 1950, p. 25:1; *NYT,* Apr. 25, 1950, p. 30:3. The Ringelmann Chart consisted of four squares with grids of varying darkness which resolved into greyish blurs of varying darkness when viewed from a distance and compared to a smoke plume. Use of the chart, which became one of the principal means for measuring smoke density in America, required one smoke inspector to hold the chart for another to observe. See Brimblecombe, *Big Smoke,* pp. 169–70.

19. *NYT,* May 3, 1950, p. 34:7; *NYT,* June 23, 1950, p. 48:1; *NYT,* June 27, 1950, p. 28:3; *NYT,* June 27, 1950, p. 31:3; *NYT,* July 16, 1950, p. 1:2; *NYT,* July 17, 1950, p. 20:3; *NYT,* Aug. 3, 1950, p. 25:6; *NYT,* Aug. 9, 1950, p. 27:2.

20. *NYT,* Mar. 18, 1951, p. 72:1; *NYT,* Mar. 21, 1951, p. 32:3; *NYT,* Apr. 3, 1951, p. 29:8.

21. *NYT,* Aug. 16, 1949, p. 25:1.

22. *NYT,* Feb. 9, 1951, p. 27:7; *NYT,* Feb. 2, 1951, p. 25:5; *NYT,* Mar. 26, 1951, p. 25:5.

23. *NYT,* May 9, 1951, p. 32:3; *NYT,* May 12, 1951, p. 24:1; *NYT,* May 16, 1951, p. 46:1; *NYT,* June 6, 1951, p. 24:3.

24. *NYT,* July 2, 1951, p. 22:3; *NYT,* June 21, 1951, p. 29:8.

25. *NYT,* July 20, 1951, p. 23:1.

26. *NYT,* July 24, 1951, p. 1:2.

27. *NYT,* July 25, 1951, p. 25:1; *NYT,* July 26, 1951, p. 23:1; *NYT,* July 27, 1951, p. 21:8; *NYT,* July 28, 1951, p. 13:1; *NYT,* Aug. 3, 1951, p. 23:5.

28. *NYT,* July 25, 1951, p. 25:1; *NYT,* July 26, 1951, p. 23:1; *NYT,* July 27, 1951, p. 21:8; *NYT,* July 28, 1951, p. 13:1; *NYT,* Aug. 3, 1951, p. 23:5.

29. *NYT,* July 25, 1951, p. 22:1; *NYT,* July 27, 1951, p. 21:8; *NYT,* July 31, 1951, p. 23:1; *NYT,* Aug. 1, 1951, p. 18:6; *NYT,* Aug. 2, 1951, p. 1:1; *NYT,* Aug. 10, 1951, p. 22:7; *NYT,* Nov. 17, 1951, p. 19:6.

30. *NYT,* Aug. 20, 1951, p. 21:8; *NYT,* Sept. 20, 1951, p. 33:8; *NYT,* Dec. 1, 1951, p. 15:6.

31. *NYT,* July 20, 1951, p. 23:1; *NYT,* July 24, 1951, p. 1:2.

32. *NYT,* Aug. 3, 1951, p. 23:5; *NYT,* Aug. 8, 1951, p. 27:5.

33. *NYT,* Aug. 17, 1951, p. 1:2; *NYT,* Aug. 28, 1951, p. 25:5; *NYT,* Sept. 6, 1951, p. 33:1; *NYT,* Oct. 10, 1951, p. 25:1; *NYT,* Oct. 11, 1951, p. 36:2; *NYT,* Oct. 11, 1951, p. 44:3; *NYT,* Oct. 16, 1951, p. 33:1; *NYT,* Oct. 19, 1951, p. 19:6; *NYT,* Oct. 25, 1951, p. 42:4.

34. *NYT,* Nov. 30, 1951, p. 1:2; *NYT,* Dec. 1, 1951, p. 15:6 and p. 12:3; *NYT,* Mar. 6, 1952, p. 33:5; *NYT,* July 18, 1952, p. 21:6; *NYT,* July 19, 1952, p. 14:2.

35. *NYT,* Jan. 16, 1952, p. 23:5.

36. Melosi, *Garbage,* pp. 47–49, 170–76, 184–87, 196–97, 217–18, and generally.

37. *NYT,* May 12, 1952, p. 24:3; *NYT,* June 4, 1952, p. 29:1.

38. *NYT,* July 24, 1952, p. 29:7; *NYT,* Aug. 15, 1951, p. 1:2; *NYT,* Aug. 1, 1952, p. 19:6.

39. *NYT,* Sept. 3, 1952, p. 31:5.

40. *NYT,* Sept. 10, 1952, p. 31:1; *NYT,* Sept. 11, 1952, p. 30:2; *NYT,* Sept. 25, 1952, p. 32:3; *NYT,* Oct. 30, 1952, p. 33:5; *NYT,* Dec. 10, 1966, p. 31:2.

41. *NYT,* July 3, 1952, p. 1:4; *NYT,* Aug. 22, 1952, p. 1:2; *NYT,* Sept. 5, 1952, p. 29:5; *NYT,* Sept. 6, 1952, p. 16:3; *NYT,* Sept. 10, 1952, p. 28:7; *NYT,* Sept. 25, 1952, p. 32:3; *New York Herald*

*Tribune,* July 3, 1952, p. 1:1, and Aug. 22, 1952, p. 1:1; New York City Council, *Proceedings, January–June 1952,* vol. 1, May 13, 1952, pp. 1025–28; June 17, 1952, pp. 1194–97.

42. *NYT,* Nov. 16, 1952, p. 1:7; *NYT,* Nov. 18, 1952, pp. 30:3 and 33:4; *NYT,* Nov. 25, 1952, p. 26:5.

43. *NYT,* June 4, 1953, p. 1:2; *NYT,* Oct. 22, 1953, p. 1:1; *NYT,* May 15, 1954, p. 17:1; *NYT,* May 8, 1956, p. 25:5; *NYT,* May 14, 1956, p. 27:7; *NYT,* May 15, 1956, p. 33:6; *NYT,* Apr. 27, 1959, p. 18:1; *NYT,* May 25, 1959, p. 32:8; *NYT,* July 19, 1959, p. 59:3; *NYT,* Aug. 8, 1959, p. 19:1.

44. *NYT,* Sept. 18, 1954, p. 17:7; *NYT,* Jan. 10, 1955, p. 28:2; *NYT,* Jan. 11, 1955, p. 19:1; *NYT,* June 21, 1955, p. 33:6; *NYT,* Nov. 14, 1955, p. 29:1; *NYT,* Feb. 4, 1956, p. 21:6; *NYT,* June 5, 1956, p. 40:7; *NYT,* Oct. 10, 1958, p. 33:1; *NYT,* Oct. 13, 1958, p. 31:3; *NYT,* Oct. 15, 1957, p.43:2; *NYT,* Sept. 20, 1960, p. 15:6; *NYT,* Feb. 6, 1961, p. 22:1; *NYT,* Dec. 17, 1963, p. 41:8.

45. *NYT,* Mar. 3, 1954, p. 29:2; *NYT,* Apr. 2, 1954, p. 17:7; *NYT,* Feb. 3, 1956, p. 25:1; *NYT,* Jan. 26, 1957, p. 21:1; *NYT,* Jan. 31, 1958, p. 23:1; *NYT,* Mar. 8, 1960, p. 29:5; *NYT,* Feb. 1, 1962, p. 20:3; *NYT,* Apr. 6, 1962, p. 26:1, 4; *NYT,* June 24, 1965, p. 22:4. The rest of New York State also lagged in pollution control expenditures. See Herman E. Hilleboe, Chairman, *A Review of Air Pollution in New York State,* p. 15.

46. *NYT,* Mar. 3, 1954, p. 29:2; *NYT,* Apr. 2, 1954, p. 17:7; *NYT,* Nov. 20, 1955, p. 134:5; *NYT,* Nov. 17, 1956, p. 20:2; *NYT,* Jan. 26, 1957, p. 21:1; *NYT,* June 9, 1958, p. 25:5; *NYT,* June 17, 1958, p. 28:2.

47. *NYT,* Mar. 4, 1960, p. 1:5; Lyle C. Fitch, Acting City Administrator, *Air Pollution Control: Organization and Operation—Report of the City Administrator.*

48. *NYT,* Feb. 19, 1957, p. 30:2; *NYT,* Oct. 13, 1958, p. 31:3.

49. *NYT,* Mar. 8, 1960, p. 29:5; *NYT,* May 26, 1966, p. 23:1.

50. *NYT,* Nov. 15, 1953, p. 1:3; *NYT,* Nov. 20, 1953, p. 1:6 and p. 22:7; *NYT,* June 19, 1954, p. 17:3; *NYT,* July 3, 1961, p. 17:4; New York State, *Act Now for Clean Air: A Program to Reduce Air Pollution in New York State,* p. 8.

51. *NYT,* Sept. 6, 1963, p. 31:5; *NYT,* Oct. 26, 1963, p. 1:1; *NYT,* May 14, 1964, p. 19:1.

52. *NYT,* May 13, 1964, p. 33:1; *NYT,* May 14, 1964, p. 19:1; *NYT,* May 20, 1964, p. 35:1; *NYT,* Dec. 2, 1964, p. 17:2; *NYT,* June 23, 1965, p. 1:5; *NYT,* Oct. 14, 1965, p. 30:4; *NYT,* Oct. 15, 1965, p. 39:6; *NYT,* Oct. 16, 1965, p. 26:6; Chris McNickle, *To Be Mayor of New York,* pp. 180–81; New York City Council, *Blueprint for Cleaner Air: Final Report of the Special Committee to Investigate Air Pollution,* pp. 1–56.

53. *NYT,* June 25, 1965, p. 1:4.

54. *NYT,* June 24, 1965, p. 22:4; *NYT,* July 30, 1965, p. 26:3; *NYT,* Sept. 2, 1965, p. 37:1; *NYT,* Sept. 4, 1965, p. 46:4; *NYT,* Sept. 14, 1965, p. 26:8; *NYT,* Oct. 27, 1965, p. 1:6.

55. *NYT,* Mar. 25, 1966, p. 1:2; *NYT,* May 4, 1966, p. 1:1; *NYT,* May 6, 1966, p. 36.

56. *NYT,* May 10, 1966, p. 1:5; *NYT,* May 12, 1966, p. 23:1; *NYT,* May 21, 1966, p. 28:1; *NYT,* July 13, 1966, p. 45:5; *NYT,* Apr. 16, 1966, p. 10:7; *NYT,* Oct. 12, 1966, p. 1:5; *NYT,* Oct. 15, 1966, p. 28:1; Norman Cousins, Chairman, *Freedom to Breathe: Report of the Mayor's Task Force on Air Pollution in the City of New York.*

57. *New York Herald Tribune,* June 23, 1965, p. 1:1; *NYT,* May 9, 1967, p. 35:2; *NYT,* Jan. 12, 1967, p. 1:4; *NYT,* Dec. 10, 1966, p. 31:2; *NYT,* May 22, 1966, p. 53:1; *NYT,* May 10, 1967, p. 33:1; *NYT,* Jan. 9, 1967, p. 33:6; *NYT,* Jan. 10, 1968, p. 20:3; *NYT,* Jan. 24, 1968, p. 47:1; *NYT,* Feb. 1, 1968, p. 1:1; *NYT,* Feb. 28, 1968, p. 58:2; *NYT,* Nov. 11, 1968, p. 27:3; *NYT,* Nov. 23, 1968, p. 39:4; Hagevik, *Decision-Making,* pp. 128–68.

58. *NYT,* Nov. 24, 1966, p. 79:6; *NYT,* Nov. 25, 1966, p. 1:8 and 36:1; *NYT,* Nov. 26, 1966, p. 1:7

and 1:8; *NYT,* Nov. 27, 1966, p. 1:1 and 1:2; *NYT,* Nov. 28, 1966, p. 1:2; *NYT,* Oct. 27, 1967, p. 56:8.

59. *NYT,* Jan. 10, 1967, p. 1:5; *NYT,* Aug. 4, 1967, p. 1:3; *NYT,* May 10, 1967, p. 33:1; *NYT,* Dec. 13, 1966, p. 24:3; *NYT,* Dec. 18, 1966, sec. IV, p. 12:2.

60. Esposito and Silverman, *Vanishing Air,* pp. 204–206, 210–13; *NYT,* Mar. 30, 1970, p. 33:1.

# Chapter 7. A Fight to the Finish

1. *NYT,* Oct. 3, 1946, p. 4:3; *NYT,* Nov. 3, 1946, sec. IV p. 12:5; *NYT,* Aug. 7, 1946, p. 26:7; *NYT,* Aug. 24, 1946, p. 10:5.

2. *NYT,* Jan. 3, 1947, p. 20:2, 6.

3. Ibid.

4. *NYT,* Jan. 3, 1947, p. 20:2; *NYT,* Jan. 6, 1947, p. 22:3; *NYT,* Jan. 8, 1947, p. 22:2; *NYT,* Jan. 10, 1947, p.20:2; *NYT,* Jan. 11, 1947, p. 18:3; *NYT,* Jan. 17, 1947, p. 22:2; *NYT,* Jan. 18, 1947, p. 14:1; *NYT,* Feb. 8, 1947, p. 16:3.

5. *NYT,* Jan. 10, 1947, p. 20:2.

6. Melosi, *Garbage,* pp. 80–82.

7. *NYT,* Jan. 3, 1947, p. 20:2.

8. *NYT,* Jan. 11, 1947, p. 18:3.

9. *NYT,* Jan. 3, 1947, p. 20:2; *NYT,* Jan. 6, 1947, p. 22:3.

10. *NYT,* Jan. 8, 1947, p. 22:2.

11. *NYT,* Jan. 17, 1947, p. 22:2.

12. *NYT,* Jan. 8, 1947, p. 22:5, 6.

13. *NYT,* Jan. 10, 1947, p. 23:3; *NYT,* Jan. 19, 1947, p. 14:1, 5; *NYT,* Feb. 4, 1947, p. 27:8.

14. *NYT,* Apr. 4, 1947, p. 5:5; *NYT,* Oct. 22, 1947, p. 31:5; *NYT,* Mar. 24, 1948, p. 22:4; *NYT,* Apr. 9, 1948, p.18:3; *NYT,* Apr. 12, 1948, p. 20:3; *NYT,* May 5, 1948, p. 24:2; *NYT,* May 10, 1948, p. 20:7.

15. *NYT,* June 3, 1948, p.27:2; *NYT,* June 15, 1948, p. 26:3; *NYT,* June 4, 1948, p. 22:3; *NYT,* June 8, 1948, p. 27:8; *NYT,* Aug. 2, 1948, p. 20:3.

16. *NYT,* Oct. 19, 1948, p. 55:4; *NYT,* Oct. 21, 1948, p. 29:7; *NYT,* Oct. 22, 1948, p. 27:1; *New York Herald Tribune,* Oct. 22, 1948, p. 18:7; *NYT,* Oct. 23, 1948, p. 14:2; *NYT,* Oct. 28, 1948, p. 28:6.

17. *NYT,* Oct. 31, 1948, p. 1:2; *NYT,* Nov. 1, 1948, p. 1:4; *New York Herald Tribune,* Oct. 31, 1948, p. 1:4.

18. *NYT,* Nov. 2, 1948, p. 28:7, 8.

19. *NYT,* Jan. 17, 1949, p. 12:3; *NYT,* Jan. 19, 1949, p. 29:8; *NYT,* Feb. 2, 1949, p. 35:1; *NYT,* Feb. 23, 1949, p. 26:3; *NYT,* Mar. 1, 1949, p. 27:1.

20. *NYT,* Feb. 22, 1949, p. 25:1; *NYT,* Feb. 23, 1949, p. 26:3.

21. *NYT,* Jan. 26, 1950, p. 30:1; *NYT,* Feb. 4, 1950, p. 14:3; *NYT,* Apr. 11, 1950, p. 30:3 and p. 26:8; *NYT,* May 3, 1950, p. 34:7; *NYT,* June 23, 1950, p. 48:1; *NYT,* Aug. 3, 1950, p 25:6; *NYT,* June 27, 1950, p. 28:3; *NYT,* June 27, 1950, p. 31:3; *NYT,* July 16, 1950, p. 1:2; *NYT,* July 17, 1950, p. 20:3.

22. *NYT,* Apr. 21, 1926, p. 14:2; *NYT,* Jan. 19, 1928, p. 14:5; *NYT,* Jan. 18, 1939, p. 21:2; *NYT,* Jan. 24, 1939, p. 22:7; *NYT,* Jan. 25, 1939, p. 23:3; *NYT,* Jan. 10, 1947, p. 23:3; *NYT,* Mar. 4, 1947, p. 51:7; *NYT,* Sept. 21, 1948, p. 26:6; *NYT,* Nov. 15, 1948, p. 27:3; *NYT,* Nov. 18, 1948, p. 26:3;

*NYT,* Nov. 19, 1948, p. 48:7; *NYT,* Nov. 28, 1948, p. 81:7; *NYT,* Mar. 23, 1949, p. 4:7.

23. *NYT,* Aug. 13, 1949, p. 10:6; *NYT,* Sept. 28, 1949, p. 26:6.

24. *NYT,* Oct. 22, 1948, p. 27:1; *NYT,* Aug. 4, 1950, p. 23:5.

25. *NYT,* Aug. 9, 1950, p. 28:3.

26. *NYT,* Jan. 11, 1951, p. 22:7; *NYT,* May 30, 1951, p. 66:1.

27. *NYT,* Jan. 11, 1951, p. 22:7; *NYT,* Jan. 17, 1951, p. 26:1.

28. *NYT,* Jan. 31, 1951, p. 18:7; *NYT,* Feb. 1, 1951, p. 24:3; *NYT,* Apr. 18, 1951, p. 30:8.

29. *NYT,* May 9, 1951, p. 32:3; *NYT,* May 30, 1951, p. 66:1.

30. *NYT,* July 31, 1951, p. 23:1; *NYT,* Aug. 1, 1951, p. 18:6; *NYT,* July 31, 1951, p. 23:1; *NYT,* Aug. 2, 1951, p. 1:1; *NYT,* Aug. 10, 1951, p. 22:7; *NYT,* Nov. 17, 1951, p. 19:6.

31. *NYT,* May 9, 1951, p. 32:2; *NYT,* July 2, 1951, p. 22:3; *NYT,* Sept. 6, 1951, p. 33:1.

32. *NYT,* July 20, 1951, p. 23:1; *NYT,* July 24, 1951, p. 1:2; *NYT,* July 25, 1951, p. 25:1; *NYT,* July 26, 1951, p. 23:1; *NYT,* July 27, 1951, p. 21:8; *NYT,* July 28, 1951, p. 13:1; *NYT,* Aug. 3, 1951, p. 23:5; *NYT,* July 25, 1951, p. 25:1; *NYT,* July 29, 1951, p. 50:8; *NYT,* July 27, 1951, p. 21:8; *New York Post,* July 25, 1951, p. 6:1; *New York Herald Tribune,* July 25, 1951, p. 3:1, July 26, 1951, p. 1:2, and Sept. 1, 1951, p. 1:2.

33. *NYT,* July 25, 1951, p. 22:1; *NYT,* July 28, 1951, p. 10:3; *NYT,* July 31, 1951, p. 20:6; *NYT,* Oct. 26, 1951, p. 22:7.

34. *NYT,* Aug. 4, 1951, p. 15:1; *NYT,* Aug. 29, 1951, p. 1:1; *NYT,* Aug. 30, 1951, p. 25:8; *NYT,* Aug. 31, 1951, p. 17:1; *NYT,* Sept. 1, 1951, p. 13:5; *NYT,* Aug. 30, 1951, p. 25:8; *NYT,* Aug. 29, 1951, p. 24:3.

35. *NYT,* July 28, 1951, p. 13:1.

36. *NYT,* Aug. 2, 1951, p. 1:1; *NYT,* Aug. 9, 1951, p. 23:5; *NYT,* Aug. 15, 1951, p. 1:2.

37. *NYT,* Aug. 15, 1951, p. 1:2.

38. *New York Herald Tribune,* July 25, 1951, p. 3:1; *NYT,* July 8, 1952, p. 51:1; *NYT,* July 23, 1952, p. 25:6; *NYT,* July 31, 1952, p. 25:6. *NYT,* Aug. 22, 1951, p. 1:1; *NYT,* Sept. 22, 1951, p. 19:4; *NYT,* Jan. 9, 1952, p. 27:2; *NYT,* Jan. 18, 1952, p. 29:8.

39. *NYT,* July 15, 1952, p. 23:5; *NYT,* July 31, 1952, p. 25:6.

40. *NYT,* Aug. 13, 1951, p. 8:5; *NYT,* Dec. 16, 1951, p. 71:5; *NYT,* May 16, 1952, p. 25:6; *NYT,* May 17, 1952, p. 13:2; *NYT,* May 26, 1952, p. 22:2; *NYT,* July 3, 1952, p. 1:4; *NYT,* Aug. 22, 1952, p. 1:2; New York City Council, *Proceedings, January–June 1952,* vol. 1, May 13, 1952, pp. 1025–28; June 17, 1952, pp. 1194–97; New York City Council, *Proceedings, July–December 1952,* vol. 2, July 2, 1952, pp. 16–17; Chris McNickle, *To Be Mayor of New York,* pp. 43, 45–48, 58–59, 63, 70.

41. *NYT,* July 3, 1952, p. 1:4; *NYT,* Aug. 22, 1952, p. 1:2; *NYT,* Sept. 5, 1952, p. 29:5; *NYT,* Sept. 6, 1952, p. 16:3; *NYT,* Sept. 10, 1952, p. 28:7; *NYT,* Sept. 25, 1952, p. 32:3; *NYT,* Nov. 16, 1952, p. 1:7; *New York Herald Tribune,* July 3, 1952, p. 1:1 and Aug. 22, 1952, p. 1:1.

42. *NYT,* Nov. 16, 1952, p. 1:7; *NYT,* Feb. 4, 1953, p. 25:2; *NYT,* Oct. 8, 1953, p. 31:1; *NYT,* Mar. 26, 1952, p. 31:5.

43. *NYT,* Mar. 26, 1952, p. 31:5.

44. *NYT,* Apr. 5, 1952, p. 20:1.

45. Ibid.

46. Regarding chemists and engineers and their role within the corporate establishment, see generally Noble, *America by Design.* For complaints about the same process being under way among physicians, see, for instance, Dr. Robert S. Mendelsohn, *Confessions of a Medical Heretic.*

47. *NYT,* Nov. 6, 1953, p. 29:1; *NYT,* Feb. 3, 1955, p. 16:5; *NYT,* Feb. 9, 1956, p. 63:2.

48. The Jersey Turnpike propellers were mentioned in the *NYT* on Aug. 27, 1954, p. 23:3 and were pictured in the *NYT,* Aug. 29, 1954, p. 1:4.

49. Brienes, "Fight against Smog," pp. 150, 170, 181–90; Krier and Ursin, *Pollution and Policy,* pp. 77–89.

50. *NYT,* Mar. 3, 1961, p. 26:6; *NYT,* May 5, 1955, p. 35:2; *NYT,* May 10, 1966, p. 1:5; *NYT,* Aug. 4, 1967, p. 1:3; *NYT,* Dec. 28, 1952, p. 12:1; *NYT,* Jan. 25, 1953, p. 33:6; *NYT,* Feb. 13, 1953, p. 3:5; *NYT,* Dec. 5, 1962, p. 55:1; *NYT,* Dec. 6, 1962, p. 1:5; *NYT,* Feb. 14, 1963, p. 4:8.

51. Regarding Citizens for Clean Air, see *NYT,* May 20, 1966, p. 49:5; *NYT,* June 17, 1966, p. 35:1; *NYT,* May 1, 1967, p. 46:2; *NYT,* May 24, 1967, p. 43:1; *NYT,* Oct. 24, 1968, p. 23:3; Constance Holden, "Hazel Henderson: Nudging Society off Its Macho Trip," *Science* 190 (Nov. 28, 1975): 862–64.

# Chapter 8. Jersey

1. Harvey Lieber, "The Politics of Air and Water Pollution Control in the New York Metropolitan Area" (Ph.D. diss., Columbia University, 1968), pp. 69–70.

2. Interstate Sanitation Commission, *Smoke and Air Pollution: New York–New Jersey,* pp. 20–21; New York State Air Pollution Control Board, *A Review of Air Pollution in New York State,* pp. 18–19.

3. *NYT,* Feb. 10, 1935, sec. 2, p. 2:5; *NYT,* Feb. 11, 1935, p. 11:7.

4. Interstate Sanitation Commission, *Smoke and Air Pollution,* pp. 21–23; New York State Air Pollution Control Board, *A Review of Air Pollution in New York State,* pp. 19–20; Lieber, "Air and Water Pollution Control," pp. 73–78; *NYT,* Jan. 18, 1950, p. 22:6 and Jan. 19, 1950, p. 26:3.

5. *NYT,* Aug. 15, 1951, p. 1:2; *NYT,* Aug. 21, 1951, p. 29:8.

6. *NYT,* Aug. 25, 1951, p. 13:2; *NYT,* Sept. 8, 1951, p. 19:4; *NYT,* Sept. 9, 1951, p. 66:4; *NYT,* Sept. 11, 1951, p. 38:3; *NYT,* Sept. 19, 1951, p. 33:1; *NYT,* Sept. 27, 1951, p. 33:1; *NYT,* Sept. 28, 1951, p. 33:5; *NYT,* Oct. 10. 1951, p. 25:1.

7. *NYT,* Jan. 12, 1952, p. 11:5; *NYT,* Jan. 23, 1952, p. 21:7; *NYT,* Feb. 2, 1952, p. 15:5; *NYT,* Feb. 19, 1952, p. 14:1; *NYT,* Feb. 26, 1952, p. 29:6; *NYT,* Mar. 7, 1952, p. 21:4; *NYT,* Apr. 5, 1952, p. 17:1; *NYT,* July 3, 1952, p. 8:3; *NYT,* Dec. 20, 1952, p. 11:5; Lieber, "Air and Water Pollution Control," pp. 79–82; James L. Sundquist, *Politics and Policy: The Eisenhower, Kennedy, and Johnson Years,* p. 332.

8. *NYT,* Apr. 8, 1948, p. 19:3; Lieber, "Air and Water Pollution Control," pp. 65–66, 113; New York State Air Pollution Control Board, *A Review of Air Pollution in New York State,* pp. 9, 12–18.

9. New Jersey Air Pollution Commission, *Report to the New Jersey Legislature,* pp. 5, 13–17, 23–28; Lieber, "Air and Water Pollution Control," pp. 67–68.

10. *NYT,* May 3, 1954, p. 27:5; *NYT,* Sept. 17, 1954, p. 46:6; New Jersey Air Pollution Control Commission, *A Progress Report of the New Jersey Air Pollution Control Commission,* pp. 4–6; William R. Bradley, Chairman, State of New Jersey Air Pollution Control Commission, "The Air Pollution Control Commission Program for New Jersey," in *Proceedings of the Forty-Ninth Annual Meeting of the Air Pollution Control Association, Buffalo, New York, May 20–24, 1956,* pp. 18–1 to 18–12; Gerald R. Moran, "The Air Pollution Control Act and Its Adminis-

tration," *Rutgers Law Review* 9, no. 4 (summer, 1955): 642, 653, 681; Lieber, "Air and Water Pollution Control," pp. 68, 211, 227–33.

11. *NYT,* Mar. 12, 1954, p. 12:5; *NYT,* Mar. 18, 1954, p. 22:4; *NYT,* May 3, 1954, p. 27:5; *NYT,* June 11, 1954, p. 25:8; *NYT,* June 22, 1954, p. 13:1; *NYT,* Aug. 18, 1954, p. 18:5; *NYT,* Sept. 8, 1954, p. 27:5; *NYT,* Sept. 10, 1954, p. 32:4; Lieber, "Air and Water Pollution Control," p. 81.

12. *NYT,* Sept. 21, 1951, p. 25:3; *NYT,* Oct. 2, 1951, p. 22:2; *NYT,* Dec. 24, 1951, p. 12:8; *NYT,* July 7, 1952, p. 41:1; *NYT,* Aug. 28, 1952, p. 26:7; *NYT,* Nov. 11, 1953, p. 30:6; *NYT,* Jan. 24, 1954, p. 69:2; *NYT,* Nov. 22, 1954, p. 25:2; *NYT,* Feb. 18, 1955, p. 19:2.

13. *NYT,* June 2, 1955, p. 59:1; *NYT,* Oct. 7, 1955, p. 14:4; *NYT,* Dec. 16, 1955, p. 25:6; *NYT,* Feb. 3, 1956, p. 25:1; *NYT,* Mar. 9, 1956, p. 15:1; *NYT,* Apr. 7, 1956, p. 23:1; *NYT,* Apr. 29, 1956, p. 58:4; *NYT,* July 24, 1956, p. 27:8; *NYT,* Aug. 4, 1956, p. 24:4; Lieber, "Air and Water Pollution Control," pp. 81–82.

14. *NYT,* Aug. 7, 1956, p. 16:6; *NYT,* Dec. 19, 1956, p. 64:5; *NYT,* Mar. 7, 1957, p. 23:1; *NYT,* May 29, 1957, p. 34:1; *NYT,* Oct. 10, 1957, p. 35:8; *NYT,* Jan. 30, 1958, p. 25:8.

15. *NYT,* Jan. 30, 1958, p. 25:8; Interstate Sanitation Commission (ISC), *Report on Air Pollution Study, 1955,* pp. 1–5; ISC, *Progress Report on Air Pollution,* pp. 1–12; ISC, *Smoke and Air Pollution,* pp. 1–16, 70–75; Lieber, "Air and Water Pollution Control," pp. 82–87.

16. *NYT,* Jan. 22, 1959, p. 24:7; *NYT,* Feb. 25, 1959, p. 33:6; *NYT,* June 29, 1959, p. 28:1; Lieber, "Air and Water Pollution Control," pp. 88–93.

17. *NYT,* Sept. 14, 1956, p. 25:8; *NYT,* Dec. 12, 1956, p. 36:1; *NYT,* May 28, 1957, p. 1:2; Lieber, "Air and Water Pollution Control," pp. 88–89.

18. *NYT,* July 9, 1961, p. 61:3; *NYT,* Jan. 24, 1965, p. 67:6; Lieber, "Air and Water Pollution Control," 351–56.

19. *NYT,* June 21, 1960, p. 35:5.

20. Lieber, "Air and Water Pollution Control," pp. 200–21.

21. *NYT,* Mar. 8, 1961, p. 35:5; *NYT,* May 3, 1966, p. 49:8; *NYT,* Jan. 6, 1967, p. 1:3; *NYT,* Jan. 10, 1967, p. 39:4; *NYT,* Jan. 29, 1967, p. 80:1; *NYT,* Apr. 25, 1967, p. 33:1; *NYT,* Apr. 28, 1967, p. 43:1; *NYT,* June 16, 1967, p. 1:4; *NYT,* June 20, 1967, p. 41:2; New Jersey Air Pollution Control Commission, *Where Does New Jersey Go from Here in Air Pollution Control? Report of the Planning Committee,* p. 13; Lieber, "Air and Water Pollution Control," pp. 202, 211–13.

22. *NYT,* Dec. 12, 1960, p. 58:5; *NYT,* Oct. 23, 1962, p. 34:3; *NYT,* Feb. 19, 1964, p. 41:8; *NYT,* Feb. 21, 1964, p. 28:2; *NYT,* Jan. 10, 1965, p. 81:5; *NYT,* July 5, 1965, p. 6:1; *New York Herald Tribune,* June 25, 1965, p. 1:1; New York City Council, *Blueprint for Cleaner Air,* p. 55.

23. *NYT,* Oct. 19, 1965, p. 36:4; *NYT,* May 2, 1966, p. 39:8; *NYT,* May 3, 1966, p. 49:8. Lieber, "Air and Water Pollution Control," pp. 445–48.

24. *NYT,* Dec. 16, 1966, p. 41:1.

25. *NYT,* Feb. 4, 1967, p. 54:4; *NYT,* Mar. 3, 1967, pp. 1:1 and 40:4; *NYT,* Mar. 3, 1967, p. 23:1; *NYT,* Mar. 28, 1967, p. 1:6; *NYT,* Apr. 25, 1967, p. 1:5; *NYT,* Apr. 28, 1967, p. 43:1; *NYT,* June 16, 1964, p. 1:4; *NYT,* June 23, 1967, p. 22:2.

26. *NYT,* Mar. 3, 1967, pp. 1:1 and 40:4.

27. *NYT,* Jan. 3, 1967, p. 11:2; *NYT,* Jan. 4, 1967, p. 1:2; *NYT,* Jan. 7, 1967, p. 1:2; *NYT,* Jan. 11, 1967, p. 26:1; Jan. 13, 1967, p. 1:8.

28. *NYT,* Jan. 12, 1967, p. 1:4; *NYT,* Mar. 5, 1967, p. 34:1.

29. *NYT,* Jan. 31, 1968, p. 45:5; Joseph F. Zimmerman, "Political Boundaries and Air Pollution Control" (master's thesis, State University of New York–Albany, 1968), pp. 27–29.

30. Air Quality Act of 1967, 81 Stat. 485; Zimmerman, "Political Boundaries," pp. 19–22.

31. *NYT,* Jan. 31, 1968, p. 45:5; *NYT,* Aug. 30, 1968, p. 66:6; *NYT,* Apr. 6, 1969, p. 55:1; *NYT,* June 4, 1969, p. 1:6; *NYT,* Aug. 17, 1969, p. 61:1; Esposito and Silverman, *Vanishing Air,* pp. 8, 135–41, 269–93, 309–10. For useful further reflection on states' inability to handle interstate air pollution, see Lewis C. Green, "State Control of Interstate Air Pollution," *Law and Contemporary Problems* 33, no. 2 (spring, 1968): 315–30.

32. Lieber, "Air and Water Pollution Control," dissertation abstract.

33. Zimmerman, "Political Boundaries," p. 23.

## Chapter 9. The Fickle Finger of Phosphate

1. Part of the information in chapter 9 was used in Scott H. Dewey's "'Is This What We Came to Florida for?': Florida Women and the Fight against Air Pollution in the 1960s," which appeared in the *Florida Historical Quarterly* 77, no. 4 (spring, 1999): 503–31. Material in chapter 9 also appeared in Scott H. Dewey, "The Fickle Finger of Phosphate: Central Florida Air Pollution and the Failure of Environmental Policy, 1957–1970," which appeared in the *Journal of Southern History* 65, no. 3 (Aug., 1999): 565–603. I gratefully acknowledge permission from the Florida Historical Society and the *Journal of Southern History* to include such material.

   Raymond F. Dasmann, *No Further Retreat: The Fight to Save Florida,* p. 52; Charlton W. Tebeau, *A History of Florida,* pp. 271–78, 409–10; Charles I. Harding, Samuel B. McKee, and Jean J. Schueneman, *A Report on Florida's Air Resources,* p. 28. For relevant background on the general patterns of often reckless industrialization and economic development in the South following the Civil War, see James C. Cobb, *Industrialization and Southern Society, 1877–1984,* and C. Vann Woodward's classic *Origins of the New South.*

2. Tebeau, *History of Florida,* pp. 181, 271–78, 416–18, 431; Michael Gannon, *Florida: A Short History,* pp. 47, 60–61, 77–85; David Nolan, *Fifty Feet in Paradise: The Booming of Florida,* p. 118; Joe A. Akerman, Jr., *Florida Cowman: A History of Florida Cattle Raising,* generally.

3. Arch Frederic Blakey, *The Florida Phosphate Industry: A History of the Development and Use of a Vital Mineral,* pp. 24–35, 56–57, 78–84; Lewis D. Harris, "The Florida Phosphate Industry and Air Pollution" (master's thesis, Florida State University, 1967), p. 12.

4. Blakey, *Florida Phosphate Industry,* pp. 90–95, 100–104; Harris, "Florida Phosphate Industry," pp. 7–11, 13–14; Harding, McKee, and Schueneman, *Florida's Air Resources,* p. 23.

5. Blakey, *Florida Phosphate Industry,* pp. 94–95, 108–109; Harding, McKee, and Schueneman, *Florida's Air Resources,* p. 22.

6. Blakey, *Florida Phosphate Industry,* pp. 1–4, 109; Harris, "Florida Phosphate Industry," p. 2.

7. Blakey, *Florida Phosphate Industry,* pp. 1–4, 109; Harris, "Florida Phosphate Industry," p. 2.

8. Blakey, *Florida Phosphate Industry,* pp. 9–12, 94–95, 108–109; Harris, "Florida Phosphate Industry," pp. 4, 27–35; Harding, McKee, and Schueneman, *Florida's Air Resources,* pp. 22–27.

9. Harris, "Florida Phosphate Industry," pp. 21, 30–33; Harding, McKee, and Schueneman, *Florida's Air Resources,* pp. 24–25; "Clean Water—Clean Air," *Florida Health Notes* 48, no. 10 (Dec., 1956): 221, in file "Florida (Polk County)," Air Pollution Engineering Branch, Correspondence, 1959–60, Records of the U.S. Public Health Service, Record Group 90, Identification Number NN3-090-91-003, National Archives, Washington, D.C. (hereafter cited as USPHS, APEB 1959–60).

10. For further information on the effects of fluorides on animal, plant, and human health, see generally National Research Council, Committee on Biologic Effects of Atmospheric Pollutants, *Fluorides;* National Research Council, Subcommittee on Fluorosis, *Effects of Fluorides in Animals;* National Research Council, Subcommittee on Health Effects of Ingested Fluoride, *Health Effects of Ingested Fluoride.*

11. Blakey, *Florida Phosphate Industry,* p. 109; Harris, "Florida Phosphate Industry," p. 40. Regarding similar fluoride damage to orchard owners in other states, see Esposito and Silverman, *Vanishing Air,* pp. 69, 183–85.

12. State and federal agricultural census data on overall livestock and citrus production in Polk County unfortunately do not reveal damage to agriculture in the phosphate belt, because as cattle and citrus operations were driven out of afflicted areas, other ranchers and orchard owners were expanding their operations in unaffected parts of the county. See Thomas D. Crocker, "Some Economics of Air Pollution Control," (Ph.D. diss., University of Missouri, 1967), pp. 55 and 64.

13. Harris, "Florida Phosphate Industry," p. 14; Harding, McKee, and Schueneman, *Florida's Air Resources,* pp. 42–44; U.S. Senate Subcommittee on Air and Water Pollution, *Clean Air: Field Hearings Held on Progress and Programs Relating to the Abatement of Air Pollution,* 88th Cong., 2d sess., p. 779.

14. Blakey, *Florida Phosphate Industry,* pp. 108–109; Harris, "Florida Phosphate Industry," pp. 38–45; U.S. Senate, *Clean Air,* p. 781, 792.

15. U.S. Senate, *Clean Air,* pp. 792–93.

16. Ibid., pp. 780–81.

17. Blakey, *Florida Phosphate Industry,* p. 110; "Clean Water—Clean Air," pp. 221–22, USPHS, APEB 1959–60; U.S. Senate, *Clean Air,* pp. 742, 808; "Polk-Hillsborough," undated report from around December 20, 1963, file "Cooperation 2—Florida," Division of Air Pollution Subject Files, 1963–64, Records of the U.S. Public Health Service, Record Group 90, Accession Number 67-A-1655, National Archives, Washington, D.C. (hereafter cited as USPHS, DAP 1963–64).

18. Harding, McKee, and Schueneman, *Florida's Air Resources,* p. 51–52; "Florida Air Pollution Control Commission," excerpt from laws of the state of Florida, dated July 2, 1958, file "721.3—to Florida," Air Pollution Medical Program, Project Records, 1955–60, Records of the U.S. Public Health Service, Record Group 90, Identification Number NN3-090-91-003, National Archives, Washington, D.C. (hereafter cited as USPHS, APMP 1955–60).

19. Harding, McKee, and Schueneman, *Florida's Air Resources,* p. 52; Blakey, *Florida Phosphate Industry,* p. 110; U.S. Senate, *Clean Air,* pp. 742, 808; Herman F. Steele to Florida Air Pollution Control Commission, Oct. 19, 1957, and attached Resolution of Florida Citrus Mutual, file "Cooperation 2—Florida," USPHS, DAP 1963–64.

20. Harding, McKee, and Schueneman, *Florida's Air Resources,* p. 52.

21. Ibid., pp. 52–53.

## Chapter 10. Conference, Conciliation, and Persuasion

1. For instance, Florida enacted a postwar law to encourage industrial development or relocation from other states by allowing waterways to be classified as "industrial streams," into which unlimited volumes of pollution could be poured. See David Helvarg, *The War against*

*the Greens,* pp. 371–79. Part of the information in chapter 10 was used in my "'Is This What We Came to Florida For?'"

2. Steele to Florida APCC, Oct. 19, 1957, and attached Resolution, file "Cooperation 2—Florida," USPHS, DAP 1963–64.

3. U.S. Public Health Service Occupational Health Program and Florida State Board of Health, *Industrial Hygiene Survey of the Phosphate Industry in Polk County, Florida,* pp. 1–2, 17; report in file "Florida Air 3-1-1," USPHS, APEB 1959–60.

4. *Tampa Morning Tribune,* Mar. 1, 1958, clipping in file "Florida Air 3-1-1," USPHS, APEB 1959–60.

5. Harry E. Seifert to Frank Tetzlaff, Mar. 5, 1958, in same folder.

6. *Lakeland Ledger,* Aug. 5, 1958, p. 1, clipping in same folder.

7. Ibid.

8. Lightfoot to Hollis, with attached "Proposed Outline of the Polk County Air Pollution Control Study," Aug. 8, 1958, in file "Florida Air 3-1-1," USPHS, APEB 1959–60; August T. Rossano, Jr., to Harry E. Seifert, Sept. 23, 1958, and Lightfoot to Hollis, Nov. 5, 1958, in same folder.

9. Durrance to Sowder, Jan. 28, 1959, in same folder.

10. Sowder to Durrance, Feb. 9, 1959, and memo from Albert V. Hardy to Sowder, Jan. 30, 1959, in same folder.

11. "Chemical Workers Call Open Meeting to Discuss Air Pollution," from newspaper of District Council No. 1 of the International Chemical Workers, Wednesday, Jan. 28, 1959, clipping in file "721.3—to Florida," USPHS, APMP 1955–60.

12. May to "U.S. Department of Public Health," Apr. 10, 1959, file "Florida—AP/61," Division of Air Pollution Subject Files, January–August 1961, Records of the U.S. Public Health Service, Record Group 90, Accession Number 65-A-0286, National Archives, Washington, D.C. (hereafter cited as USPHS, DAP 1961–62); Howell to "U.S. Department of Public Health," Apr. 10, 1959, in same folder.

13. "Trip Report—Tampa, Florida—March 30th–April 3, 1959," memo from August T. Rossano, Jr., to Arthur C. Stern, Apr. 16, 1959, file "Florida Air 3-1-1," USPHS, APEB 1959–60.

14. Arthur C. Stern to Albert V. Hardy, May 20, 1960, file "Florida Air 3-1-1," USPHS, DAP 1961–62; "Air Pollution—Florida," memo from Roy O. McCaldin, May 25, 1960; David G. Stephan to Wilson T. Sowder, June 26, 1960; "State-Wide Air Pollution Survey," memo from David B. Lee to Sowder, July 1, 1960, all in same folder. See also Harding, McKee, and Schueneman, *Florida's Air Resources.*

15. U.S. Senate, *Clean Air,* pp. 813–14.

16. Jones to Smathers, Aug. 4, 1962, file "Cooperation 2—Florida," USPHS, DAP 1963–64; Smathers to Jones, Aug. 23, 1962; memo from A. Rego, Sept. 6, 1962; Richard A. Prindle to Smathers, Sept. 18, 1962, all in same folder.

17. Blakey, *Florida Phosphate Industry,* p. 111; Crocker, "Economic Aspects," p. 242; Jane H. May to United States Senator Edmund S. Muskie, Feb. 23, 1964, in U.S. Senate: Senate Office (625-5), Edmund S. Muskie Collection, Edmund S. Muskie Archives, Bates College, Lewiston, Maine (hereafter cited as Muskie Collection); memo from Vernon G. MacKenzie, June 7, 1963, file "Cooperation 2—Florida," USPHS, DAP 1963–64; David B. Lee to MacKenzie, June 18, 1963, and memo from R. J. Anderson, July 24, 1963, in same folder.

18. McLean to U.S. Department of Health, Education, and Welfare, June 7, 1963, file "Cooperation 2—Florida," USPHS, DAP 1963–64.

19. Lightfoot to Florida APCC, Aug. 13, 1963, file "OCC: Florida Air Pollution Commission," National Center for Air Pollution Control, Subject Files, 1967–68, Records of the U.S. Public Health Service, Record Group 90, Accession Number 70-A-4011, National Archives, Washington, D.C. (hereafter cited as USPHS, NCAPC 1967–68).

20. Lightfoot to Bryant, Oct. 12, 1963, in same folder.

21. David B. Lee to Lightfoot, Nov. 1, 1963, and Lightfoot to Bryant, Feb. 10, 1964, in same folder.

22. Lightfoot to Kennedy, Feb. 11, 1964, and Vernon G. MacKenzie to Lightfoot, Mar. 30, 1964, in same folder.

23. Huff to Smathers, Dec. 26, 1963, file "Cooperation 2—Florida," USPHS, DAP 1963–64.

24. Smathers to "Office of the Chief, Office of Air Pollution, Public Health Service," undated (Jan., 1964); Charles D. Yaffe to Smathers, Jan. 23, 1964.; memo from Austin N. Heller, Dec. 20, 1963, all in same folder.

25. U.S. Senate, *Clean Air,* pp. 721–33.

26. Ibid., pp. 741–46.

27. Ibid., pp. 746–47.

28. Ibid. pp. 748–49, 754–61.

29. Ibid., pp. 777–79.

30. Ibid., pp. 780–88.

31. Ibid., pp. 788–90, 798–800.

32. Ibid., pp. 792, 795–96, 800–803.

33. Ibid., pp. 807–16.

34. Ibid., pp. 806, 819–22.

35. Ibid., pp. 819–20.

36. Ibid., p. 823.

37. Ibid., pp. 822, 825–26.

38. Ibid., pp. 804–805, 826–27.

39. Ibid., pp. 817–19, 834–35.

40. *Tampa Tribune,* undated clipping in U.S. Senate: Senate Office (1005-8), Muskie Collection; May to Muskie, Feb. 23, 1964, and Howell and S. Opal Howell to Muskie, Feb. 20, 1964, in same folder.

41. Letter from the Howells to Muskie; Taylor to Muskie, Mar. 16, 1964, in same folder.

42. "Courts May Decide Florida's Phosphate Industry Pollution Issue," *Air/Water Pollution Report* 2, no. 12 (Monday, June 15, 1964), in U.S. Senate: Senate Office (99-4), Muskie Collection.

43. Blakey, *Florida Phosphate Industry,* pp. 110–11; Crocker, "Economic Aspects," pp. 244, 252; Harris, "Florida Phosphate Industry," pp. 55–58.

44. *Transcript: Proceedings of the Hearing Before the Florida Air Pollution Control Commission, Lakeland, Florida,* Sept. 17, 1965, pp. 16–19, 20–35.

45. Lightfoot to Vernon G. MacKenzie, May 14, 1966, file "OCC: Florida Air Pollution Commission," USPHS, NCAPC 1967–68; Lightfoot to Florida State Board of Health and Florida APCC, Feb. 11, 1966, in same folder.

46. "Minutes—Meeting of the Florida Air Pollution Control Commission, Tampa, Florida," Apr. 15, 1966, pp. 2–9, in same folder.; Harriett Lightfoot to Vernon G. MacKenzie, May 14, 1966, and MacKenzie to Lightfoot, May 27, 1966, in same folder.

47. Lightfoot to the Florida Air Pollution Control Commission, Apr. 15, 1966, included as Addendum Number 8 in "Minutes—Meeting of the Florida Air Pollution Control Commission," Apr. 15, 1966; Huffstutler to Lightfoot, May 31, 1966, in same folder; W. R. Lamb to the Florida State Board of Health and the Florida Air Pollution Control Commission, June 3, 1966, Addendum Number 2 in "Minutes—Meeting of the Florida Air Pollution Control Commission, Lakeland, Florida," June 3, 1966, in same folder. State officials and industrialists praised Huffstutler, however. See Blakey, *Florida Phosphate Industry,* p. 113.

48. "Transcript: Proceedings of Hearings—Possible Effect of Fluorides on Citrus—before the Florida Air Pollution Control Commission, Lakeland, Florida," June 2–3, 1966, pp. 2–8, file "OCC: Florida Air Pollution Commission," USPHS, NCAPC 1967–68. See also "Phosphate Board Opposes Citrus 'Fluoride Standard,'" unidentified, undated newspaper clipping (probably from the *Sarasota Herald-Tribune,* June 7, 1966), in same folder.

49. "Transcript: Possible Effect of Fluorides on Citrus," pp. 8–21; "Exhibit Number 1" (Statement of Mrs. W. A. Dobbs), in same folder.

50. "Transcript: Possible Effect of Fluorides on Citrus," pp. 21–103. Murray's cross-examination takes up pp. 59–103.

51. Ibid., pp. 103–104.

52. Ibid., pp. 113–54.

53. Ibid., pp. 155–68.

54. Ibid., pp. 168–88, 191–205, and 207–43 generally.

55. For a good, even-handed discussion of the politics of science with regard to environmental issues, see the chapter on the topic in Hays, *Beauty, Health, and Permanence,* pp. 329–62.

56. "Transcript: Proceedings of Hearings—Sulfur Oxide Emissions—before the Florida Air Pollution Control Commission, Lakeland, Florida," June 3, 1966, pp. 1–50, in file "OCC: Florida Air Pollution Commission," USPHS, NCAPC 1967–68; Clyde Burnett, "State Health Chief Proposes Emission Controls for Acid Plants," unidentified, undated newspaper clipping (evidently from the *Sarasota Herald-Tribune,* June 7, 1966), in same folder.

57. "Air Pollution Control Dealt Severe Setback," unidentified, undated newspaper clipping (probably from *Tampa Tribune* or *Lakeland Ledger,* late January or February, 1967), U.S. Senate: Senate Office (822-9), Muskie Collection.

58. Ibid.

59. Ibid.

60. Arnold to President Johnson, Feb. 20, 1967, U.S. Senate: Senate Office (822-9), Muskie Collection.

61. Blakey, *Florida Phosphate Industry,* p. 112; Florida State Department of Pollution Control, *Air Pollution Control in Florida,* pp. 5–6.

62. Blakey, *Florida Phosphate Industry,* pp. 111–12; Harris, "Florida Phosphate Industry," pp. 57–60.

63. Harris, "Florida Phosphate Industry," pp. 66–68, 83; Florida State Department of Pollution Control, *Air Pollution Control in Florida,* pp. 3–4.

64. Crocker, "Economic Aspects," pp. 199–204, 214–16; Harris, "Florida Phosphate Industry," pp. 87–93; Blakey, *Florida Phosphate Industry,* pp. 112–13. Crocker states that Vincent D. Patton and Kay K. Huffstutler of the state air pollution control program both told him that state authorities always favored the land-purchase approach.

# Chapter 11. The Perils of Federalism

1. Material in chapter 11 first appeared in the *Journal of Southern History* 65, no. 3 (Aug., 1999): 565–603. I wish to thank the *Journal of Southern History* for granting permission to include such material. Southern states such as Florida were particularly sensitive about states' rights and federal intervention during the 1950s and 1960s due to the growing challenge of the civil rights movement against segregation.

2. "Visit to Florida Board of Health, Division of Industrial Hygiene," memo from Harry Heimann, M.D., Apr. 17, 1957, file "Florida Air 3-1-1," USPHS, APEB 1959–60; "Division of Special Health Services—Air Pollution Medical Program: Trip Report," memo from Harry Heimann, Apr. 25, 1957, file "721.3—to Florida," USPHS, APMP 1955–60.

3. U.S. Public Health Service Occupational Health Program and Florida State Board of Health, *Industrial Hygiene Survey of the Phosphate Industry in Polk County, Florida*, pp. 1, 4, 5, 17, and generally; in file "721.3–to Florida," USPHS, APMP 1955–60.

4. "Trip Report—Lakeland, Florida—July 24–26, 1957," memo from C. Stafford Brandt to Arthur C. Stern, July 31, 1957, in file "Florida Air 3-1-1," USPHS, APEB 1959–60.

5. Ibid. Again, cattle raising had been conducted extensively and profitably for many years in western Polk County.

6. Ibid.

7. "Trip Report (A. C. Stern and C. S. Brandt)—Tampa, Florida—February 27–28, 1958," memo from Arthur C. Stern to Harry G. Hanson, Mar. 7, 1958, file "Florida Air 3-1-1," USPHS, APEB 1959–60; see also "Agenda for February 28, 1958," in same folder.

8. Brandt's comments about shoddy care of livestock were later challenged by animal nutritionist Dr. George K. Davis of the University of Florida's Agricultural Experiment Station, a veterinary expert who rejected claims of poor care and felt quite certain about fluoride poisoning. See letter from Brandt to Davis, Mar. 25, 1958, ibid. Regarding the technocratic assumptions of American scientists and engineers during the 1950s and early 1960s, see generally Rachel Carson, *Silent Spring*.

9. See Hays, *Beauty, Health, and Permanence,* pp. 329–62.

10. Harry E. Seifert to Frank Tetzlaff, Mar. 5, 1958, file "Florida Air 3-1-1," USPHS, APEB 1959–60; Tetzlaff to Seifert, Mar. 10, 1958; memo from Wilton M. Fisher, M.D., Apr. 15, 1958; Wilson T. Sowder, M.D., to Fisher, May 6, 1958; Fisher to Sowder, May 21, 1958, all in same folder.

11. Letter and livestock inspection report from Dr. Norman L. Garlick, D.V.M., June 2, 1958, pp. 4–7, in same folder.

12. Dewell to Harry G. Hensen [*sic*], Jan. 26, 1959, in same folder.

13. Draft letter from Harry G. Hanson to John H. Dewell, Feb. 2, 1959, file "721.3—to Florida," USPHS, APMP 1955–60; Hanson to Dewell, Feb. 9, 1959, and memo from Richard A. Prindle, Feb. 4, 1959, in same folder.

14. Prindle to John H. Dewell, Feb. 27, 1959, in same folder.

15. Julian C. Durrance to Wilson T. Sowder, Jan. 28, 1959, file "Florida Air 3-1-1," USPHS, APEB 1959–60; memo from A. L. Chapman, Feb. 9, 1959, in same folder; memo from Richard A. Prindle, Jan. 29, 1959, file "721.3—to Florida," USPHS, APMP 1955–60; memo from Norman F. Gerrie, Feb. 12, 1959, and Prindle to Wilson T. Sowder, Feb. 16, 1959, in same folder.

16. Draft "Statement," undated, but from early February, 1959, file "721.3—to Florida," USPHS, APMP 1955–60.

17. Ibid.

18. Harry Seifert to Frank Tetzlaff, Feb. 24, 1959, file "Florida Air 3-1-1," USPHS, APEB 1959–60; August T. Rossano, Jr., to Seifert, Mar. 6, 1959, in same folder.

19. "Trip Report—Tampa, Florida—March 30th–April 3, 1959," memo from Rossano to Arthur C. Stern, Apr. 16, 1959, in same folder.

20. *Lakeland Ledger,* Thursday, Apr. 2, 1959, and *Tampa Tribune,* Friday, Apr. 3, 1959, clippings in same folder.

21. Scott H. Dewey, "The Fickle Finger of Phosphate: Central Florida Air Pollution and the Failure of Environmental Policy, 1957–1970," *Journal of Southern History* 65, no. 3 (Aug., 1999): 589–93.

22. For sharp criticism of the federal air pollution control program's persistent research focus through the late 1960s, see generally Esposito and Silverman, *Vanishing Air.*

23. Memo from Ralph C. Graber, July 9, 1959, file "Florida Air 3-1-1," USPHS, APEB 1959–60.

24. Wilson T. Sowder to W. H. Aufranc, July 22, 1959, file "Florida Air 3-1-1," USPHS, DAP 1961–62; H. B. Cottrell, M.D., to Sowder, Aug. 13, 1959, in same folder; "Report on a Review of Certain Air Pollution Activities of the Florida State Board of Health," undated but from early September 1959, file "Florida Air 3-1-1," USPHS, APEB 1959–60; memo from Jean J. Schueneman, Sept. 24, 1959, in same folder.

25. "Report on a Review of Certain Air Pollution Activities," file "Florida Air 3-1-1," USPHS, APEB 1959–60; letter to Frank Tetzlaff from unidentified correspondent with the Health Department of Hamilton County, Florida, Nov. 2, 1959, in same folder; Blakey, *Florida Phosphate Industry,* p. 113.

26. See, for example, Thomas F. Williams to Donald S. McLean, Aug. 8, 1963, file "Cooperation 2—Florida," USPHS, DAP 1963–64; Vernon G. MacKenzie to Harriett Lightfoot, Mar. 30, 1964, and May 27, 1966, file "OCC: Florida Air Pollution Commission," USPHS, NCAPC 1967–68.

27. Lee to Emil C. Jensen, Dec. 26, 1963, file "Cooperation 2—Florida," USPHS, DAP 1963–64. Regarding the Clean Air Act of 1963 and its enforcement provisions, see Ripley, "Congress and Clean Air," pp. 224–78.

28. Arthur C. Stern to Albert V. Hardy, May 20, 1960, file "Florida Air 3-1-1," USPHS, DAP 1961–62; memo from Roy O. McCaldin, May 25, 1960; memo from Vernon R. Hanson, July 6, 1960; memo from Ralph C. Graber, July 11, 1960, with attached, undated "Cooperative Project Agreement"; memo from Jean J. Schueneman, Dec. 8, 1960, with attached preliminary report, all four memos in same folder; Harding, McKee, and Schueneman, *Florida's Air Resources;* memo from Paul A. Humphrey, June 27, 1961, file "Florida—AP/61," USPHS, DAP 1961–62; memo from K. L. Johnson, Sept. 1, 1961, in same folder.

29. Brandt to Arthur C. Stern, Dec. 7, 1960, file "Florida Air 3-1-1," USPHS, APEB 1959–60; Stern to Huffstutler, Dec. 29, 1960, in same folder.

30. Memo from MacKenzie, June 7, 1963, file "Cooperation 2—Florida," USPHS, DAP 1963–64; David B. Lee to MacKenzie, June 18, 1963, and memo from R. J. Anderson, July 24, 1963, in same folder.

31. Brandt to MacKenzie, July 5, 1963, and MacKenzie to Brandt, July 18, 1963, in same folder.

32. Memo from Austin N. Heller, Dec. 20, 1963, in same folder.

33. "Minutes—Meeting of the Florida Air Pollution Control Commission, Tampa, Florida," Apr. 15, 1966, pp. 2–3, 7–9, file "OCC: Florida Air Pollution Commission," USPHS, NCAPC 1967–68; memo from S. Smith Griswold to Vernon G. MacKenzie, July 1, 1966, and memo from Dr. E. Blomquist, Arthur C. Stern, and Ralph C. Graber to Vernon G. MacKenzie, July 6, 1966, in same folder; memo from James A. Anderegg for the record, undated (apparently 1966), file "OCC-F," USPHS, NCAPC 1967–68.

## Chapter 12. The Three-Thousand-Mile-Long Sewer

1. Porter to Nicoll, May 17, 1966, U.S. Senate: Senate Office (329-2), Muskie Collection. Regarding the disorganized and disunited state of the federal air pollution control program during the 1960s, Clarence and Barbara Davies note that "there was little coordination among the component units of the division. The enforcement people did not talk to the technical assistance people, the research program was not tied to the control program, the grants were distributed without much thought being given to the operating needs of the agency." See Davies and Davies, *Politics of Pollution* 2e, p. 105.

2. Krier and Ursin, *Pollution and Policy,* pp. 89–91; *NYT,* Mar. 1, 1953, p. 64:1; *NYT,* Apr. 5, 1965, p. 1:2; Schiller, "The Los Angeles Smog," p. 564.

3. Louis C. McCabe, Chairman, *Air Pollution: Proceedings of the United States Technical Conference on Air Pollution,* pp. 109–14; Stanley Scott and John F. McCarty, *Air Pollution Control: 1951,* Legislative Problems, No. 7, p. 10; University of California–Berkeley Bureau of Public Administration, *Air Pollution Control,* p. 18.

4. *NYT,* May 15, 1957, p. 37:4; *NYT,* Oct. 1, 1967, p. 79:3.

5. Quotes from Assembly of the State of California, Subcommittee on Air Pollution and Radiation Protection, *Air Pollution, Its Health Effects, and Its Control,* pp. 20–27. For examples of books sounding the alarm about air pollution during the 1960s, see Howard R. Lewis, *With Every Breath You Take;* Donald E. Carr, *The Breath of Life;* Edward Edelson and Fred Warshofsky, *Poisons in the Air;* John C. Esposito and Larry J. Silverman, Project Directors, *Vanishing Air: The Ralph Nader Study Group Report on Air Pollution.*

6. *NYT,* June 8, 1966, p. 18:4; *NYT,* June 21, 1966, p. 14:3.

7. Esposito and Silverman, *Vanishing Air,* pp. 10–14; Barry Commoner, *Science and Survival,* p. 28.

8. Davies, *Politics of Pollution,* pp. 125–30; Ripley, "Congress and Clean Air," pp. 226–27.

9. Davies, *Politics of Pollution,* pp. 125–30; Ripley, "Congress and Clean Air," pp. 226–27; *NYT,* Dec. 7, 1970, p. 1:1.

10. See, for example, *NYT,* Apr. 9, 1965, p. 38:6; letter from Leonard Levine regarding phone conversation with Cecil M. Pepperman, Aug. 17, 1965, file "OCC-'K'-AP/65," box 53, Division of Air Pollution Correspondence, 1965–66, Records of the U.S. Public Health Service, Record Group 90, Identification Number NN3-090-91-004, National Archives, Washington, D.C. (hereafter cited as USPHS, DAP 1965–66); letter from Steven Szablewski, Aug. 8, 1966, file "OCC-'J'-AP/65," box 53; Thomas F. Williams to Marie Storlazzi, Dec. 9, 1966, file "OCC-'C'-AP/1965," box 42; letters of Oct. 27, Nov. 7, and Dec. 12, 1966, file "OCC-'T'-AP/1965," box 48, all in same collection.

11. Letter from Kenneth R. Brown, Dec. 2, 1966, file "EL Device—General, part 1/5—AP 66," box 20, in USPHS, DAP 1965–66; Irwin Auerbach to Brown, Dec. 13, 1966, in same folder;

Auerbach to Representative Dan H. Kuykendall of Tennessee, Aug. 22, 1967, file "IF 11-1—Congressional Inquiries, July–September/AP 67," Carton 9, USPHS, NCAPC 1967–68.

12. Kreutzer to Muskie, July 28, 1969, U.S. Senate: Senate Office (764-10), Muskie Collection.

13. True to Muskie, undated (probably 1969), U.S. Senate: Senate Office (625-11), Muskie Collection.

14. Owen to Muskie, Mar. 24, 1969, U.S. Senate: Senate Office (765-1), Muskie Collection.

15. Randy McLaughlin, "Talleyrand Pollution Fighters Win Hearing," *Jacksonville Journal,* Oct. 9, 1963, newspaper clipping in U.S. Senate: Senate Office (625-5), Muskie Collection.

16. Ibid.; Belcher to Florida Governor Farris Bryant, Dec. 5, 1963, in U.S. Senate: Senate Office (625-5), Muskie Collection; Belcher to Muskie, Jan. 2, 1964, and Mrs. Joseph C. McGuffy to Muskie, Jan. 6, 1964, in same folder.

17. Kai to Muskie, Sept. 21, 1966, U.S. Senate: Senate Office (595-1), and Martinez to Muskie, May 11, 1971, U.S. Senate: Senate Office (1504-7), both in Muskie Collection.

18. For more detailed information regarding pre-1970s union environmentalism, see Scott H. Dewey, "Working for the Environment: Organized Labor and the Origins of Environmentalism in the United States, 1948–1970," *Environmental History* 3, no. 1 (Jan., 1998): 45–63.

19. Snyder, "Death-Dealing Smog," pp. 117–39 and generally; H. H. Schrenk et al., *Air Pollution in Donora, Pennsylvania: Epidemiology of the Unusual Smog Episode of October, 1948;* Krier and Ursin, *Pollution and Policy,* pp. 102–104; Harry Heimann, "Effects on Human Health," in World Health Organization, *Air Pollution,* pp. 165–72.

20. Krier and Ursin, *Pollution and Policy,* pp. 104–105; Bernstein, "Air Pollution Control," pp. 66–68; Dyck, "Federal Air Pollution Control," pp. 21–22.

21. Bernstein, "Air Pollution Control," pp. 53–99; Dyck, "Federal Air Pollution Control," pp. 28–45; Krier and Ursin, *Pollution and Policy,* pp. 106–11; Ripley, "Congress and Clean Air," pp. 229–35; Sundquist, *Politics and Policy,* pp. 331–33.

22. Bernstein, "Air Pollution Control," pp. 53–99; Dyck, "Federal Air Pollution Control," pp. 28–45; Krier and Ursin, *Pollution and Policy,* pp. 106–11; Ripley, "Congress and Clean Air," pp. 229–35; Sundquist, *Politics and Policy,* pp. 331–33. The Kuchel bill became P.L. 84-159 when signed into law on July 14, 1955. Bernstein's dissertation offers the fullest discussion of the passage and implementation of the 1955 federal air pollution control act.

23. Davies, *Politics of Pollution,* pp. 50–53; Dyck, "Federal Air Pollution Control," pp. 45–58; Krier and Ursin, *Pollution and Policy,* pp. 169–73; Ripley, "Congress and Clean Air," pp. 235–78; Sundquist, *Politics and Policy,* pp. 351–55. Ripley offers the most thorough discussion of the political maneuvering in Congress and the federal executive branch leading to the passage of the Clean Air Act (P.L. 88-206).

24. Davies, *Politics of Pollution,* pp. 52–53; Krier and Ursin, *Pollution and Policy,* pp. 172–73; Ripley, "Congress and Clean Air," pp. 235–78; Sundquist, *Politics and Policy,* pp. 351–55.

25. Davies, *Politics of Pollution,* pp. 50, 53; Dyck, "Federal Air Pollution Control," pp. 63–76; Krier and Ursin, *Pollution and Policy,* pp. 173–74; Sundquist, *Politics and Policy,* pp. 367–71.

26. Davies, *Politics of Pollution,* pp. 54–55; Krier and Ursin, *Pollution and Policy,* pp. 179–80.

27. Air Quality Act of 1967, 81 Stat. 485; Davies, *Politics of Pollution,* pp. 54–58; Dyck, "Federal Air Pollution Control," pp. 77–124; Krier and Ursin, *Pollution and Policy,* pp. 180–81, 183–84. Dyck offers perhaps the fullest discussion of the process of passing the Air Quality Act of 1967.

28. The quote opening the paragraph is from the first edition of Davies's *Politics of Pollution*

(p. 58), published before the Clean Air Act Amendments of 1970. Regarding the continuing problems of efforts at interstate cooperation and abatement during the 1960s, see Esposito and Silverman, *Vanishing Air,* pp. 113–18, 121–29, 135–41, and Davies, *Politics of Pollution,* pp. 134–35, 138–40.

29. Esposito and Silverman, *Vanishing Air,* pp. 258–94, 310 (quote from p. 289).

30. Krier and Ursin, *Pollution and Policy,* pp. 199–205; Davies and Davies, *Politics of Pollution* 2e, pp. 52–56, 107–14; Shabecoff, *Fierce Green Fire,* pp. 129–31; Scheffer, *Shaping of Environmentalism,* pp. 143–44. Statutory citations for the National Environmental Policy Act of 1969 are P.L. 91-190 and 83 Stat. 852; for the Clean Air Act Amendments of 1970, P.L. 91-604.

31. Davies and Davies, *Politics of Pollution* 2e, pp. 103–105; Bernstein, "Air Pollution Control," pp. 22–23, 42–43, 101–104; Dyck, "Federal Air Pollution Control," pp. 43–44.

32. Bernstein, "Air Pollution Control," pp. 22–60, 91–93, 101–104, 111–26, 140–60; Dyck, "Federal Air Pollution Control," pp. 27, 30, 34–35, 43–44, 51, 63–65.

33. Esposito and Silverman, *Vanishing Air,* pp. 23–25, 51–62, 182–89.

34. Manufacturing Chemists' Association, Air Pollution Abatement Committee, Subcommittee on Legislation Principles, *A Rational Approach to Air Pollution Legislation,* pp. 3–12.

35. Ibid.

36. Bernstein, "Air Pollution Control," pp. 56 note 35, 82–86.

37. Bernstein, "Air Pollution Control," pp. 86–87, 235–37; Ripley, "Congress and Clean Air," p. 277; Davies, *Politics of Pollution,* pp. 90–96.

38. Copy of advertisement, letter from Norman Cousins to Senator Edmund S. Muskie, Sept. 26, 1966, and letter from Muskie to Cousins, Nov. 10, 1966, all in U.S. Senate: Senate Office (595-1), Muskie Collection.

39. *NYT,* May 19, 1971, p. 93:1; *NYT,* May 23, 1971, p. 59:1; *NYT,* May 28, 1971, p. 66:1.

40. Esposito and Silverman, *Vanishing Air,* pp. 26–33, 51–64; *NYT,* June 8, 1971, p. 45:3.

41. *NYT,* Feb. 9, 1965, p. 1:4 and 27:1–5; *NYT,* July 25, 1965, p. 51:4; Sundquist, *Politics and Policy,* p. 371. For further examples of the spread of ecological concepts among American scientists and in the media, see also *NYT,* Apr. 20, 1965, p. 41:8; *NYT,* Oct. 28, 1965, p. 45:8; *NYT,* Mar. 15, 1966, p. 19:1.

42. *NYT,* Dec. 13, 1966, p. 24:3; *NYT,* Dec. 18, 1966, sec. IV, p. 12:2. For the full text of Gardner's speech, see U.S. Department of Health, Education, and Welfare, National Center for Air Pollution Control, Office of Legislative and Public Affairs, *Proceedings: The Third National Conference on Air Pollution, Washington, D.C., December 12–14, 1966.*

43. *NYT,* Dec. 18, 1966, sec. IV, p. 12:2.

44. Sale, *Green Revolution,* pp. 11, 24–25; Shabecoff, *Fierce Green Fire,* pp. 111–21; Scheffer, *Shaping of Environmentalism,* pp. 124–25; *NYT,* Jan. 23, 1970, p. 1:8 ff.; *NYT,* Feb. 11, 1970, p. 1:8 ff.

45. Bernstein, "Air Pollution Control," pp. 235, 238–39; Melosi, *Coping with Abundance,* pp. 277–93.

46. Protesting workers quoted in Richard Kazis and Richard L. Grossman, *Fear at Work: Job Blackmail, Labor and the Environment,* p. 4. Watt quoted in Sale, *Green Revolution,* p. 102. Regarding corporate or popular anti-environmentalism, see generally David Helvarg, *The War against the Greens;* Carl Deal, *The Greenpeace Guide to Anti-Environmental Organizations,* pp. 6–22; John C. Stauber and Sheldon Rampton, "'Democracy' for Hire: Public Relations and Environmental Movements," *Ecologist* 25, no. 5 (Sept.–Oct., 1995): 173–80. Regarding union anti-environmentalism, see Kazis and Grossman, *Fear at Work,* pp. ix–xi, 1–5

ff.; Heinrich Siegman, *The Conflicts between Labor and Environmentalism in the Federal Republic of Germany and the United States,* pp. 23–27; Frederick H. Buttel, Charles C. Geisler, and Irving W. Wiswall, *Labor and the Environment: An Analysis of and Annotated Bibliography on Workplace Environmental Quality in the United States,* pp. 4–15; Hays, *Beauty, Health, and Permanence,* pp. 297–300, 302–303. Regarding the critique of environmentalists as overprivileged elitists or irrational fanatics, see Tucker, *Progress and Privilege;* Wildavsky, "Aesthetic Power," in *America's Changing Environment,* ed. Revelle and Landsberg, pp. 147–60; Bailey, *Eco-Scam;* Kaufman, *No Turning Back;* Ray, *Environmental Overkill.* Along with Rush Limbaugh, the arch-conservative, libertarian, anti-communist newsletter the *Spotlight* regularly fulminates against environmentalists with varying degrees of factual accuracy, as does the somewhat more moderate, populist-conservative *Middle-American News.*

47.  Barry Stavro, "A Hidden Mountain Treasure in the City," *Los Angeles Times,* May 19, 1998, p. A-1; Marc Stirdivant, "Battle for the Verdugos is about Quality of Life," *Foothill Leader,* May 27, 1998, p. A-9. Regarding environmentalism as protection of quality of life, see generally Hays, *Beauty, Health, and Permanence.*

48.  Shabecoff, *Fierce Green Fire,* pp. 203–30; Sale, *Green Revolution,* pp. 48–56; Edmund S. Muskie, "Why the U.S. Needs a Strong Clean Air Act," Lewiston (Maine) *Sun-Journal/Sunday,* Sept. 17, 1995, p. 2D. Regarding the defensive posture of the environmental movement in recent decades, see generally Dowie, *Losing Ground.*

49.  For reflections on the achievements and shortcomings of the environmental movement, see Shabecoff, *Fierce Green Fire,* pp. 266 ff. For commentary and criticism on moderate, "Third Wave" environmentalists who seek cooperation rather than conflict with industrialists, see Dowie, *Losing Ground,* pp. 105–24; Sale, *Green Revolution,* pp. 83–84; Shabecoff, *Fierce Green Fire,* pp. 257–61.

50.  Regarding the environmental impacts of global consumerism, see generally Durning, *How Much Is Enough?*

# Bibliography

## Archival and Manuscript Sources

Documents of the Air Pollution Engineering Branch, the Air Pollution Medical Program, the Division of Air Pollution, and the National Center for Air Pollution Control, in records of the United States Public Health Service, Record Group 90, National Archives and Records Administration, Washington, D.C.; College Park, Md.; and Suitland, Md.

Municipal Archives of the City of New York.

Papers of Los Angeles County Supervisor Kenneth F. Hahn, Henry E. Huntington Library, San Marino, Calif.

Papers of United States Senator Edmund S. Muskie, U.S. Senate: Senate Office, Edmund S. Muskie Collection, Edmund S. Muskie Archives, Bates College, Lewiston, Me.

## Other Sources

Aborn, Richard A., and Carl E. Axelrod. "Current Legislation: State Air Pollution Control Legislation." *Boston College Industrial and Commercial Law Review* 9, no. 3 (spring, 1968): 712–56.

Adler, Cy A. *Ecological Fantasies, or Death from Falling Watermelons: A Defense of Innovation, Science, and Rational Approaches to Environmental Problems.* New York: Green Eagle Press, 1973.

"The Advocates: Should the Sale of the Internal Combustion Engine Be Banned in California by 1975?" Transcript of "The Advocates" episode on Los Angeles public television station KCET, 7:00–8:00 P.M., Pacific Standard Time, October 5, 1969.

Air Pollution Foundation. *Final Report.* San Marino, Calif.: 1961.

Akerman, Joe A., Jr. *Florida Cowman: A History of Florida Cattle Raising.* Kissimmee: Florida Cattlemen's Association, 1976.

Allaby, Michael. *Air: The Nature of Atmosphere and the Climate.* New York: Facts on File, 1992.

Allison, Oscar Hugh. "Raymond R. Tucker: The Smoke Elimination Years, 1934–1950." Ph.D. dissertation, St. Louis University, 1978.

American Industrial Hygiene Association, Northern California Section, California State Department of Public Health. *Proceedings: First Northern California Air Pollution Symposium.* Berkeley: September 7, 1956.

American Petroleum Institute. *Memorandum on Conference on Air Pollution Research.* New York: December 19, 1959.

Anderson, Walt, ed. *Politics and Environment: A Reader in Ecological Crisis.* Pacific Palisades, Calif.: Goodyear Publishing Company, 1970.

Ashby, Eric, and Mary Anderson. *The Politics of Clean Air.* Oxford: Clarendon Press, 1981.

*The Automobile and Air Pollution: A Panel Discussion* (Transcript). Berkeley: Academic Publishing, January 26, 1970.

Ayres, Robert U. "Air Pollution in Cities." In *Politics and Environment: A Reader in Ecological Crisis,* ed. Walt Anderson. Pacific Palisades, Calif.: Goodyear Publishing Company, 1970.

Bach, Wilfrid. *Atmospheric Pollution.* New York: McGraw-Hill Book Company, 1972.

Bailey, Ronald. *Eco-Scam: The False Prophets of Ecological Apocalypse.* New York: St. Martin's Press, 1993.

Barbour, Ian G. *Technology, Environment, and Human Values.* New York: Praeger Publishers, 1980.

Bates, David V. *A Citizen's Guide to Air Pollution.* Montreal: McGill–Queen's University Press, 1972.

Battan, Louis J. *The Unclean Sky: A Meteorologist Looks at Air Pollution.* Garden City, N.J.: Anchor Books, 1966.

Beaver, Sir Hugh E. C. "The Growth of Public Opinion." In *Problems and Control of Air Pollution: Proceedings of the First International Congress on Air Pollution,* New York City, March 1–2, 1955, ed. Frederick S. Mallette, pp. 1–11. New York: Reinhold Publishing Corp., 1955.

Beckman, Arnold O. et al. *Report of Special Committee on Air Pollution Made to Governor Goodwin J. Knight's Air Pollution Control Conference.* Los Angeles: December 5, 1953.

Bernstein, Shawn. "The Rise of Air Pollution Control as a National Political Issue: A Study of Issue Development." Ph.D. dissertation, Columbia University, 1982.

Blakey, Arch Frederic. *The Florida Phosphate Industry: A History of the Development and Use of a Vital Mineral.* Cambridge, Mass.: Harvard University Press, 1973.

Bradley, William R., Chairman, New Jersey State Air Pollution Control Commission. "The Air Pollution Control Commission Program for New Jersey." In Proceedings of the Forty-Ninth Annual Meeting of the Air Pollution Control Association, Buffalo, N.Y., May 20–24, 1956, pp. 18-1 to 18-12. Pittsburgh: Air Pollution Control Association, 1956.

Bramwell, Anna. *Ecology in the Twentieth Century: A History.* New Haven, Conn.: Yale University Press, 1989.

———. *The Fading of the Greens: The Decline of Environmental Politics in the West.* New Haven, Conn.: Yale University Press, 1994.

Breslow, Lester. *Air Pollution: Effects Reported by California Residents.* Berkeley: California State Department of Public Health, 1958.

Brienes, Marvin. "The Fight against Smog in Los Angeles, 1943–1957." Ph.D. dissertation, University of California–Davis, 1975.

Brimblecombe, Peter. *The Big Smoke: A History of Air Pollution in London since Medieval Times.* London: Methuen, 1987.

Brimblecombe, Peter, and Catherine Bowler. "Air Pollution in York 1850–1900." In *The Silent Countdown: Essays in European Environmental History,* ed. Peter Brimblecombe and Christian Pfister, pp. 182–95. New York: Springer-Verlag, 1990.

Brimblecombe, Peter, and Christian Pfister, eds. *The Silent Countdown: Essays in European Environmental History.* New York: Springer-Verlag, 1990.

Brüggemeier, Franz-Josef. "The Ruhr Basin 1850–1980: A Case of Large-Scale Environmental Pollution." In *The Silent Countdown: Essays in European Environmental History,* ed. Peter Brimblecombe and Christian Pfister, pp. 210–27. New York: Springer-Verlag, 1990.

Bugler, Jeremy. *Polluting Britain: A Report.* Harmondsworth, U.K.: Penguin Books, 1972.

Bullard, Robert D., ed. *Unequal Protection: Environmental Justice and Communities of Color.* San Francisco: Sierra Club Books, 1994.

Burton, Ian, Robert W. Kates, and Gilbert F. White. *The Environment as Hazard.* New York: Oxford University Press, 1978.

Bush, A. F. et al. *The Effects of Engine Exhaust on the Atmosphere When Automobiles Are Equipped with Afterburners.* Los Angeles: University of California–Los Angeles, Department of Engineering, 1962.

Buttel, Frederick H., Charles C. Geisler, and Irving W. Wiswall. *Labor and the Environment: An Analysis of and Annotated Bibliography on Workplace Environmental Quality in the United States.* Westport, Conn.: Greenwood Press, 1984.

Cadle, Richard D., and Henry C. Wohlers. "Smog Lore." *Air Repair: Journal of the Air Pollution and Smoke Prevention Association of America* 1, no. 4 (May, 1952): 30–35.

Cagin, Seth, and Philip Dray. *Between Earth and Sky: How CFCs Changed Our World and Endangered the Ozone Layer.* New York: Pantheon Books, 1993.

Caldwell, Lynton Keith. *Environment: A Challenge for Modern Society.* Garden City, N.Y.: Natural History Press, 1970.

Caldwell, Lynton Keith, Lynton R. Hayes, and Isabel M. MacWhirter. *Citizens and the Environment: Case Studies in Popular Action.* Bloomington: Indiana University Press, 1976.

California Air Resources Board. *Reduction of Air Pollution by the Use of Natural Gas or Liquefied Petroleum Gas Fuels for Motor Vehicles.* Sacramento: March, 1970.

———. Technical Advisory Committee. *A Rational Program for Control of Lead in Motor Gasoline.* Sacramento: March 18, 1970.

California Air Resources Board, Los Angeles County Air Pollution Control District, and Western Oil and Gas Association. *Gasoline Modification: Its Potential as an Air Pollution Control Measure in Los Angeles County.* Los Angeles: November, 1969.

California State Assembly Committee on Transportation and Commerce. *A Series of Interim Hearings on Air Pollution Control.* Sacramento: 1966.

California State Assembly Interim Committee on Governmental Efficiency and Economy. *Report of the Investigation into the Control of Smog.* Sacramento: April 1, 1949.

———. *Report of the Investigation into the Control of Smog.* Sacramento: July, 1949.

———. *Study and Analysis of the Facts Pertaining to Air Pollution Control in Los Angeles County.* Sacramento: March 5, 1953.

California State Assembly Interim Committee on Public Health, Subcommittee on Air Pollution. *Motor Vehicle Created Air Pollution: A Control Program for California.* Sacramento: December, 1960.

California State Assembly Interim Committee on Public Health, Subcommittee on Air Pollution and Radiation Protection. *Air Pollution, Its Health Effects, and Its Control.* Sacramento: March, 1959.

California State Assembly Interim Committee on Public Health, Subcommittee on the Health Effects of Smog. *Air Pollution and the Public Health.* Sacramento: March 20, 1957.

California State Assembly Interim Fact-Finding Committee on Water Pollution. *Interim Report on Air Pollution Research.* Sacramento: April 13, 1950.

California State Chamber of Commerce. *Proceedings of the Second Southern California Conference on Elimination of Air Pollution: Twelve Months' Progress—and the Year Ahead.* Los Angeles: November 14, 1956.

———. *Proceedings of the Third Southern California Conference on Elimination of Air Pollution: Moving Closer to Clean Air, Featuring If's, And's and But's of Automobile Exhaust Control.* Los Angeles: November 21, 1957.

———. *Proceedings of the Valley Air Pollution Conference,* Stockton, Calif.: January 31, 1958.

California State Department of Public Health. *California Standards for Ambient Air and Motor Vehicle Exhaust.* Berkeley: December 4, 1959.

————. *Clean Air for California: Initial Report of the Air Pollution Study Project.* San Francisco: March, 1955.

————. *Lead in the Environment and Its Effects on Humans.* Berkeley: March, 1967.

————. *Progress Report: Air Pollution.* Berkeley: July 17, 1962.

California State Motor Vehicle Pollution Control Board. *Miscellaneous Publications* (collection), vol. 1. Berkeley: various dates.

————. *Our Exhausted Air: California's Fight against Automobile Smog—The Story of the California Motor Vehicle Pollution Control Board.* Sacramento: June, 1965.

California State Office of Emergency Services. *Community Emergency Plan: San Francisco.* Sacramento: undated [1960s].

Calvert, Seymour, and Harold M. Englund, eds. *Handbook of Air Pollution Technology.* New York: John Wiley and Sons, 1984.

Campbell, Rex R., and Jerry L. Wade, eds. *Society and Environment: The Coming Collision.* Boston: Allyn and Bacon, 1972.

Carlin, Alan P., and George E. Kocher. *Environmental Problems: Their Causes, Cures, and Evolution, Using Southern California Smog as an Example.* Santa Monica, Calif.: Rand Corporation, Report Number R-640-CC/RC, May, 1971.

Carr, Donald E. *The Breath of Life.* New York: W. W. Norton and Company, 1965.

Carson, Rachel. *Silent Spring.* Boston: Houghton Mifflin, 1962.

Carter, F. W., and D. Turnock, D., eds. *Environmental Problems in Eastern Europe.* New York: Routledge, 1993.

Cassell, Eric J. "The Health Effects of Air Pollution and Their Implications for Control." *Law and Contemporary Problems* 33, no. 2 (spring, 1968): 197–216.

Chambers, Leslie A. "Classification and Extent of Air Pollution Problems." In *Air Pollution* 3e, edited by Arthur C. Stern, 1:1–33. New York: Academic Press, 1976.

Chase, Anthony R. "Assessing and Addressing Problems Posed by Environmental Racism." *Rutgers Law Review* 45 (1993): 335–69.

Chass, Robert L., and Edward S. Feldman. "Tears for John Doe." *Southern California Law Review* 27, no. 4 (July, 1954): 349–72.

Christy, William G. "Area Air Pollution Surveys." In *Proceedings of the Forty-Eighth Annual Meeting of the Air Pollution Control Association,* Detroit, Michigan, 1955, pp. 12-1 to 12-10. Pittsburgh: Air Pollution Control Association, 1955.

City College of the City University of New York, School of Engineering and Architecture. *Proceedings of the Three-State Conference on Air Resource Management.* New York: May, 1967.

City of Cincinnati et al. *The Bridge to Clean Air.* Cincinnati: undated [probably 1965].

Clapp, Brian William. *An Environmental History of Britain since the Industrial Revolution.* New York: Longman Publishing, 1994.

Clarkson, Diana, and John T. Middleton. *The California Control Program for Motor Vehicle Created Air Pollution.* Berkeley: California State Motor Vehicle Pollution Control Board, 1961.

Clay, Grady. "Copper Basin Cover-up." *Landscape Architect* (July, 1983): 49–55 ff.

Cobb, James C. *Industrialization and Southern Society, 1877–1984.* Lexington: University Press of Kentucky, 1984.

Commoner, Barry. *Science and Survival.* New York: Viking Press, 1966.

————. *The Closing Circle: Nature, Man and Technology.* New York: Alfred A. Knopf, 1971.

————. *Making Peace with the Planet.* New York: Pantheon Books, 1990.

Costain, W. Douglas, and James P. Lester. "The Evolution of Environmentalism." In *Environmental Politics and Policy: Theories and Evidence* 2e, ed. James P. Lester, pp. 15–38. Durham, N.C.: Duke University Press, 1995.

Cousins, Norman, Chairman. *Freedom to Breathe: Report of the Mayor's Task Force on Air Pollution in the City of New York*. New York: Office of the Mayor of the City of New York, June 20, 1966.

———. *Evaluation of the Department of Air Resources since the Completion of the Report of the Mayor's Task Force on Air Pollution in 1966*. City of New York, October, 1969.

Cowan, Thomas A. "Air Pollution Control in New Jersey." *Rutgers Law Review* 9, no. 4 (summer, 1955): 607–33.

Crandall, Robert W. *Controlling Industrial Pollution: The Economics and Politics of Clean Air*. Washington, D.C.: Brookings Institution, 1983.

Crenson, Matthew A. *The Un-Politics of Air Pollution: A Study of Non-Decisionmaking in the Cities*. Baltimore: Johns Hopkins University Press, 1971.

Crist, Raymond E. "The Citrus Industry in Florida." *American Journal of Economics and Sociology* 15, no. 1 (October, 1995): 1–11.

Crocker, Thomas D. "Some Economics of Air Pollution Control." Ph.D. dissertation, University of Missouri, 1967.

Cronon, William. *Changes in the Land: Indians, Colonists, and the Ecology of New England*. New York: Hill and Wang, 1983.

———. *Nature's Metropolis: Chicago and the Great West*. New York: W. W. Norton and Company, 1991.

———, ed. *Uncommon Ground: Rethinking the Human Place in Nature*. New York: W. W. Norton and Company, 1988.

Currie, David P. "Motor Vehicle Air Pollution: State Authority and Federal Pre-Emption." *Michigan Law Review* 68, no. 6 (May, 1970): 1083–1102.

Daines, Robert H., Ida Leone, and Eileen Brennan. "The Effect of Fluorine on Plants as Determined by Soil Nutrition and Fumigation Studies." In *Air Pollution: Proceedings of the United States Technical Conference on Air Pollution,* Louis C. McCabe, Chairman, pp. 97–105. New York: McGraw-Hill Book Company, 1952.

Daines, Robert H., Ida A. Leone, and Eileen Brennan. *Air Pollution as It Effects Agriculture in New Jersey*. New Jersey Agricultural Experiment Station Bulletin 794. New Brunswick: Rutgers University, March, 1965.

Darnovsky, Marcy. "Stories Less Told: Histories of US Environmentalism." *Socialist Review* 22, no. 4 (October–December, 1992): 11–54.

Dasmann, Raymond F. *No Further Retreat: The Fight to Save Florida*. New York: Macmillan, 1971.

———. *The Destruction of California*. New York: Collier Books, 1971.

Davenport, S. J., and G. G. Morgis. *Air Pollution: A Bibliography*. U.S. Department of Interior, Bureau of Mines, Bulletin 537. Washington, D.C.: U.S. Government Printing Office, 1954.

Davies, J. Clarence III. *The Politics of Pollution*. New York: Pegasus, 1970.

Davies, J. Clarence III, and Barbara S. Davies. *The Politics of Pollution* 2e. Indianapolis: Bobbs-Merrill Company, 1975.

De Bell, Garrett. *The Environmental Handbook, Prepared for the First National Environmental Teach-In*. New York: Ballantine Books, 1970.

Deal, Carl. *The Greenpeace Guide to Anti-Environmental Organizations*. Berkeley: Odonian Press, 1993.

Degler, Carl N. *At Odds: Women and the Family in America from the Revolution to the Present.* New York: Oxford University Press, 1980.

Dewey, Scott H. "'The Antitrust Case of the Century': Kenneth F. Hahn and the Fight against Smog." *Southern California Quarterly* 81, no. 3 (fall, 1999): 341–76.

———. "The Fickle Finger of Phosphate: Central Florida Air Pollution and the Failure of Environmental Policy, 1957–1970." *Journal of Southern History* 65, no. 3 (Aug., 1999): 565–603.

———. "'Is This What We Came to Florida for?': Florida Women and the Fight against Air Pollution in the 1960s," which appeared in the *Florida Historical Quarterly* 77, no. 4 (spring, 1999): 503–31.

———. "Part Cause, Part Cure: The Changing Relationship between Air Pollution and Aviation in the United States, 1927–1973." In *The Meaning of Flight in the 20th Century: Proceedings of the National Aerospace Conference at Wright State University, Dayton, Ohio, October 1998,* pp. 344–54. Dayton: Wright State University, 1999.

———. "Working for the Environment: Organized Labor and the Origins of Environmentalism in the United States, 1948–1970," *Environmental History* 3, no. 1 (Jan., 1998): 45–63.

Ditzel, Paul. *Smog in Perspective* (a series of four articles). Los Angeles: Automobile Club of Southern California, January, 1970.

Doty, Robert Adam. "Life Cycle Theories of Regulatory Agency Behavior: The Los Angeles Air Pollution Control District." Ph.D. dissertation, University of California–Riverside, June, 1978.

Doty, Robert Adam, and Leonard Levine. *Profile of an Air Pollution Controversy: Volume II.* Riverside, Calif.: Clean Air Now, 1973.

Dowie, Mark. *Losing Ground: American Environmentalism at the Close of the Twentieth Century.* Cambridge, Mass.: MIT Press, 1995.

Durning, Alan Thein. *How Much Is Enough? The Consumer Society and the Future of the Earth.* New York: W. W. Norton and Company, 1992.

Dyck, Robert Gilkey. "Evolution of Federal Air Pollution Control Policy." Ph.D. dissertation, University of Pittsburgh, 1971.

Easterbrook, Gregg. *A Moment on the Earth: The Coming Age of Environmental Optimism.* New York: Viking–Penguin Books USA, 1995.

Edelman, Sidney. "The Law of Federal Air Pollution Control." *Journal of the Air Pollution Control Association* (October, 1966): 523–25.

Edelson, Edward, and Fred Warshofsky. *Poisons in the Air.* New York: Pocket Books, 1966.

Efron, Edith. *The Apocalyptics: Cancer and the Big Lie—How Environmental Politics Controls What We Know about Cancer.* New York: Simon and Schuster, 1984.

Eigen, Lewis D., and Jonathan P. Siegel, eds. *The Macmillan Dictionary of Political Quotations.* New York: Macmillan, 1993.

Eller, Ronald D. *Miners, Millhands, and Mountaineers: The Industrialization of the Appalachian South, 1880–1930.* Knoxville: University of Tennessee Press, 1982.

Elsom, Derek. *Atmospheric Pollution: Causes, Effects and Control Policies.* London: Basil Blackwell, 1987.

Epstein, Samuel S. *The Politics of Cancer.* San Francisco: Sierra Club Books, 1978.

Esposito, John C., and Larry J. Silverman, Project Directors. *Vanishing Air: The Ralph Nader Study Group Report on Air Pollution.* New York: Grossman Publishers, 1970.

Evelyn, John. *Fumifugium, or the Inconveniencie of the Aer and the Smoak of London Dissipated:*

*Together with Some Remedies Humbly Proposed.* Reprinted in *The Smoak of London: Two Prophecies,* ed. James P. Lodge, Jr., pp. 1–29. Elmsford, N.Y.: Maxwell Reprint Co., 1969.

Faith, W. L. *Air Pollution Control.* New York: John Wiley and Sons, 1959.

———. *Combustion and Smog: The Relation of Combustion Processes to Los Angeles Smog.* Los Angeles: Air Pollution Foundation, September, 1954.

Faith, W. L., and Arthur A. Atkisson, Jr. *Air Pollution* 2e. New York: John Wiley and Sons, 1972.

Farber, Seymour M. *The Air We Breathe: A Study of Man and His Environment.* Springfield, Ill.: Thomas Publishing Co., 1961.

Fawcett, Jeffry. *The Political Economy of Smog in Southern California.* New York: Garland Publishing, 1990.

Feshbach, Murray, and Alfred Friendly, Jr. *Ecocide in the USSR: Health and Nature under Siege.* New York: Basic Books, 1992.

First National City Bank of New York Economics Department. *Profile of a City.* New York: McGraw-Hill Book Company, 1972.

Fitch, Lyle C., Acting City Administrator. *Air Pollution Control: Organization and Operation—Report of the City Administrator.* New York: Office of the Mayor, City of New York, March 1, 1960.

Flanagan, Maureen A. "The City Profitable, the City Livable: Environmental Policy, Gender, and Power in Chicago in the 1910s." *Journal of Urban History* 22, no. 2 (January, 1996): 163–90.

Flavin, Christopher, and Odil Tunali. *Climate of Hope: New Strategies for Stabilizing the World's Atmosphere.* Worldwatch Paper 130. Washington, D.C.: Worldwatch Institute, June, 1996.

Flick, Carlos. "The Movement for Smoke Abatement in 19th-Century Britain," *Technology and Culture* 21 no. 1 (January, 1980): 29–50.

Florida Phosphate Council. *Phosphate: Florida's Hidden Blessing—Mineral of Life.* Lakeland, 1966.

Florida State Department of Pollution Control. *Air Pollution Control in Florida.* Tallahassee, July, 1973.

Friends of the Earth. *How Green is Britain?* London: Hutchinson Radius, 1990.

Fromson, Jeffrey. "A History of Federal Air Pollution Control." *Ohio State Law Journal* 30, no. 3 (summer, 1969): 516–36.

Fuller, Jack Richard. "Politics, Pollution and Human Values." Ph.D. dissertation, University of Washington, 1972.

Gannon, Michael. *Florida: A Short History.* Gainesville: University Press of Florida, 1993.

Gardner, John W. "Statement of Honorable John W. Gardner, Secretary of Health, Education, and Welfare," *Hearings before the Committee on Interstate and Foreign Commerce, House of Representatives on HR 9509 and S. 780.* Washington, D.C.: Department of Health, Education, and Welfare, 1967.

General Motors Corporation. *Progress of Power: A General Motors Report on Vehicular Power Systems.* Warren, Mich.: General Motors Technical Center, May 7–8, 1969.

Gerhardt, Paul H. "Incentives to Air Pollution Control." *Law and Contemporary Problems* 33, no. 2 (spring, 1968): 358–68.

Goedhard, Neil. *Report No. 6: Basic Statistics of the Los Angeles Area.* Los Angeles: Air Pollution Foundation, January, 1955.

Goldman, Marshall I. *The Spoils of Progress: Environmental Pollution in the Soviet Union.* Cambridge, Mass.: MIT Press, 1972.

Goldsmith, John R. *Health Effects of Motor Vehicle Exhaust.* Los Angeles: California State Motor Vehicle Pollution Control Board, August 11, 1961.

Goldsmith, John R., and Lars T. Friberg. "Effects of Air Pollution on Human Health." In *Air Pollution* 3e, ed. by Arthur C. Stern, 2:458–511. New York: Academic Press, 1976.

Gottlieb, Robert. *Forcing the Spring: The Transformation of the American Environmental Movement.* Washington, D.C., and Covelo, Calif.: Island Press, 1993.

Grad, Frank P. et al. *The Automobile and the Regulation of Its Impact on the Environment.* Norman: University of Oklahoma Press, 1975.

Graebner, William. "Hegemony through Science: Information Engineering and Lead Toxicology, 1925–1965." In *Dying for Work: Workers' Safety and Health in Twentieth-Century America,* ed. David P. Rosner and Gerald R. Markowitz, pp. 140–59. Bloomington: Indiana University Press, 1987.

Grant, Wyn. *Autos, Smog and Pollution Control: The Politics of Air Quality Management in California.* Brookfield, Vt.: Edward Elgar Publishing Co., 1995.

Green, Lewis C. "State Control of Interstate Air Pollution." *Law and Contemporary Problems* 33, no. 2 (spring, 1968): 315–30.

Greenberg, Michael R. *Urbanization and Cancer Mortality: The United States Experience, 1950–1975.* New York: Oxford University Press, 1983.

Greenburg, Leonard et al. "Report of an Air Pollution Incident in New York City, November 1953." *Public Health Reports* (January, 1962): 7–16.

Greenburg, Leonard, and Morris B. Jacobs. "Sulphur Dioxide in New York City Atmosphere." *Industrial and Engineering Chemistry* 48, no. 9 (September, 1956): 1517–21.

Greenwood, D. A., R. A. Call, J. V. Hales, and J. P. Kesler. *Progress Report of a Research Project on the Effect of Atmospheric Fluorides on Man.* Washington, D.C.: National Institutes of Health Division of Research Grants, September 25, 1956.

Grinder, Robert Dale. "The Anti-Smoke Crusades: Early Attempts to Reform the Urban Environment, 1893–1918." Ph.D. dissertation, University of Missouri–Columbia, 1973.

———. "The Battle for Clean Air: The Smoke Problem in Post–Civil War America." In *Pollution and Reform in American Cities, 1870–1930,* ed. Martin Melosi, pp. 83–103. Austin: University of Texas Press, 1980.

Griswold, S. Smith. *Study and Recommendations on Air Pollution Control in the County of Los Angeles.* Los Angeles: County Board of Supervisors, October 1, 1954.

Gruber, Charles W. *Air Pollution Control in Cincinnati.* Cincinnati: City of Cincinnati Department of Safety, 1964.

Haagen-Smit, Arie J., C. E. Bradley, and M. M. Fox. "Ozone Formation in Photochemical Oxidation of Organic Substances." *Industrial and Engineering Chemistry* 45, no. 9 (September, 1953): 2086–89.

Hagevik, George H. *Decision-Making in Air Pollution Control: A Review of Theory and Practice, with Emphasis on Selected Los Angeles and New York City Management Experiences.* New York: Praeger Publishers, 1970.

———. "Legislating for Air Quality Management: Reducing Theory to Practice." *Law and Contemporary Problems* 33, no. 2 (spring, 1968): 369–98.

Hahn, Kenneth F. *Smog: A Factual Record of Correspondence between Kenneth Hahn and Automobile Companies* 6e. Los Angeles: County Board of Supervisors, 1970.

Halliday, E. C. "A Historical Review of Atmospheric Pollution." In World Health Organization, *Air Pollution,* pp. 39–48. New York: Columbia University Press, 1961.

Hardin, Garrett. "The Tragedy of the Commons." *Science* 162, no. 3859 (December 13, 1968): 1243–48.

Harding, Charles I. *Final Progress Report: Greater Jacksonville Air Pollution Control Program.* Gainesville: University of Florida College of Engineering, Bioenvironmental Engineering Department, August, 1966.

Harding, Charles I. et al. *Clearing the Air in Jacksonville.* Jacksonville: Air Improvement Committee of Greater Jacksonville, October, 1967.

Harding, Charles I., Samuel B. McKee, and Jean J. Schueneman. *A Report on Florida's Air Resources.* Jacksonville: Florida State Board of Health, February, 1961.

Harris, Lewis D. "The Florida Phosphate Industry and Air Pollution." Master's thesis, Florida State University, August, 1967.

Harris, Louis, and Associates. *The Public's View of Environmental Problems in the State of New York: Survey Conducted for the New York State Council of Environmental Advisers,* May, 1971.

Hays, Samuel P. *Beauty, Health, and Permanence: Environmental Politics in the United States, 1955–1985.* New York: Cambridge University Press, 1987.

———. *Conservation and the Gospel of Efficiency: The Progressive Conservation Movement, 1890–1920.* Cambridge, Mass.: Harvard University Press, 1959.

———. "The Structure of Environmental Politics since World War II." *Journal of Social History* 14, no. 4 (1981): 719–38.

Heifetz, Ruth. "Women, Lead, and Reproductive Hazards: Defining a New Risk." In *Dying for Work: Workers' Safety and Health in Twentieth-Century America,* ed. David P. Rosner and Gerald R. Markowitz, pp. 160–74. Bloomington: Indiana University Press, 1987.

Heimann, Harry. "Effects on Human Health." In World Health Organization, *Air Pollution,* pp. 159–220. New York: Columbia University Press, 1961.

Helvarg, David. *The War against the Greens.* San Francisco: Sierra Club Books, 1994.

Hill, Gladwin. "The Politics of Pollution." *Nation* (October 11, 1965): 220–23.

Hilleboe, Herman E., Chairman. *A Review of Air Pollution in New York State.* Albany: New York State Air Pollution Control Board, July, 1958.

———. *Rules to Prevent New Air Pollution.* Albany: New York State Air Pollution Control Board, 1962.

Hoffman, George A. *Los Angeles Smog Control.* Los Angeles: University of California–Los Angeles, Institute of Government and Public Affairs, April, 1966.

Holden, Constance. "Hazel Henderson: Nudging Society off Its Macho Trip." *Science* 190 (November 28, 1975): 862–64.

Holmes, Joseph A., Edward C. Franklin, and Ralph A. Gould. *Report of the Selby Smelter Commission, with Accompanying Papers.* Department of the Interior, United States Bureau of Mines, Bulletin 98. Washington, D.C.: Government Printing Office, 1915.

Hoy, Suellen M. "'Municipal Housekeeping': The Role of Women in Improving Urban Sanitation Practices, 1880–1917." In *Pollution and Reform in American Cities, 1870–1930,* ed. Martin Melosi, pp. 173–98. Austin: University of Texas Press, 1980.

Hurley, Andrew. "Challenging Corporate Polluters: Race, Class, and Environmental Politics in Gary, Indiana, since 1945," *Indiana Magazine of History* 88, no 4 (December 1992), pp. 273–302.

Interstate Sanitation Commission. *Progress Report on Air Pollution.* New York, January, 1957.

————. *Report on Air Pollution Study, 1955.* New York, 1955.

————. *Smoke and Air Pollution: New York–New Jersey.* New York, February, 1958.

Jackson and Moreland, Division of United Engineers and Constructors. *New York City's Air Pollution Problem: Another Look.* Summary Report, Survey for American Petroleum Institute. Boston: Jackson and Moreland–American Petroleum Institute, May, 1967.

Johnson, Lamar. *Beyond the Fourth Generation.* Gainesville: University Press of Florida, 1974.

Johnstone, H. F. *Technical Aspects of the Los Angeles Smog Problem: A Report Prepared for the Los Angeles County Air Pollution Control District.* Los Angeles: County Board of Supervisors, May 15, 1948.

Jones, Charles O. *Clean Air: The Policies and Politics of Pollution Control.* Pittsburgh: University of Pittsburgh Press, 1975.

*Journal of the Air Pollution Control Association,* various editions.

Kaufman, Wallace. *No Turning Back: Dismantling the Fantasies of Environmental Thinking.* New York: Basic Books, 1994.

Kavaler, Lucy. "How Los Angeles Women are Fighting Smog—and Winning." *Family Circle,* Sept., 1968, pp. 55, 99–100.

Kazis, Richard, and Richard L. Grossman. *Fear at Work: Job Blackmail, Labor and the Environment.* New York: Pilgrim Press, 1982.

Kehoe, Terrance. "'You Alone Have the Answer': Lake Erie and Federal Water Pollution Control Policy, 1960–1972." *Journal of Policy History* 8, no. 4 (1996): 440–69.

Kennedy, Harold W. *The History, Legal and Administrative Aspects of Air Pollution Control in the County of Los Angeles: Report Submitted to the Board of Supervisors of the County of Los Angeles, May 9, 1954.* Los Angeles: 1954.

Kennedy, Harold W., and Martin E. Weekes. "Control of Automobile Emissions: California Experience and the Federal Legislation." *Law and Contemporary Problems* 33, no. 2 (spring, 1968): 297–314.

Klein, Maury, and Harvey A. Kantor. *Prisoners of Progress: American Industrial Cities, 1850–1920.* New York: Macmillan, 1976.

Komarov, Boris. *The Destruction of Nature in the Soviet Union.* White Plains, N.Y.: M. E. Sharpe, 1980.

Krier, James E., and Edmund Ursin. *Pollution and Policy: A Case Essay on California and Federal Experience with Motor Vehicle Air Pollution, 1940–1975.* Berkeley: University of California Press, 1977.

Kuhn, Thomas S. *The Structure of Scientific Revolutions* 2e. Chicago: University of Chicago Press, 1970.

La Gumina, Salvatore John. *New York at Mid-Century: The Impellitteri Years.* Westport, Conn.: Greenwood Press, 1992.

Lagarias, John S. "The Story of the Air Pollution Control Association: Seventy-Five Years of Growth." *Journal of the Air Pollution Control Association* 32, no. 1 (January, 1982): 31–43.

Lamare, Judith Louise. "Urban Mass Transportation Politics in the Los Angeles Area: A Case Study in Metropolitan Policy-Making." Ph.D. dissertation, University of California–Los Angeles, 1973.

Largent, Edward J. "The Effects of Air-borne Fluorides on Livestock." In *Air Pollution: Proceedings of the United States Technical Conference on Air Pollution,* Louis C. McCabe, Chairman, pp. 64–72. New York: McGraw-Hill Book Company, 1952.

Layton, Edwin T., Jr. *The Revolt of the Engineers: Social Responsibility and the American Engineering Profession.* Cleveland: Press of Case Western Reserve University, 1971.

Lehrer, Tom. *Too Many Songs by Tom Lehrer.* New York: Pantheon Books, 1981.

Leighton, Philip A. "Man and Air in California." Unpublished paper presented at Statewide Conference on Man in California, Sacramento, California, January 27, 1964.

Lewis, Howard R. *With Every Breath You Take: The Poisons of Air Pollution, How They Are Injuring Our Health, and What We Must Do About Them.* New York: Crown Publishers, 1965.

Lester, James P., ed. *Environmental Politics and Policy: Theories and Evidence* 2e. Durham: Duke University Press, 1995.

Lieber, Harvey. "The Politics of Air and Water Pollution Control in the New York Metropolitan Area." Ph.D. dissertation, Columbia University, 1968.

Lillard, Richard. *The Great Forest.* New York: Alfred A. Knopf, 1948.

Linsky, Benjamin. *Fifty Years Backward and Forward in Air Pollution Control.* San Francisco: Bay Area Air Pollution Control District, 1957.

List, E. J. *Energy Use in California: Implications for the Environment.* Pasadena: California Institute of Technology, Environmental Quality Laboratory, December, 1971.

Long, Harold W. *A Study of the Sulfur Oxide Ambient Air Concentration in Polk County, Florida, December 1964–October 1965.* Jacksonville: Florida State Board of Health, Bureau of Sanitary Engineering, April, 1966.

Los Angeles County Air Pollution Control District. *Annual Report, 1954–1955.* Los Angeles, 1955.

————. *APCD: History and Function.* Los Angeles: undated [approximately 1977].

————. *Crossing the Smog Barrier: A Factual Account of Southern California's Fight against Air Pollution.* Reprinted from the Newspapers of Los Angeles County. Los Angeles, 1957.

————. *Profile of Air Pollution Control in Los Angeles County.* Los Angeles, January, 1969.

————. *Testimony of S. Smith Griswold, Dr. Leslie A. Chambers, and Hoyt R. Crabaugh before the Select Committee on Small Business of the House of Representatives at Los Angeles, California, May 18 and 19, 1956.* Los Angeles, 1956.

Los Angeles County Medical Association, Los Angeles County Tuberculosis and Health Association. *Physicians' Environmental Health Survey: A Poll of Medical Opinion.* Los Angeles, May, 1961.

Los Angeles County Office of Air Pollution Control. *First Annual Report of Office of Air Pollution Control, 1945–1946.* Los Angeles: County Board of Supervisors, September 30, 1946.

*Los Angeles Times,* various editions.

Low, Robert A., Chairman. *Air Pollution in New York City: An Interim Technical Report of the Special Committee to Investigate Air Pollution.* New York: City Council of New York, June 22, 1965.

Ludwig, John H. "Air Pollution Control Technology: Research and Development on New and Improved Systems." *Law and Contemporary Problems* 33, no. 2 (spring, 1968): 217–26.

Lunche, Robert G. et al. *Air Pollution Engineering in Los Angeles County.* Los Angeles: Los Angeles County Air Pollution Control District, July 1, 1966.

Maddox, John. *The Doomsday Syndrome.* New York: McGraw-Hill Book Company, 1972.

Mallette, Frederick S., ed. *Problems and Control of Air Pollution: Proceedings of the First International Congress on Air Pollution,* New York City, March 1–2, 1955. New York: Reinhold Publishing Corp., 1955.

Manufacturing Chemists' Association, Air Pollution Abatement Committee, Subcommittee on

Legislation Principles. *A Rational Approach to Air Pollution Legislation.* Washington, D.C.: Manufacturing Chemists' Association, February, 1952.

Martin, Robert, and Lloyd Symington. "A Guide to the Air Quality Act of 1967." *Law and Contemporary Problems* 33, no. 2 (spring, 1968): 239–74.

Commonwealth of Massachusetts. *Report Submitted by the Legislative Research Council Relative to Air Pollution in the Metropolitan Boston Area.* Boston, February 5, 1960.

McCabe, Louis C., Chairman. *Air Pollution: Proceedings of the United States Technical Conference on Air Pollution.* New York: McGraw Hill Book Company, 1952.

McCarthy, Kathleen D. "Parallel Power Structures: Women and the Voluntary Sphere." In *Lady Bountiful Revisited: Women, Philanthropy, and Power,* ed. Kathleen D. McCarthy, pp. 1–29. New Brunswick: Rutgers University Press, 1990.

McCluney, Ross, ed. *The Environmental Destruction of South Florida: A Handbook for Citizens.* Coral Gables: University of Miami Press, 1971.

McGill, John T. *Report to the Supreme Court of the United States in the Cause, Number 1, Original, October Term, 1914,* The State of Georgia, Complainant, vs. The Tennessee Copper Company and The Ducktown Sulphur, Copper & Iron Company, Limited, *on Motion to Enter a Final Decree Against the Ducktown Sulfur, Copper & Iron Company.* Nashville: January 1, 1916.

McNickle, Chris. *To Be Mayor of New York: Ethnic Politics in the City.* New York: Columbia University Press, 1993.

Meetham, A. R. et al. *Atmospheric Pollution: Its History, Origins and Prevention* 4e. New York: Pergamon Press, 1981.

Melosi, Martin V. *Coping with Abundance: Energy and Environment in Industrial America.* Philadelphia: Temple University Press, 1985.

———. *Garbage in the Cities: Refuse, Reform, and the Environment, 1880–1980.* College Station: Texas A&M University Press, 1981.

———. "The Place of the City in Environmental History." *Environmental History Review* 17, no. 1 (spring, 1993): 1–23.

———, ed. *Pollution and Reform in American Cities, 1870–1930.* Austin: University of Texas Press, 1980.

Mendelsohn, Robert S. *Confessions of a Medical Heretic.* New York: Warner Books, 1979.

Merchant, Carolyn. "Women of the Progressive Conservation Movement, 1900–1916." *Environmental Review* 8, no. 1 (spring, 1984): 57–85.

Mieck, Ilja. "Luftverunreinigung und Imissionsschutz in Frankreich und Preussen zur Zeit der frühen Industrialisierung [Air pollution and emissions control measures in France and Prussia during the period of early industrialization]." *Technikgeschichte* 48, no. 3 (1981): 239–51.

Mills, Clarence A. *Air Pollution and Community Health.* Boston: Christopher Publishing House, 1954.

Moran, Gerald R. "The Air Pollution Control Act and Its Administration." *Rutgers Law Review* 9, no. 4 (summer, 1955): 640–81.

Mowrey, Marc, and Tim Redmond. *Not in Our Backyard: The People and Events that Shaped America's Modern Environmental Movement.* New York: William Morrow and Company, 1993.

Mumford, Lewis. *The City in History.* New York: Harcourt, Brace Jovanovich, 1961.

Murray, Keith A. "The Trail Smelter Case: International Air Pollution in the Columbia Valley," *BC Studies,* no. 15 (autumn, 1972): 68–82.

National Research Council Board on Agriculture and Renewable Resources, Committee on Animal Nutrition, Subcommittee on Fluorosis. *Effects of Fluorides in Animals*. Washington, D.C.: National Academy of Sciences, 1974.

National Research Council Commission on Life Sciences, Board on Environmental Studies and Toxicology, Committee on Toxicology, Subcommittee on Health Effects of Ingested Fluoride. *Health Effects of Ingested Fluoride*. Washington, D.C.: National Academy Press, 1993.

National Research Council Division of Medical Sciences, Committee on Biologic Effects of Atmospheric Pollutants. *Fluorides*. Washington, D.C.: National Academy of Sciences, 1971.

Neiburger, Morris. *Report No. 11: Visibility Trend in Los Angeles*. Los Angeles: Air Pollution Foundation, September, 1955.

Nelson, Edwin E. *Hydrocarbon Control for Los Angeles by Reducing Gasoline Volatility*. Detroit: International Automotive Engineering Congress, January 13–17, 1969.

New Jersey Air Pollution Commission. *Report to the New Jersey Legislature on Air Pollution in New Jersey and Recommendations for Its Abatement*. Trenton, March 19, 1952.

New Jersey Air Pollution Control Commission. *Air Pollution Control in New Jersey: Second Progress Report*. Trenton: New Jersey State Department of Health, December, 1961.

———. *A Progress Report of the New Jersey Air Pollution Control Commission, Including a Report of the Department of Health, Air Sanitation Program, for the Period July 1, 1956 to October 31, 1957*. Trenton: New Jersey State Department of Health, 1958.

———. *Where Does New Jersey Go from Here in Air Pollution Control? Report of the Planning Committee*. Trenton: State of New Jersey, July 18, 1966.

New Jersey State Legislature. *A Public Hearing before the Joint Legislative Committee on Air and Water Pollution and Public Health*, various locations, March 7–17, 1967, vols. 1–4. Trenton, 1967.

New York City Council. *Blueprint for Cleaner Air: Final Report of the Special Committee to Investigate Air Pollution*. New York: December, 1965.

———. *Proceedings of the Council of the City of New York*, various editions, 1948–52.

New York City Environmental Protection Administration, Department of Air Resources. *Air Pollution Implementation Manual for a High Air Pollution Alert and Warning System*. New York, October 1, 1968.

New York State. *Act Now for Clean Air: A Program to Reduce Air Pollution in New York State*. Albany, 1966.

New York State Air Pollution Control Board. *A Review of Air Pollution in New York State*. Albany, July, 1958.

*New York Times*, various editions.

Noble, David F. *America by Design: Science, Technology, and the Rise of Corporate Capitalism*. New York: Alfred A. Knopf, 1977.

Nolan, David. *Fifty Feet in Paradise: The Booming of Florida*. New York: Harcourt Brace Jovanovich, 1984.

O'Fallon, John E. "Deficiencies in the Air Quality Act of 1967." *Law and Contemporary Problems* 33, no. 2 (spring, 1968): 275–94.

Pancheri, Giovanni. "Industrial Atmospheric Pollution in Italy." In *Problems and Control of Air Pollution: Proceedings of the First International Congress on Air Pollution*, New York City, March 1–2, 1955, ed. Frederick S. Mallette, pp. 252–63. New York: Reinhold Publishing Corp., 1955.

Parker, Albert. "Air Pollution Legislation: Standards and Enforcement." In World Health Organization, *Air Pollution*, pp. 365–80. New York: Columbia University Press, 1961.

Perkins, Henry C. *Air Pollution*. New York: McGraw-Hill Book Company, 1974.

Petulla, Joseph M. *American Environmental History* 2e. Columbus, Ohio: Merrill Publishing Company, 1988.

Planning and Conservation League. *Prospects for Clean Air: A PCL Position Paper*. Sacramento, 1973.

Platt, Harold L. "Invisible Gases: Smoke, Gender, and the Redefinition of Environmental Policy in Chicago, 1900–1920." *Planning Perspectives* 10, no. 1 (January, 1995): 67–97.

Pollack, Lawrence W. "Legal Boundaries of Air Pollution Control: State and Local Legislative Purpose and Techniques." *Law and Contemporary Problems* 33, no. 2 (spring, 1968): 331–57.

Poppendiek, H. F. et al. *A Report on an Atmospheric Investigation in the Los Angeles Basin*. Los Angeles: University of California Departments of Engineering and Meteorology, June, 1948.

Primack, Joel, and Frank von Hippel. *Advice and Dissent: Scientists in the Political Arena*. New York: Basic Books, 1974.

Rackleff, Robert B. *Close to Crisis: Florida's Environmental Problems*. Tallahassee: New Issues Press, 1972.

Ray, Dixy Lee, and Lou Guzzo. *Environmental Overkill: Whatever Happened to Common Sense?* Washington, D.C.: Regnery Gateway, 1993.

———. *Trashing the Planet: How Science Can Help Us Deal with Acid Rain, Depletion of the Ozone, and Nuclear Waste (among Other Things)*. Washington, D.C.: Regnery Gateway, 1990.

Ridker, Ronald G. *Economic Costs of Air Pollution: Studies in Measurement*. New York: Frederick A. Praeger, Publishers, 1967.

Ripley, Randall B. "Congress and Clean Air: The Issue of Enforcement, 1963." In *Congress and Urban Problems*, ed. Frederic N. Cleaveland, pp. 224–78. Washington, D.C.: Brookings Institution, 1969.

City of Riverside, California, Office of Disaster Preparedness. *Air Pollution–Smog: An Existing Health Hazard, an Imminent Disaster*. A Resolution, Report and Proposed Emergency Plan from the Mayor and Council of the City of Riverside to the Governor of the State of California. Riverside, June, 1972.

Roan, Sharon. *Ozone Crisis: The 15-Year Evolution of a Sudden Global Emergency*. New York: John Wiley and Sons, 1989.

Roberts, Thomas Raymond. "Motor Vehicle Air Pollution Control in California: A Case Study in Political Unresponsiveness." Baccalaureate honors thesis, Harvard College, March, 1969.

Rolle, Andrew. *Los Angeles: From Pueblo to City of the Future* 2e. San Francisco: MTL, 1995.

Rose, Chris. *The Dirty Man of Europe: The Great British Pollution Scandal*. London: Simon and Schuster, 1990.

Rosen, Christine Meisner. "Businessmen against Pollution in Late Nineteenth Century Chicago." *Business History Review* 69, no. 3 (autumn, 1995): 351–97.

Rosenbaum, Walter A. *Environmental Politics and Policy* 3e. Washington, D.C.: Congressional Quarterly, 1995.

Rosner, David P., and Gerald R. Markowitz. "'A Gift of God'?: The Public Health Controversy over Leaded Gasoline during the 1920s." In *Dying for Work: Workers' Safety and Health in Twentieth-Century America*, ed. David P. Rosner and Gerald R. Markowitz, pp. 121–39. Bloomington: Indiana University Press, 1987.

Sacramento Regional Area Planning Commission. *The Air Pollution Threat: Summary of Technical Papers*. Sacramento, October 7, 1969.

Sale, Kirkpatrick. *The Green Revolution: The American Environmental Movement, 1962–1992.* New York: Hill and Wang, 1993.

San Francisco Bay Area Air Pollution Control District. *Air Pollution and the San Francisco Bay Area,* 4e. San Francisco, October, 1969.

———. *Miscellaneous Publications, 1957–1964.* San Francisco, 1964.

San Francisco Bay Area Council. *A Voluntary Program for Air Pollution Control in the San Francisco Bay Area: Initial Report.* San Francisco, January, 1953.

Schachter, Esther Roditti, Project Director. *New York City Air Pollution Control Code.* New York: New York City Department of Air Pollution Control, December 31, 1963.

Scheffer, Victor B. *The Shaping of Environmentalism in America.* Seattle: University of Washington Press, 1991.

Schiller, Ronald. "The Los Angeles Smog." *National Municipal Review* 44, no. 11 (December, 1955): 558–64.

Schlesinger, Arthur M., Jr. *The Cycles of American History.* Boston: Houghton Mifflin Company, 1986.

Schramm, E. "Experts in the Smelter Smoke Debate." In *The Silent Countdown: Essays in European Environmental History,* ed. Peter Brimblecombe and Christian Pfister, pp. 196–209. New York: Springer-Verlag, 1990.

Schrenk, H. H., Harry Heimann, George D. Clayton, W. M. Gafafer, and Harry Wexler. *Air Pollution in Donora, Pennsylvania: Epidemiology of the Unusual Smog Episode of October, 1948.* Preliminary Report/Public Health Bulletin Number 306. Washington, D.C.: U.S. Public Health Service Bureau of State Services, Division of Industrial Hygiene, 1949.

Schultz, Stanley K., and Clay McShane. "Pollution and Political Reform in Urban America: The Role of the Municipal Engineer." In *Pollution and Reform in American Cities, 1870–1930,* ed. Martin V. Melosi, pp. 155–72. Austin: University of Texas Press, 1980.

Schulze, Richard H. "The 20-Year History of the Evolution of Air Pollution Control Legislation in the U.S.A." *Atmospheric Environment* 27B, no. 1 (1993): 15–22.

Schuurmans, Rick, ed. *A History of Air Pollution Control in Chattanooga and Hamilton County.* Chattanooga: Air Pollution Control Bureau, 1991.

Scorer, R. S. *Pollution in the Air: Problems, Policies and Priorities.* Boston: Routledge and Kegan Paul, 1973.

Scott, Anne Firor. "Women's Voluntary Associations: From Charity to Reform," In *Lady Bountiful Revisited: Women, Philanthropy, and Power,* ed. Kathleen D. McCarthy, pp. 30–62. New Brunswick: Rutgers University Press, 1990.

Scott, Stanley, and John F. McCarty. *Air Pollution Control: 1951.* Legislative Problems, No. 7. Berkeley: University of California Bureau of Public Administration, May, 1951.

Selcraig, Bruce. "Border Patrol." *Sierra* 79, no. 3 (May–June, 1994): 58–74, 79–81.

Shabecoff, Philip. *A Fierce Green Fire: The American Environmental Movement.* New York: Hill and Wang, 1993.

Shaw, Alfred J. *Preliminary Inventory of Air Pollution Emissions in Hillsborough County, Florida, 1966.* Tampa: Hillsborough County Health Department, Office of Environmental Engineering, June, 1968.

Siegman, Heinrich. *The Conflicts between Labor and Environmentalism in the Federal Republic of Germany and the United States.* Aldershot, U.K.: Gower Publishing Company, 1985.

Simmons, Ian Gordon. *Environmental History: A Concise Introduction.* Cambridge, Mass.: Blackwell Publishers, 1993.

Sims, Harold Lahn. "The Emergence of Air Pollution as a Political Issue in Southern California, 1940–1970." Ph.D. dissertation, University of California–Riverside, 1973.

Smilor, Raymond W. "Toward an Environmental Perspective: The Anti-Noise Campaign, 1893–1932." In *Pollution and Reform in American Cities, 1870–1930,* ed. Martin V. Melosi, pp. 135–51. Austin: University of Texas Press, 1980.

Smith, Duane A. *Mining America: The Industry and the Environment, 1800–1980.* Lawrence: University of Kansas Press, 1987.

Smith, V. Kerry. *The Economic Consequences of Air Pollution.* Cambridge, Mass.: Ballinger Publishing Company, 1976.

Snyder, Lynne Page. "'The Death-Dealing Smog over Donora, Pennsylvania': Industrial Air Pollution, Public Health, and Federal Policy, 1915–1963." Ph.D. dissertation, University of Pennsylvania, 1994.

———. "'The Death-Dealing Smog over Donora, Pennsylvania': Industrial Air Pollution, Public Health Policy, and the Politics of Expertise, 1948–1949." *Environmental History Review* 18, no. 1 (spring, 1994): 117–39.

Spelsberg, Gerd. *Rauchplage: Hundert Jahre Saurer Regen* [Smoke plague: A hundred years of acid rain]. Aachen, Germany: Alano Verlag, 1984.

Stanford Research Institute. *The Smog Problem in Los Angeles County: A Report by Stanford Research Institute on Studies to Determine the Nature and Causes of Smog.* Los Angeles: Western Oil and Gas Institute, January, 1954.

Stauber, John C., and Sheldon Rampton. "'Democracy' for Hire: Public Relations and Environmental Movements." *Ecologist* 25, no. 5 (September–October, 1995): 173–80.

Stead, Frank M. "Background Paper: Pollution or Purity? Air, Water, Solid Wastes." Unpublished paper presented at Conference on Bay Area Regional Organization, University of California–Berkeley, September 14, 1968.

Stern, Arthur C., ed. *Air Pollution* 3e. 5 vols. New York: Academic Press, 1976.

———. "History of Air Pollution Legislation in the United States." *Journal of the Air Pollution Control Association* 32, no. 1 (January, 1982): 44–61.

Sundquist, James L. *Politics and Policy: The Eisenhower, Kennedy, and Johnson Years.* Washington, D.C.: Brookings Institution, 1968.

Tarr, Joel A. "Changing Fuel Use Behavior and Energy Transitions: The Pittsburgh Smoke Control Movement, 1940–1950." In *The Search for the Ultimate Sink: Urban Pollution in Historical Perspective,* ed. Joel A. Tarr, pp. 227–61. Akron: University of Akron Press, 1996.

———. "Railroad Smoke Control: The Regulation of a Mobile Pollution Source." In *The Search for the Ultimate Sink: Urban Pollution in Historical Perspective,* ed. Joel A. Tarr, pp. 262–83. Akron: University of Akron Press, 1996.

———. *The Search for the Ultimate Sink: Urban Pollution in Historical Perspective.* Akron: University of Akron Press, 1996.

Tebeau, Charlton W. *A History of Florida.* Coral Gables: University of Miami Press, 1971.

TeBrake, William H. "Air Pollution and Fuel Crises in Preindustrial London, 1250–1650," *Technology and Culture* 16, no. 3 (July, 1975): 337–59.

Theophrastus. *On Stones.* Ed. and trans. D. E. Eichholz. Oxford: Clarendon Press, 1965.

Tucker, Raymond R. *The Los Angeles Smog Report.* Los Angeles: Times-Mirror Co., 1947.

Tucker, William. *Progress and Privilege: America in the Age of Environmentalism.* Garden City, N.Y.: Anchor Press, 1982.

U.S. Department of Health, Education, and Welfare. *Proceedings: National Conference on Air Pollution, November 18–20, 1958.* Washington, D.C.: U.S. Government Printing Office: 1959.

———. *Proceedings: National Conference on Air Pollution, December 10–12, 1962.* Washington, D.C.: U.S. Government Printing Office: 1963.

———. *Topical Summary of Consultation Relative to Air Pollution Discharged in the State of New Jersey and Alleged to Endanger the Health and Welfare of Persons in New York State.* Washington, D.C., 1965.

U.S. Department of Health, Education, and Welfare, National Center for Air Pollution Control, Office of Legislative and Public Affairs. *Proceedings: The Third National Conference on Air Pollution, Washington, D.C., December 12–14, 1966.* Washington, D.C.: U.S. Department of Health, Education, and Welfare, 1966.

U.S. Department of Health, Education, and Welfare, Robert H. Taft Sanitary Engineering Center. *Symposium: Air over Cities.* SEC Technical Report A62-5. Cincinnati, 1962.

U.S. House of Representatives, Committee on Interstate and Foreign Commerce. *Air Pollution: Hearings before a Subcommittee of the Committee on Interstate and Foreign Commerce, United States House of Representatives, on HR 3507, HR 4061, HR 4415, HR 4750,* 88th Cong., 1st sess., 1963.

———. *Air Quality Act of 1967: Hearings before a Subcommittee of the Committee on Interstate and Foreign Commerce on HR 9509, S. 780,* 90th Cong., 1st sess., 1967.

U.S. House of Representatives, Committee on Merchant Marine and Fisheries, Subcommittee on Fisheries and Wildlife Conservation. *Environmental Quality: Hearings before a Subcommittee of the Committee on Merchant Marine and Fisheries,* 91st Cong., 1st sess., May–June, 1969.

U.S. President's Science Advisory Committee, Environmental Pollution Panel. *Restoring the Quality of Our Environment.* Washington, D.C.: White House, November, 1965.

U.S. Public Health Service. *Air Pollution in Donora, Pa.: Epidemiology of the Unusual Smog Episode of October 1948.* Washington, D.C., 1949.

———, Division of Air Pollution. *The Effect of Air Pollution.* Washington, D.C.: U.S. Department of Health, Education, and Welfare, 1966.

U.S. Senate Committee on Public Works, Subcommittee on Air and Water Pollution. *Air Pollution Control: Hearings before the Subcommittee on Air and Water Pollution of the Senate Committee on Public Works,* 88th Cong., 1st sess. Washington, D.C., 1963.

———. *Clean Air: Field Hearings Held on Progress and Programs Relating to the Abatement of Air Pollution,* 88th Cong., 2d sess. Washington, D.C., 1964.

———. *Air Pollution Control: Hearings before the Subcommittee on Air and Water Pollution of the Senate Committee on Public Works,* 89th Cong., 1st sess. Washington, D.C., 1965.

———. *Air Pollution: Hearings before the Subcommittee on Air and Water Pollution of the Senate Committee on Public Works,* 89th Cong., 2d sess. Washington, D.C., 1966.

———. *Air Pollution: Hearings before the Subcommittee on Air and Water Pollution of the Senate Committee on Public Works,* 90th Cong., 1st sess. Washington, D.C., 1967.

———. *Air Pollution: Hearings before the Subcommittee on Air and Water Pollution of the Senate Committee on Public Works Relating to the Health Effects of Air Pollution,* 90th Cong., 2d sess. Washington, D.C., 1968.

———. *Air Pollution: Hearings before the Subcommittee on Air and Water Pollution of the Senate Committee on Public Works,* 91st Cong., 1st sess., St. Louis, 1969. Washington, D.C., 1969.

———. *Air Pollution: Hearings before the Subcommittee on Air and Water Pollution of the Senate Committee on Public Works,* 91st Cong., 2d sess. Washington, D.C., 1970.

————. *Resource Recovery Act of 1969: Hearings before the Subcommittee on Air and Water Pollution of the Senate Committee on Public Works,* 91st Cong., 1st sess., Detroit, October 28, 1969.

————. *Steps Toward Cleaner Air: A Report from the Special Subcommittee on Air and Water Pollution.* Washington, D.C.: U.S. Government Printing Office, 1964.

————. *A Study of Pollution: Air.* Staff Report to the Committee on Public Works, United States Senate. Washington, D.C.: U.S. Government Printing Office, 1963.

University of California Air Pollution Research Center. *The Polluted Air: Smog Research at the University of California.* Special Report to the Regents. Riverside, 1967.

University of California–Berkeley Bureau of Public Administration. *Air Pollution Control.* Sacramento: California State Assembly, January, 1955.

Utah Legislative Council Air Pollution Advisory Committee. *Air Resources of Utah.* Salt Lake City, June, 1962.

Vanderver, Timothy A., Jr., ed. in chief et al. *Clean Air Law and Regulation.* Washington, D.C.: Bureau of National Affairs, 1992.

Venezia, Ronald A. *Air Pollution in Hillsborough County, Florida, 1962.* Jacksonville: Florida State Board of Health, Bureau of Sanitary Engineering, March, 1963.

Wagner, Richard. *Environment and Man.* New York: W. W. Norton and Company, 1971.

Warner, Sam Bass, Jr. *The Urban Wilderness: A History of the American City.* Berkeley: University of California Press, 1995.

Watson, Ann Y., Richard R. Bates, and Donald Kennedy, eds. *Air Pollution, the Automobile, and Public Health.* Washington, D.C.: National Academy Press, 1988.

Weale, Albert, Timothy O'Riordan, and Louise Kramme. *Controlling Pollution in the Round: Change and Choice in Environmental Regulation in Britain and Germany.* London: Anglo-German Foundation for the Study of Industrial Society, 1991.

Weidner, Helmut. *Clean Air Policy in Great Britain: Problem-Shifting as Best Practicable Means.* Berlin: Edition Sigma Bohn, 1987.

Welsh, Gene B., and Thomas E. Kreichelt. *Clean Air for Chattanooga.* Cincinnati: U.S. Public Health Service, Robert A. Taft Sanitary Engineering Center, July 1964.

Wiebe, Robert H. *The Search for Order, 1877–1920.* New York: Hill and Wang, 1967.

Wildavsky, Aaron. "Aesthetic Power or the Triumph of the Sensitive Minority over the Vulgar Masses: A Political Analysis of the New Economics." In *America's Changing Environment,* ed. Roger Revelle and Hans H. Landsberg, pp. 147–60. Boston: Houghton Mifflin Company, 1970.

Williams, Charles R. "Air Pollution from Fluorides." In *Proceedings of the Forty-Ninth Annual Meeting of the Air Pollution Control Association,* Buffalo, New York, May 20–24, 1956, pp. 17-1 to 17-10. Pittsburgh: Air Pollution Control Association, 1956.

Williams, James D., and Norman G. Edmiston. *An Air Resource Management Plan for the Nashville Metropolitan Area.* Cincinnati: United States Department of Health, Education, and Welfare, Public Health Service, Division of Air Pollution, 1965.

Wolozin, Harold. "The Economics of Air Pollution: Central Problems." *Law and Contemporary Problems* 33, no. 2 (spring, 1968): 227–38.

Wood, Samuel E. *Air Pollution Control.* Sacramento: California State Assembly, January, 1955.

Woodward, C. Vann. *The Origins of the New South.* Baton Rouge: Louisiana State University Press, 1951.

World Health Organization. *Air Pollution.* New York: Columbia University Press, 1961.

Wrigley, E. Anthony. *Continuity, Chance and Change: The Character of the Industrial Revolution in England.* Cambridge: Cambridge University Press, 1988.

Young, Louise B. *Sowing the Wind: Reflections on the Earth's Atmosphere.* New York: Prentice-Hall Press, 1990.

Zimmerman, Joseph F. "Political Boundaries and Air Pollution Control." Master's thesis, State University of New York–Albany, 1968.

Zimmerman, Percy W. "Effects on Plants of Impurities Associated with Air Pollution." In *Air Pollution: Proceedings of the United States Technical Conference on Air Pollution,* Louis C. McCabe, Chairman, pp. 127–33. New York: McGraw-Hill Book Company, 1952.

# Index

Adams, Henry, 141
AFL-CIO, 75, 236. *See also* labor organizations
Ainsworth, Ed, 87–88
air pollutants. *See* pollutants
air pollution as bipartisan issue, 106, 115–16, 131, 242–43; bipartisan opposition to control, 163
air pollution as federal issue, 12, 236–44. *See also* federal government
air pollution as international issue, 29, 237, 253–54
air pollution as interstate issue, 11–12, 28, 228, 240–42; in Greater St. Louis, 226–28; in metropolitan New York, 158–72
air pollution as state/local issue, 5, 7, 11, 24, 26, 37, 237–39, 241; countywide control agencies, 24, 38, 43–44, 118, 159, 181–82; inadequacy of state/local control efforts, 158, 165, 167–68, 172, 190, 194–96, 228, 231; multicounty control approach, 55, 62, 108, 181–82; state laws, 24, 37, 43–44, 62–65, 73–74, 77–78, 80, 159, 161–68, 180–81, 207–208, 231
air pollution control agencies, 31, 38, 41, 55–56, 159, 226–27, 228, 231; in California, 41–42, 46–47, 73–74, 77, 108; federal, 27, 44–45, 217; in Florida, 181–82, 202, 207–208. *See also* Florida Air Pollution Control Commission; Florida State Board of Health; Los Angeles County Air Pollution Control District (APCD); New York City Bureau of Smoke Control (BCS); New York City Department of Air Pollution Control; Polk County Air Pollution Control District (APCD); United States Public Health Service (PHS)
Air Pollution Control Association, 230. *See also* civic groups

air pollution control budgets and funding: at federal level, 238–40; in Florida, 186–88, 219; generally, at state or local level, 231; in Los Angeles County, 44, 57, 128; in New York City, 127–30, 132, 139, 142–44
air pollution control officials: favoring strong enforcement, 44–46, 50–53, 57–58, 121–22, 148–51, 243–44; favoring voluntary cleanup, 118, 129, 134, 140, 161–63, 187; favoring weaker enforcement, education, and persuasion, 26, 115–16, 120, 122, 127, 129, 134, 157, 163, 165, 216, 220, 238–39, 243–44; public relations and public relation problems, 51–52, 87–88, 91, 96, 109, 122; seen as too cozy with industry, 50, 72–73, 122, 197
air pollution control techniques, 10–11, 17–18; fuel regulation, 119, 128, 130; fuel substitution, 10, 18, 31–32, 54; zoning, 52–53. *See also* air pollution control technology
air pollution control technology, 9–10, 13, 25–26, 31–32, 59–66, 118–19, 155; mandatory use of, 10, 19, 31–32, 44–46, 50–53, 63–64, 75, 131. *See also* automotive emissions control devices; science and technology
air pollution damage and costs: accidents, 19, 23; aesthetic harm, 137; costs estimated, 25, 131, 137–39, 229; crop damage, 47–48, 161, 163, 178–80, 189, 195, 200–205, 212–13, 229, 278n 12; deaths, 19, 23; harm to commercial flower growers, 48, 178–79, 195, 222–23; livestock damage, 179–80, 190, 195–97, 212–13, 218–19, 229, 278n 12; low visibility, 19, 23–24, 164; property damage, 17–18, 24, 31, 131, 138, 164, 188–89, 197, 200, 229; statistically elevated death rates, 17, 133, 170, 230; threat to

Boston, 21, 133, 229, 231
Bowen, Floyd, 197–99, 203
Bowron, Fletcher, 84, 94
Brandt, C. Stafford, 211–14, 222–23, 244
Breslow, Lester, 230
Brewer, R. S., 204
Brienes, Marvin, 97
Brimblecombe, Peter, 17, 19
Britain, 16–20; Alkali Act, 19–20; British
    Clean Air Act of 1956, 20. *See also* air pol-
    lution incidents and disasters; London
Brooklyn, 115, 122–23, 147
Bronx, 141
Brown, Edmund G., Sr., 95
Brown, Kenneth R., 232–33
Bryant, Farris, 191–92
Bureau of Smoke Control (BSC). *See* New
    York City Bureau of Smoke Control
    (BSC)
Burney, Leroy E., 91, 195, 219, 230
business and industry: control of research,
    29–30, 61–62, 245; relocation threats or
    worries, 31, 167, 169, 171, 197–99, 200; re-
    search as delay, 30, 41, 64, 116, 198, 244–
    45; support for pollution control, 24, 25,
    86, 87, 95, 117, 149, 184; voluntary pollu-
    tion control, 20, 54, 115
business and industry resistance to air pollu-
    tion control, 8–9, 11; in auto industry, 47,
    49–50, 64, 68–69; in England, 19; in
    Florida, 196–201, 203–204, 206; in Los
    Angeles, 41–44, 47, 49–50, 54–55, 64, 86,
    88–89, 96; in New Jersey, 162, 165, 167; in
    New York City, 116, 119, 122–24, 126, 128,
    132, 142, 145–46; in postwar United
    States, 238, 244–48; in prewar United
    States, 26, 31–32; standard business and
    industry arguments, 86, 88–89, 115, 167,
    245–47
business associations: in Florida, 184, 197; na-
    tional, 50, 59, 77, 79, 242, 245–46; in
    New York City, 117, 122, 142, 147; South-
    ern California, 44, 48, 54–55, 85–86, 89,
    95–96. *See also* Chamber of Commerce
Byrne, William H., 117–18, 120, 128, 144

Cabrillo, Juan Rodriguez, 38
California, State of: Air Resources Board
    (ARB), 73–74, 77, 108; Assembly Com-
    mittee on Air and Water Pollution, 44–
    45; Department of Motor Vehicles, 63, 73;
    Department of Public Health, 63–64, 71,
    73, 91, 230; Highway Patrol, 63, 73, 78;
    Senate Committee on Highways and
    Transportation, 70, 104
California, University of, 63, 95; Berkeley
    campus, 47; Los Angeles campus, 105;
    Riverside campus, 105; state agricultural
    experiment station at Riverside, 48, 55,
    204, 229
California Institute of Technology, 47, 48, 88,
    93
California Manufacturers Association, 96
California Motor Vehicle Pollution Control
    Board (MVPCB): and antitrust suit, 77;
    averaging policy, 68, 71–72; enabling leg-
    islation, 63–65, 72–73, 98; feud with LA
    County APCD, 71–72; mistakes, 69–70;
    replaced by California ARB, 73–74; seen
    as too close to industry, 69, 72–73, 98;
    and used–car crankcase devices, 102–104
Campbell, John M., 49
Canada, 29, 237
Capehart, Homer, 237, 244, 246
carbon monoxide, 53, 64, 71, 76–77, 128, 133,
    155–56
Carnegie, Andrew, 25
Carson, Rachel, 231, 248–49
Carter, Stanley Charles, 234
catalytic converters. *See* automotive emis-
    sions control devices
cattle. *See* livestock
Celebrezze, Anthony S., 75
Chamber of Commerce: in Long Beach, Ca-
    lif., 96; in Los Angeles, 85–86, 89, 95; in
    New Jersey, 162; in New York State, 117;
    in Pasadena, Calif., 85; in Pittsburgh, 25;
    United States, 246. *See also* business asso-
    ciations
Chandler, Norman and Dorothy, 87
Chase, Edna Woolman, 144

women, 25, 87, 94, 97–98, 124, 140–41, 143–48, 153–54, 191, 201–202, 233–35, 260*n* 32; among working class, 217, 233–36; in Britain, 20; claim of clean air as a basic right, 192, 194; in Florida, 180, 183–87, 189–202, 207, 234; in minority communities, 234–35; in New Jersey, 166, 170; in New York City, 117, 135–49, 151–52, 156; in postwar United States, 232–37; in pre-1940s United States, 24–27; public resistance to air pollution control, 20, 31, 49, 64, 69, 88, 102–104, 109, 124–27, 132; in Southern California, 83–88, 93–98, 109–10; on Staten Island, 164, 170. *See also* civic groups

Public Health Service (PHS). *See* United States Public Health Service (PHS)

Progressive era, 12, 137, 140

Providence, 229

Pure Air Committee, 93. *See also* civic groups

Pure Air Council of Southern California, 93. *See also* civic groups

Queens, N.Y., 115, 120, 122, 136, 142, 146–47

Randolph, Jennings, 195, 199, 240–41

Ray, John H., 161

Reagan, Ronald W.: as California governor, 74; as president, 252

Reavis, Mrs. C. Frank, 144

Reitz, Herman J., 204

Ribicoff, Abraham, 67

Ringelmann Chart, 118–19, 271*n* 18

Riverside County, Calif., 102, 108

Roberts, Kenneth A., 74, 238–39

Robinson, Elizabeth, 145–48, 153–54, 156

Rockefeller, Nelson, 128, 133, 168–69

Rome, 16, 33

Roosevelt, Franklin D., 159

Rossano, August, 218–21, 227

Royce, Stephen, 87, 93

Sargent, Frederick, 92

St. Louis: Division of Smoke Regulation, 31; early smoke problems and abatement

efforts, 23–26, 37; example for New York and other cities, 33, 119, 137; interstate air pollution, 158, 172, 227, 236; local business resistance, 31, 86, 89; successful 1930s control campaign, 31–33, 58, 87, 109, 119

Salt Lake City, 27

San Bernardino County, Calif., 38, 102, 108

San Diego, 229

San Francisco, 27, 229; San Francisco Bay Area, 47; San Francisco Bay Area Air Pollution Control District, 55

Sao Paulo, 229

Scheele, Leonard A., 163

Schenck, Paul F., 67; Schenck Act, 67, 74

Schueneman, Jean J., 220–22, 226

science and technology: engineers, 25–26, 136; faith in, to solve air pollution and other problems, 13, 25–26, 44, 61, 98–101, 109, 118, 127, 134, 152, 155, 220, 222; limited early scientific understanding of air pollution, 9, 46, 155; political impacts or use of science, 9–10, 30, 203–206, 210–11, 214, 247–48; scientific professionalism as quietism, 9–10, 41, 126–27, 135–36, 153–54, 157, 210–14, 220, 225–27, 244; scientific research on air pollution, 17–18, 44, 46–50, 61–62, 127, 129, 155, 237; scientists and technicians close to business and industry, 8–9, 26, 136, 153–54, 203–206. *See also* air pollution control technology; automobiles; business and industry; health concerns; United States Public Health Service (PHS)

Seattle, 229

Seifert, Harry, 186, 218–20

Shafer, Raymond P., 169–70

Sharkey, Joseph, 114, 130; demands stricter enforcement, 119–21, 148; struggles to establish BSC, 114–17, 141–44; writes bills to regulate incinerators, 125–26; writes bill to replace BSC with DAPC, 122, 151–52

Slade, Afton, 107

Smathers, George, 189–90, 193

Smelters, 27–28

Smith, Chesterfield, 203–204, 206